Inside Rhinoceros®

Ron K. C. Cheng

ONWORD PRESS
THOMSON LEARNING

Australia   Canada   Mexico   Singapore   United Kingdom   United States

Inside Rhinoceros®
By Ron K. C. Cheng

Publisher:
Alar Elken

Executive Editor:
Sandy Clark

Acquisitions Editor:
James Gish

Managing Editor:
Carol Leyba

Development Editor:
Daril Bentley

Editorial Assistant:
Jaimie Wetzel

Executive Marketing Manager:
Maura Theriault

Executive Production Manager:
Mary Ellen Black

Production Manager:
Larry Main

Manufacturing Coordinator:
Betsy Hough

Technology Project Manager:
David Porush

Cover Design:
Cammi Noah

Trademarks
Rhinoceros is a registered trademark of Robert McNeel & Associates, Inc.

CD-ROM:
Copyright © 2001 Robert McNeel and Associates. All rights reserved.

Copyright © 2002 by OnWord Press. OnWord Press is an imprint of Thomson Learning
SAN 694-0269

Printed in Canada
10 9 8 7 6 5 4 3 2 1

For permission to use material from this text, contact us by
Tel : 1-800-730-2214
Fax: 1-800-730-2215
www.thomsonrights.com

For more information, contact:
OnWord Press
An imprint of Thomson Learning
Box 15-015
Albany, New York 12212-15015

Or find us on the World Wide Web at
http://www.onwordpress.com

All rights reserved. No part of this work covered by the copyright hereon may be reproduced or used in any form or by any means—graphic, electronic or mechanical, including photocopying, recording, taping, Web distribution or information storage and retrieval systems—without the written permission of the publisher.

Library of Congress Cataloging-in-Publication Data is available for this title.

ISBN: 0-7668-5437-X

NOTICE TO THE READER

Publisher does not warrant or guarantee any of the products described herein or perform any independent analysis in connection with any of the product information contained herein. Publisher does not assume, and expressly disclaims, any obligation to obtain and include information other than that provided to it by the manufacturer.

The reader is expressly warned to consider and adopt all safety precautions that might be indicated by the activities herein and to avoid all potential hazards. By following the instructions contained herein, the reader willingly assumes all risks in connection with such instructions.

The publisher makes no representation or warranties of any kind, including but not limited to, the warranties of fitness for particular purpose or merchantability, nor are any such representations implied with respect to the material set forth herein, and the publisher takes no responsibility with respect to such material. The publisher shall not be liable for any special, consequential, or exemplary damages resulting, in whole or part, from the readers' use of, or reliance upon, this material.

■■ About the Author

Ron K. C. Cheng leads the Product Design Unit of the Industrial Center of The Hong Kong Polytechnic University, where he is involved in computer-aided design and the development of computer-based learning materials. He is also the developer and manager of the instructional web site *http://pdu.ic.polyu.edu.hk*. Ron is author of eight Autodesk Press training guides, including books on AutoCAD, Mechanical Desktop, and Autodesk Inventor. You may contact the author via his e-mail address: *icrcheng@polyu. edu.hk*.

■■ Acknowledgments

This book never would have been realized without the contributions of many individuals. The author and the publisher are grateful to technical reviewer Jerry Hambly at McNeel for thoughtful suggestions and help regarding both the text and the companion CD-ROM content.

Several people at OnWord Press/Thomson Learning deserve special mention: acquisitions editor James Gish, developmental editor Daril Bentley, managing editor Carol Leyba, editorial assistant Jaimie Wetzel, manufacturing coordinator Betsy Hough, and cover designer Cammi Noah.

Contents

Chapter 1: An Introduction to Rhinoceros 1
   Objectives 1
   Overview 1
   Computer Modeling 1
   Rhino Functions 4
   Starting Rhino 8
   Summary 25
   Review Questions 26

Chapter 2: Wireframe Modeling and Curves: Part 1 27
   Objectives 27
   Overview 27
   Wireframe Modeling Concepts 28
   Curves 30
   Rhino Curve Construction Tools 34
   Summary 82
   Review Questions 82

Chapter 3: Wireframe Modeling and Curves: Part 2 83
   Objectives 83
   Overview 83
   More on Curves 83
   Summary 149
   Review Questions 149

Chapter 4: Wireframe Modeling and Curves: Part 3 151
   Objectives 151
   Overview 151
   Analyzing Curves 151
   Curve Construction Projects 162
   Summary 198
   Review Questions 198

Chapter 5: Surface Modeling: Part 1 199
   Objectives 199
   Overview 199
   Surface Modeling Concepts 200
   Rhino Surface Modeling Tools 206
   Summary 252
   Review Questions 253

Contents

Chapter 6: Surface Modeling: Part 2 255
Objectives 255
Overview 255
More Surface Modeling Tools 255
Summary 302
Review Questions 302

Chapter 7: Surface Modeling: Part 3 303
Objectives 303
Overview 303
Final Surface Modeling Tools 303
Surface Modeling Projects 320
Summary 349
Review Questions 350

Chapter 8: Solid Modeling 351
Objectives 351
Overview 351
Solid Modeling Concepts 352
Rhino Solid Modeling Tools 355
Solid Modeling Projects 386
Summary 409
Review Questions 410

Chapter 9: Consolidation 411
Objectives 411
Overview 411
NURBS Surface Modeling 412
The Construction Plane 436
Import and Export 453
Use of Polygon Meshes 455
Modeling Projects 472
Summary 485
Review Questions 486

Chapter 10: Presentation 487
Objectives 487
Overview 487
Shaded and Wireframe Viewport Display 487
Rendering 492
2D Drawing 508
Dimensioning and Annotation 510
Presentation Projects 511
Summary 514
Review Questions 515

Index 517

Introduction

Rhinoceros (or Rhino, as it is popularly known) is a Windows-based 3D surface modeling program fast gaining in popularity. Rhino is based on the popular NURBS (non uniform rational B-spline) mathematics. NURBS is a computer modeling paradigm that makes it easy to build curved surfaces and organic, free-form shapes, which are the rule rather than the exception in today's design world.

Using Rhino tools, you construct, edit, transform, and analyze curves, surfaces, and solids. Because curves alone have limited application in computerized design and manufacturing, you use Rhino curve tools mainly for building frameworks for subsequent construction of surfaces and solids. It is these underlying frameworks that allow models constructed with Rhino to have maximum utility in the real world of manufactured designs.

In addition to its use as a modeler, Rhino is used to construct polygon meshes capable of coping with various applications of rendered solids and surfaces, as well as to output rendered images and 2D engineering drawings of 3D modeled objects. Used in conjunction with the Flamingo rendering tool, Rhino also provides you with the ability to produce photorealistic images as final output. All of these capabilities are covered in this book.

■■ Audience and Prerequisites

This book is designed for students, designers, and engineers who wish to gain a general understanding of computer modeling concepts and to learn how to construct free-form organic shapes that address both upstream and downstream design/manufacturing considerations. These objectives are achieved in the context of learning to use Rhino and its associated Flamingo rendering tool as a computer modeling system for industrial and engineering design.

Computer modeling experience and a basic knowledge of design concepts are always helpful in learning any new software program. However, there really are no prerequisites in either case in regard to using this book, which explains basic design concepts, techniques, and considerations as it teaches you from the beginning of the learning curve how to use Rhino.

Philosophy and Approach

To bridge the theoretical and software-oriented approaches to computer modeling, this book provides a balanced presentation combining theory, concepts, and tutorials. Theory is relatively useless without hands-on experience, and vice versa. This book begins with an overview of the three basic types of computer models and the Rhino interface. The book then addresses the limitations of wireframe models, discusses various types of curves and underscores why learning how to construct curves is important, and shows you how to construct curves using Rhino, including the use of digitizers.

Subsequent chapters continue to pursue this logical progression from the conceptual to the practical, and from the basic to the more complex. Following the delineation of wireframe models and curves, the book continues with an in-depth examination of surface modeling via discussion and hands-on tutorials that show you step by step how to construct surfaces using Rhino.

Following surface modeling, the book provides detailed discussion of the various types of solid modeling methods, the forms in which Rhino represents solids, and the techniques and tools used to construct solids in Rhino. To consolidate what you have learned to this point, a latter chapter reviews curves, surfaces, and solids together. Finally, the book addresses rendering, drawing output, and outputting to various file formats.

Content

This book is written to Release 2 of Rhinoceros, although the concepts, techniques, and the majority of exercises covered in this book do not require the use of Release 2. Each chapter of this book includes an overview of the chapter's topic and content, a section on concepts, a section on methodology, an exercise project or projects, a chapter summary, and review questions. Chapter 1 introduces computer modeling methods, as well as an examination of the Rhino user interface, including its key functional components.

Chapters 2 through 4 examine the limitations of wireframe modeling, introduce the use of curves in surface and solid modeling, and explore general methods of working with curves. After learning wireframe modeling and curves, in chapters 5 through 7 you learn more detailed surface modeling concepts, as well as how to construct, edit, transform, and analyze surfaces using Rhino.

Chapter 8 covers various solid modeling methods, as well as the manner in which Rhino represents solids in the computer. Chapter 9 consolidates what you learned in chapters 2 through 8. Finally, Chapter 10 addresses two main presentation methods: using shaded and rendered images, and producing 2D engineering drawings.

■■ Features and Conventions

Italic font in regular text is used to distinguish certain command names, code elements, file names, directory and path names, user input, and similar items. Italic is also used to highlight terms and for emphasis.

The following is an example of the monospaced font used for code examples (i.e., command statements) and computer/operating system responses, as well as passages of programming script.

```
var myimage = InternetExplorer ? parent.
cell : parent.document.embeds[0];
```

The following are the design conventions used for various "working parts" of the text. In addition to these, you will find that the text incorporates many exercises and examples.

> **NOTE:** *Information on features and tasks that requires emphasis or that is not immediately obvious appears in notes.*
>
> **TIP:** *Tips on command usage, shortcuts, and other information aimed at saving you time and work appear like this.*

■■ About the Companion CD-ROM

The companion CD-ROM found at the back of this book contaoms a trial version of Rhinoceros Release 2.

Chapter 1

An Introduction to Rhinoceros

■■ Objectives

The goals of this chapter are to introduce the concepts of 3D computer modeling, to outline the key functions of Rhinoceros (commonly called Rhino), and to familiarize you with the Rhino user interface. After studying this chapter, you should be able to:

❐ Explain the principles of computer modeling
❐ Describe the key functions of Rhino
❐ Use the Rhino user interface

■■ Overview

To facilitate the process of designing a product or system, you use various media to capture your design ideas. Initially, you use sketches to represent the subject of the design. To elaborate on this design, you use models. Models can be physical or digital. Making a physical model can often be time consuming, or simply not feasible. To create a digital model, you use computer technology and computer-aided design applications. This chapter provides an overview of computer modeling. You will also learn the basic functions, user interface, and basic operation methods of Rhino.

■■ Computer Modeling

The purpose of constructing a computer model of an object is to represent it in the computer in digital form in order to facilitate design, analysis, and downstream computerized operations. Using computer-aided design and rendering applications, you represent the object's geometry, texture, and color. To make better use of the computer and computer-aided design application, you need to know the various ways models are represented in the computer, the types of modeling tools available, and the techniques for using these tools. There are three ways to represent a 3D object in the computer: as a wireframe model, as a surface model, or as a solid model.

Wireframe Model

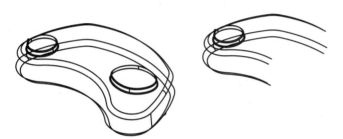

In the history of computer modeling, the 3D wireframe model is the earliest type of 3D model. It is the most primitive type of 3D object in the computer. In essence, a wireframe model is a set of unassociated curves assembled in 3D space. The curves serve only to give the pattern of a 3D object. There is no relationship between the curves. Therefore, the model does not have any surface information or volume

Fig. 1-1. Wireframe model and its cutaway view.

information. It has only data that describe the edges of the 3D object. Because of the limited information provided by the model, the use of wireframe models is very confined. Figure 1-1 shows a wireframe model of a joypad and the same model in a cutaway view.

Surface Model

A surface is a mathematical expression represented as a thin sheet without thickness. A surface model is a set of surfaces assembled in 3D space to represent a 3D object. When compared to a 3D wireframe model, a surface model has in addition to edge data information on the contour and silhouette of the 3D object. Surface models are typically used in computerized manufacturing systems and to generate photo-realistic rendering or animation. Figure 1-2 shows a surface model of a car body, a section across the model, and a rendered image of the sectioned model.

Fig. 1-2. Surface model, a section across the model, and a rendered image of the sectioned model.

Solid Model

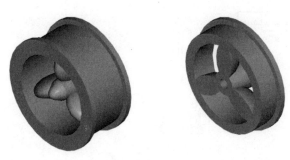

Fig. 1-3. Solid model and section across the model.

In regard to information, a 3D solid model is superior to the other two models because a solid model in a computer is a complete representation of the object. It integrates mathematical data that includes surface and edge data as well as data on the volume of the object the model describes. In addition to visualization and manufacturing, solid modeling data is used in design calculation. Figure 1-3 shows a solid model of a wheel and a section across the model.

Assembly Model

To facilitate evaluation and visualization of how various component parts of a product or system can or should be put together, you assemble individual components in an assembly model. Figure 1-4 shows the assembly of a toggle clamp.

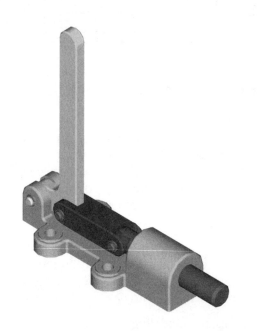

Fig. 1-4. Toggle clamp assembly.

Engineering Drawing

To represent a 3D object in a 2D drawing sheet, you use an orthographic engineering drawing. If you already have a 3D computer model, you use the computer to generate orthographic views of the model. Figure 1-5 shows a 2D engineering drawing generated from the assembly of the toggle clamp.

Chapter 1: An Introduction to Rhinoceros

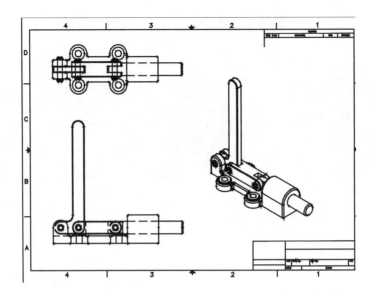

Fig. 1-5. 2D drawing generated from the toggle clamp assembly.

Downstream and Upstream Operations

Creating 3D models and producing 2D drawings from 3D models is not necessarily the end of the computer modeling process. In a computerized manufacturing system, for example, the same computer model can and often should be designed to be used in all downstream operations, such as finite element analysis, rapid prototyping, computer numerical control (CNC) machining, and computerized assembly.

To enhance illustration of the 3D object, you construct renderings and animations. Because these operations may be done using different types of computer applications, and because each application may use a unique type of data format, the computer modeling system must enable the conversion of the 3D computer model into various file formats.

On the other hand, the computer modeling system must enable the opening of various types of file formats so that computer models constructed in other systems can be used for further elaboration of the design. Figure 1-6 shows a rapid prototyping machine making a 3D object, and figure 1-7 shows the CNC machining of a 3D object.

■■ Rhino Functions

Rhino is a 3D computer modeling application that enables you to construct six types of objects: points, curves, NURBS surfaces, polysurfaces, solids, and polygon meshes. To facilitate downstream computerized

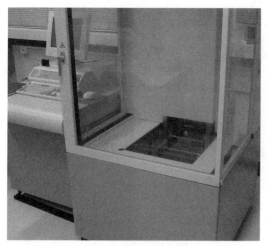

Fig. 1-6. Rapid prototyping machine making a rapid prototype from a 3D model.

Fig. 1-7. CNC machining of a free-form object.

operations and to reuse existing computer models constructed using some other computer application, you can export Rhino models to various file formats and import various file formats into Rhino. You will learn more about this later in the book.

Curves and Points

Wireframe models by themselves have limited utility in design and manufacture because they are simply a set of unassociated curves. However, curves and points are required in many surface and solid construction operations. Therefore, you need to learn how to construct curves and points for the purpose of making surfaces and solids. Using Rhino, you construct points and various types of 3D curves. To construct free-form surfaces, NURBS (non-uniform rational B-spline) curves are required. You will learn about NURBS curves in Chapters 2 through 4.

3D Surfaces

There are two basic ways to represent a surface in the computer: using a NURBS surface to exactly represent the surface or using a polygon mesh to approximate the surface. By joining a set of contiguous NURBS surfaces, you create a polysurface. You will learn about NURBS surfaces and polysurfaces in Chapters 5 through 7, and about polygonal meshes in Chapter 9. Figure 1-8 shows a NURBS surface, and figure 1-9 shows a polygonal mesh.

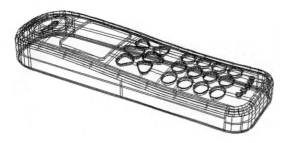 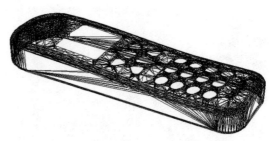

Fig. 1-8. NURBS surface model of a mobile phone casing.

Fig. 1-9. Polygonal mesh model of the same mobile phone casing.

3D Solids

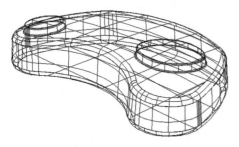

A Rhino solid is a closed-loop polysurface (i.e., a set of joined surfaces with no gap or opening). You construct Rhino solids in two basic ways: directly, using the solid modeling tools, or by converting a set of contiguous NURBS surfaces into a solid volume by joining them. You will learn more about solid modeling in Chapter 8. Figure 1-10 shows a Rhino solid.

Fig. 1-10. Rhino solid.

Rendering

You output photo-realistic renderings from surfaces, polysurfaces, and polygon meshes. You will learn about rendering in Chapter 10. Figure 1-11 shows the rendering of a mobile phone casing.

Fig. 1-11. Rendering of a mobile phone casing.

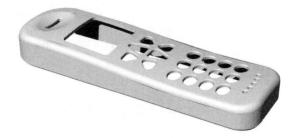

2D Drawing

To illustrate a 3D object on a 2D sheet, you typically generate a 2D drawing from the computer model of the 3D object and add appropriate dimensions and annotations to the drawing. You will learn about generation of 2D drawings from computer models in Chapter 6.

Rhino Functions

Import and Export

Many file types can be imported into and opened (a single operation) in Rhino to make reuse of existing engineering designs constructed via other computer design applications possible. To enable and facilitate engineering data exchange in which computer models you have constructed using Rhino can be used in other applications, you can save Rhino files as various file formats. The types of files you can open and/or save in Rhino are outlined in table 1-1.

Table 1-1: File Formats Rhino Can Open and/or Save

Open File Format	Save As File Format
IGES (*.igs, *.iges)	IGES (*.igs, *.iges)
STEP (*.stp, *.step)	STEP (*.stp, *.step)
—	WaveFront (*.obj)
AutoCAD DWG (*.dwg)	AutoCAD DWG (*.dwg)
AutoCAD DXF (*.dxf)	AutoCAD DXF (*.dxf)
3D Studio (*.3ds)	3D Studio (*.3ds)
—	ACIS (*.sat)
—	Parasolid (*.x_t)
LightWave (*.lwo)	LightWave (*.lwo)
Adobe Illustrator (*.ai)	Adobe Illustrator (*.ai)
Raw Triangles (*.raw)	Raw Triangles (*.raw)
—	POV Ray Mesh (*.pov)
—	Moray UDO (*.udo)
Sculptura Files (*.scn, *.3do)	—
STL (*.stl)	STL (*.stl)
—	VRML (*.wrl)
—	Windows MetaFile (*.wmf)
—	RenderMan (*.rib)
—	Comma separated value (*.csv)
VDA (*.vda)	VDA (*.vda)
AGLib (*.ag)	AGLib (*.ag)

Starting Rhino

Now that you are familiar with some rudiments of the Rhino environment, let's try a bit of hands-on examination. Start Rhino by selecting the Rhinoceros 2.0 icon from your desktop. In the application window, you will find five major areas: standard Windows title bar, pull-down menu, command line area, graphics area, and status bar. In addition, there are a number of toolbars. (See figure 1-12.) These components are described in the sections that follow.

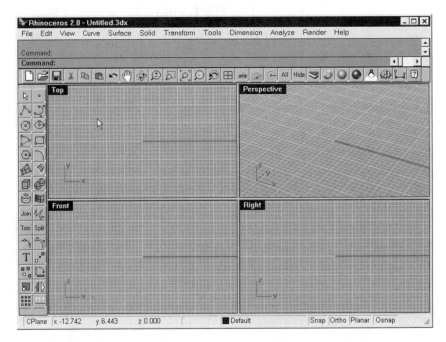

Fig. 1-12. Rhino application window.

Standard Windows Title Bar

At the top of the application, there is the standard Windows title bar. This title bar functions no differently and contains nothing different than the basic Windows title bar.

Pull-down Menu

Below the standard Windows title bar is the pull-down menu, which contains twelve menu options: File, Edit, View, Curve, Surface, Solid, Transform, Tools, Dimension, Analyze, Render, and Help. The functions of these options are outlined in table 1-2.

Starting Rhino

Table 1-2: Pull-down Menu Options and Their Functions

Pull-down Menu Option	Function
File	For working on files and templates
Edit	For editing curves, surfaces, and solids
View	For manipulating display settings and establishing construction planes
Curve	For constructing wires
Surface	For constructing NURBS surfaces
Solid	For constructing solids
Transform	For transforming objects you have constructed
Tools	Provides various types of useful tools
Dimension	For constructing 2D drawings and adding annotations
Analyze	Helps you analyze objects you have constructed
Render	For shading and rendering
Help	Provides useful help information

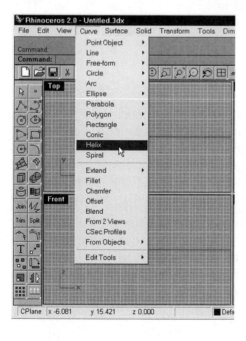

To run a command, you select an item from a pull-down menu or cascading menu and left-click.

1 Select Curve > Helix. (See figure 1-13.) The Helix command from the Curve pull-down menu is activated.

2 Select Solids > Sphere > 3 Points. (See figure 1-14.) The Sphere3Pt command from the Sphere cascading menu of the Solid pull-down menu is used to construct a three-point sphere.

Fig. 1-13. Pull-down menu for constructing a helix.

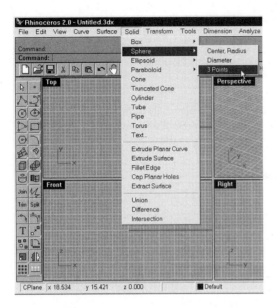

Command Line Area

Below the pull-down menu is the command line area, which provides a place for textual interaction. Here you run any command by typing the command name or alias of the command and then pressing the Enter key or the space bar. After a command is run, further prompts or instructions will appear in this area or in any associated pop-up dialog boxes. Command names are not case sensitive, and therefore you can use any combination of small and capital letters to specify a command name. Figure 1-15 shows the command line area.

Fig. 1-14. Cascading menu for constructing a three-point sphere.

Fig. 1-15. Command line area.

Graphics Area

The graphics area is where you construct your model. After you select a command, you select a location in the graphics area for constructing any number of types of geometric objects. This area can be divided into any number of viewports, which may be docked or floating. Figure 1-16 shows a four-viewport configuration (i.e., Top, Front, Right, and Perspective viewports).

Initially, the number of viewports and their orientation are determined by the viewport setting of the template file you use to start a new file. To use a template, you start a new file by selecting New from the File pull-down menu. This accesses the Template File dialog box, in which you select a template. (See figure 1-17.)

To set the viewport configuration while working on a file, you use the Viewport Layout cascading menu of the View pull-down menu or the Viewport Layout toolbar. (See figure 1-18.)

Starting Rhino

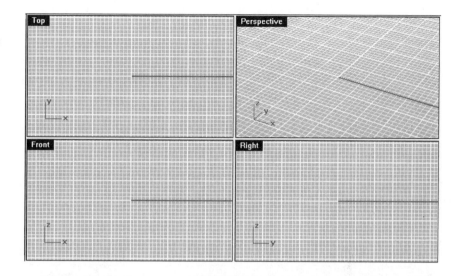

Fig. 1-16. Graphics area showing a four-viewport configuration.

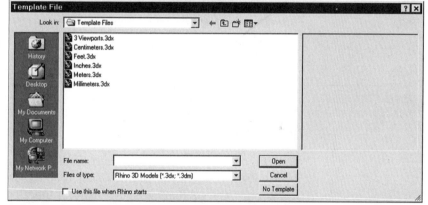

Fig. 1-17. Template File dialog box.

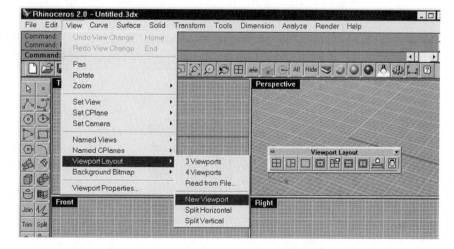

Fig. 1-18. Setting the viewport configuration.

Status Bar

At the bottom of the application window is the status bar. It shows the location of the cursor marker, the current layer, and the state of the drawing aids. It contains eight panes, the functions of which are outlined in table 1-3.

Table 1-3: Status Bar Panes and Their Functions

Pane	Function
World/Cplane	Toggles the coordinate display to show either world coordinates or construction plane coordinates
Coordinate Display	Shows the coordinates of the mouse pointer
Distance	Displays the distance from the mouse pointer to the last picked point
Layer	Shows the current layer
Snap	Constrains the cursor movement to the specified snap intervals
Ortho	Constrains the cursor movement to be orthogonal
Planar	Constrains the cursor movement to be parallel to the current construction plane
Osnap	Toggles the display of the Osnap dialog box

Toolbars

Buttons on toolbars represent commands in a graphical way. To run a command, you select a button on a toolbar. Because there are many commands and toolbars, displaying all of them would take up the entire screen display. Therefore, only the Standard (figure 1-19) and Main (figure 1-20) toolbars are displayed by default.

Fig. 1-19. Standard toolbar.

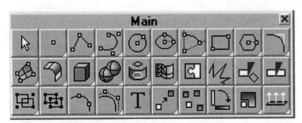

Fig. 1-20. Main toolbar.

To find out which toolbars are available, and to display them on the screen, you use the Toolbar command by typing the command name at the command line area or by selecting Tools > Toolbar Layout > Edit from the pull-down menu. In the Toolbars

Starting Rhino

dialog box (figure 1-21), place a check mark in the box beside a toolbar option to display that toolbar.

Command Interaction

To summarize, there are three ways to run a Rhino command.

- Select an item from the pull-down menu or the cascading menu.
- Type a command at the command line area.
- Select a button on the toolbar.

Left-click and Right-click

Normally, your pointing device (mouse) has two buttons (left and right). You use the left button to select an item from the pull-down menu, a button on a toolbar, or a check box in a dialog box, as well as to specify a location on the graphics area.

Depending on where you place your cursor, right-clicking has different effects. Some toolbars have two commands sharing a single button. For example, two commands (Zoom Extents and Zoom Extents All Views), shown in figure 1-22, share a single button on the Standard toolbar. Left-clicking the button brings out the Zoom Extents command and right-clicking the button brings out the Zoom Extents All Views command.

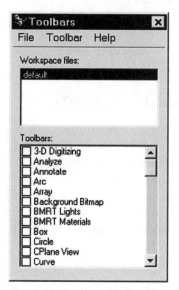

Fig. 1-21. Toolbars dialog box.

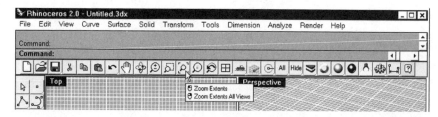

Fig. 1-22. Zoom Extents and Zoom Extents All Views commands.

When you place the cursor over the command line area and right-click, it brings out the command history (the most recent one being placed at the top). You repeat any previous command by selecting the command you want to repeat from the pop-up menu and left-clicking. Initially, the command history is blank. (See figure 1-23.)

When you place the cursor over the graphics area, you right-click in two ways. Simply right-clicking repeats the last command. Right-clicking, holding down the mouse button for a while to wait for a special symbol to appear, and releasing the button brings out the context menu. The content of the menu depends on the current context. (See figure 1-24.)

14 CHAPTER 1: An Introduction to Rhinoceros

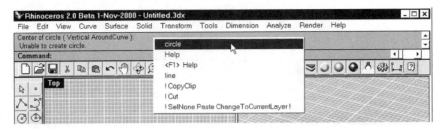

Fig. 1-23. Command history.

Fig. 1-24. Context menu.

Construction Plane Concept

When you construct a geometric object, you select a command by using the pull-down menu, toolbar, or command line area. You then select a location. To specify a location using the pointing device, you select a location in one of the viewports. To precisely specify a point, you key in a set of coordinates at the command line area. (You will learn the use of coordinate systems in the next section.)

When you pick a point in one of the viewports, you select a point on an imaginary construction plane corresponding to the selected viewport. If you pick a point in the Top viewport, you select a point on a construction plane parallel to the Top viewport and passing through the origin. Similarly, if you pick a point in the Front viewport, you select a point on the construction plane parallel to the Front viewport and passing through the origin. To construct two circles on two different construction planes, perform the following steps.

1. Select Curve > Circle > Center, Radius, or the *Circle/Center, Radius* option from the Circle toolbar.

   Command: Circle

2. Move the cursor over the Top viewport.

   To depict that the viewport is active, the label of the viewport at the top left corner is highlighted and the labels in the other viewports are grayed out. (See figure 1-25.)

3. Select a point to specify the center location.
4. Select a point to specify a point on the circumference. A circle is constructed on a construction plane parallel to the Top viewport.
5. Press the Enter key to repeat the Circle command.

Starting Rhino

Fig. 1-25. A circle being constructed on the construction plane corresponding to the top view.

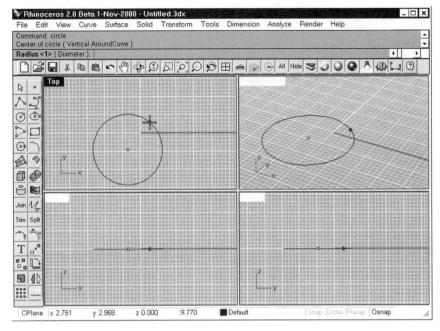

6 Move the cursor over the Front viewport.

Now the label of the Front viewport is highlighted. You are working on a construction plane parallel to the Front viewport.

7 Select a point to specify the center location. (See figure 1-26.)

Fig. 1-26. Circle being constructed on the front construction plane.

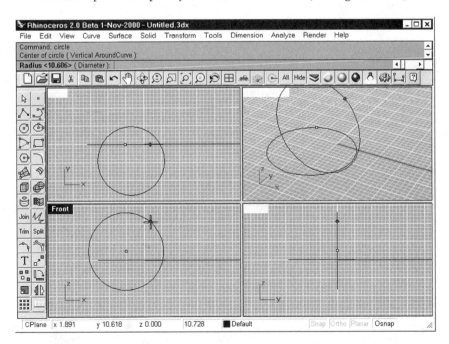

8 Select a point to indicate a point on the circumference.

A circle is constructed on a construction plane parallel to the Front viewport.

To summarize, the orientation of the objects you construct depends on the active construction plane, and the active construction plane depends on which viewport you select. In a four-viewport display, there are three construction planes. With the exception of the Perspective viewport, which has the same construction plane as the top view, each viewport has a construction plane parallel to the viewport. In addition to the default construction planes, you construct new construction planes by using one of the commands in the Set CPlane cascading menu of the View pull-down menu. (See figure 1-27.) You will learn how to set up construction planes in Chapter 9.

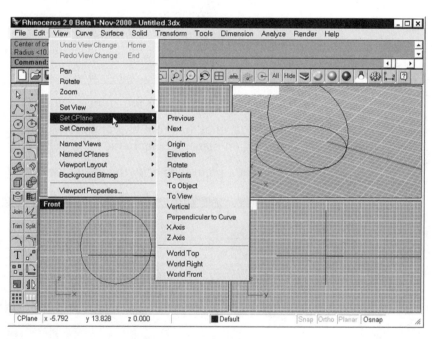

Fig. 1-27. Commands for manipulating the construction plane.

Coordinate Systems

There are two coordinate systems: the construction plane coordinate system (corresponding to the active viewport) and world coordinate system (independent of the active viewport). The construction plane coordinate system corresponds to the construction plane. In each viewport, there are a red line and a green line depicting, respectively, the X and Y axes of the construction plane in the viewport. The Z direction is perpendicular to the construction plane.

Starting Rhino

In addition to the red and green lines, there is a world axes icon at the lower left corner. The world axes icon is in the shape of a tripod. The lines on the tripod depict the absolute X, Y, and Z axes of the world coordinate system.

To specify a location at the command line area, you use a Cartesian coordinate system or a polar coordinate system. In a Cartesian coordinate system, you specify the X, Y, and Z values separated by a comma. If you specify only the X and Y values, the Z value is assumed to be zero. In a polar coordinate system, you specify the distance and angular values separated by a less than (<) sign, which stands for "angle." If you specify a point relative to the last selected point, you prefix the coordinate with the letter *r*. If you want to use the world coordinate, you prefix the coordinate with the letter *w*. Coordinate systems that operate under Rhino, and what each specifies, are summarized in table 1-4.

Table 1-4: Coordinate Systems and Their Functions

Coordinate System	Example	Specifies
Construction plane Cartesian system	2,3	A point 2 units in the X direction and 3 units in the Y direction from the origin
	2,3,4	A point 2 units in the X direction, 3 units in the Y direction, and 4 units in the Z direction from the origin
Construction plane polar system	2<45	A point 2 units at an angle of 45 degrees from the origin
Relative construction plane Cartesian system	r2,3	A point 2 units in the X direction and 3 units in the Y direction from the last reference point
Relative construction plane polar system	r4<60	A point 4 units at an angle of 60 degrees from the last reference point
World Cartesian system	w3,5	A point 3 units in the absolute X direction and 5 units in the absolute Y direction from the absolute origin, regardless of the location of the current construction plane
World relative Cartesian system	wr3,6	A point 3 units in the absolute X direction and 6 units in the absolute Y direction from the last reference point, regardless of the location of the current construction plane
World polar system	w4<30	A point 4 units at an angle of 30 degrees on the absolute XY plane from the absolute origin, regardless of the location of the current construction plane
World relative polar system	wr5<45	A point 5 units at an angle of 45 degrees on the absolute XY plane from a reference point, regardless of the location of the current construction plane

The Concept of Layers

The term *layer* originates from manual drafting. It refers to overlay of clear transparent sheets. It is a grouping mechanism in which groups of objects are drawn on different transparent sheets. By removing or overlaying the sheets, you control the objects to be shown on a drawing.

In computer-aided design, layers are not physical sheets. You might say they are conceptual layers. You set up layers in a file and place objects on different layers. By turning layers on or off, you control the display of the objects on the screen. In addition to on and off, you can lock a layer so that objects placed on the layer can be seen and selected but cannot be manipulated (such as moving or erasing). To try editing a layer, perform the following steps.

1 Select Edit > Layers > Edit Layers, or select Edit Layers on the Standard toolbar. (See figure 1-28.)

Command: Layer

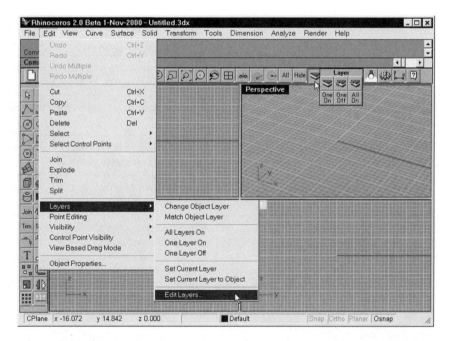

Fig. 1-28. Edit Layers command.

2 In the Edit Layers dialog box (figure 1-29), you add new layers, delete existing layers, set visibility of layers, lock objects in layers, and define the color and material properties of objects placed in a layer.

Starting Rhino

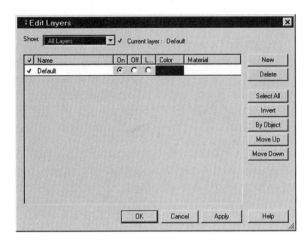

Fig. 1-29. Edit Layers dialog box.

3 To set the color of objects in a layer, select a layer and then click in the Color column of the Edit Layers dialog box. (See figure 1-29.)

4 In the Select Color dialog box (figure 1-30), select a color by selecting a color in the color swatch or specifying the color code in terms of the color's hue, saturation, and value (i.e., red, green, and blue, respectively).

5 In addition to color, you set the material of the object by selecting the layer and then the material column of the Edit Layers dialog box, which accesses the Material Properties dialog box. (See figure 1-31.) Material concerns the display of objects when the display is rendered. You will learn more about material in Chapter 10.

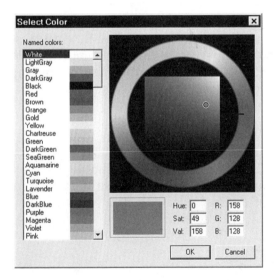

Fig. 1-30. Select Color dialog box.

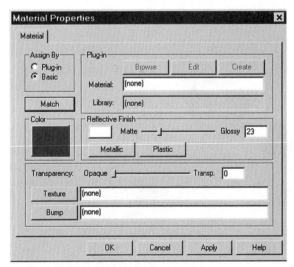

Fig. 1-31. Material Properties dialog box.

Drawing Aids

Rhino offers several drawings aids: grid mesh, snapping to grid, planar mode, elevation mode, ortho mode, and object snap. These aids are described in the sections that follow.

Grid Mesh

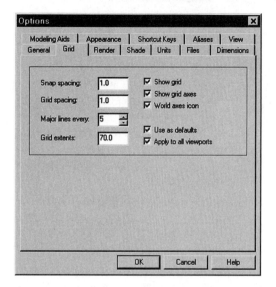

Grid mesh of known spacing on the screen gives you a sense of the actual size of the current viewport. To control the spacing of the grid mesh, use the Options command by selecting Tools > Options and then selecting the Grid tab in the Options dialog box. (See figure 1-32.)

In the Grid tab of the Options dialog box, you control the display and set the spacing of the grid lines. In addition, you control the display of the world axes icon. To quickly turn on or off the grid mesh in a viewport, press the F7 key.

Fig. 1-32. Grid tab of the Options dialog box.

Snap to Grid

The grid display shown in the viewport is for visual reference only. Without the use of further aids, it is virtually impossible to select these points precisely using the pointing device. To restrict the movement of the cursor so that it will stop only at the grid intervals, you use the Snap command, or select or deselect the Snap button on the status bar.

Planar Mode

Because the active construction plane depends on the current working viewport, you may get an unexpected outcome if you change the active viewport in the middle of an active command. To restrict the current construction plane to that of the last selected point, you use planar mode via the Planar command, or by selecting or deselecting the Planar button on the status bar.

Elevator Mode

To move the cursor perpendicularly to the current construction plane, hold down the Ctrl key and pick, and then drag the mouse.

Ortho Mode

To restrict cursor movement in a specified angular direction, you select or deselect Ortho on the status bar. To set the angular intervals, you use the Modeling Aids tab of the Options dialog box. (See figure 1-33.)

Starting Rhino

Fig. 1-33. Modeling Aids tab of the Options dialog box.

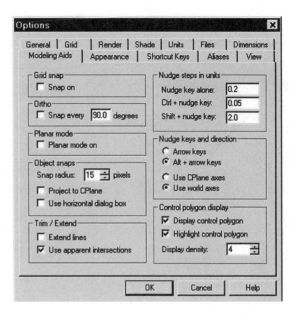

Object Snap

To help locate the cursor to selected features of existing geometric objects, you use object snap. Features you snap to are end, near, point, midpoint, center, intersection, perp, tan, quad, and knot.

There are two ways to set object snap. You can temporarily set object snap mode (i.e., for each snap to a feature) by specifying the snap mode before you select an object. You set persistent object snap by selecting Osnap on the status bar to use the Snap command. (See figure 1-34.)

Fig. 1-34. Osnap dialog box.

Help System

The help system contains several categories. These are Help Topics, Frequently Asked Questions, and Introduction to Rhino, which are described in the sections that follow.

Help Topics

By selecting Help Topics from either the Standard toolbar or Help pull-down menu, you gain access to the comprehensive Help dialog box. (See figure 1-35.) Here you will find all the information you need.

Fig. 1-35. Help dialog box.

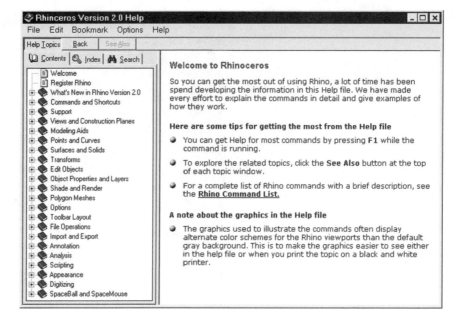

Frequently Asked Questions

Selecting Frequently Asked Questions from the Help pull-down menu brings you to the web site *http://rhino3d.com/faq/*. (See figure 1-36.)

Fig. 1-36. Frequently asked questions.

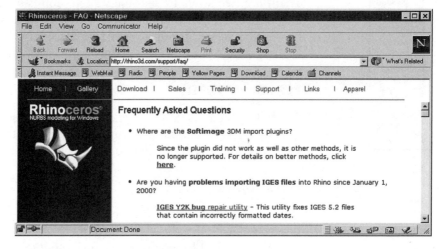

Introduction to Rhino

Select Getting Started from the Help pull-down menu to view the Introduction to Rhino dialog box. (See figure 1-37.)

Starting Rhino

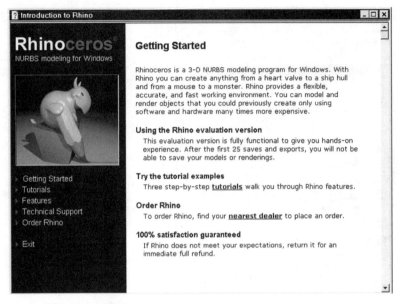

Fig. 1-37. Introduction to Rhino dialog box.

System Settings

By now, you should have a general understanding of Rhino's user interface and various drawing aids. Before you proceed to the other chapters to learn how to construct various types of objects, spend some time learning the meaning of various settings in the Options dialog box. These options are discussed in the sections that follow.

General Setting

The General tab contains eight fields: Mouse group select, Popup menu, Command lists, Undo, Default surface isoparm density, Dragging, Middle mouse button, and Right mouse button. (See figure 1-38.)

Shortcut Keys

Using the Keyboard tab, you can assign shortcut keys to the commands you use most frequently. (See figure 1-39.)

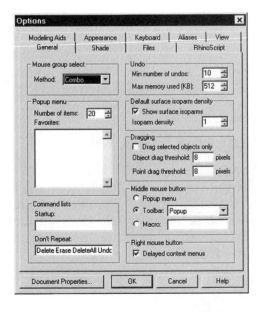

Fig. 1-38. General tab.

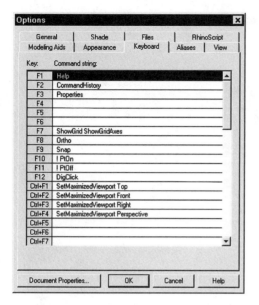

Fig. 1-39. Keyboard tab.

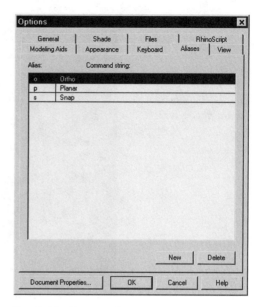

Fig. 1-40. Aliases tab.

Command Aliases

The Aliases tab enables you to set an alias for sets of command strings. (See figure 1-40.)

File Location

To set the template file and auto-save file location, you use the Files tab. (See figure 1-41.)

Appearance

In the Appearance tab, you set the color and appearance of the user interface. (See figure 1-42.)

File Properties Dialog and Units Tab

Before you start constructing a model, you should check the units of measurement so that the model you construct is compatible with any upstream and downstream operations. Select Files > Properties to display the Document Properties dialog box. Then select the Units tab. (See figure 1-43.) Here, you set the units of measurement and the tolerance of the model. In addition, you set the display tolerance

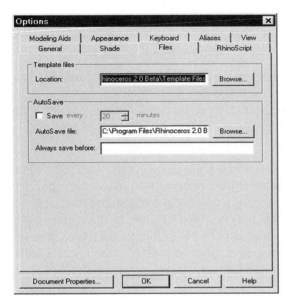

Fig. 1-41. Files tab.

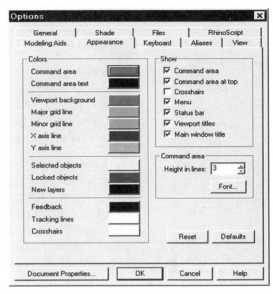

Fig. 1-42. Appearance tab.

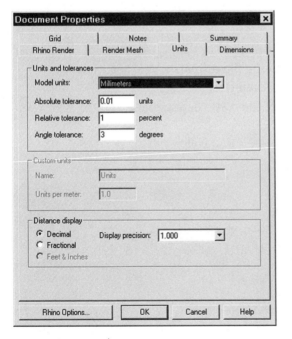

Fig. 1-43. Units tab.

■■ Summary

In this chapter you learned that computer modeling is a representation of an object in the computer. You also learned that there are three types of computer models: wireframe, surface, and solid. A wireframe model is a set of unassociated curves in 3D space that does not have any surface or volume information. A surface model is a set of assembled surfaces and is best suited to representing free-form objects. A solid model is the most comprehensive of the three types, containing information about the vertices, edges, surfaces, and volume of the object it represents.

Rhino is a 3D computer modeling tool. You use it to construct points, curves, surfaces, polysurfaces, solids, and polygon meshes. In addition, you output photo-realistic renderings, 2D drawings, and file formats of various types for downstream computerized operations. You also reuse upstream computer models by opening various file formats.

The Rhino user interface contains five major areas: standard Windows title bar, pull-down menu, command line area, graphics area, and status bar. In addition, there are a number of toolbars. To use a Rhino command, you use the pull-down menu, command line area, or toolbar. The graphics area is where you construct computer models.

When you use the mouse to select a point in one of the viewports in the graphics area, you select a point on a construction plane corresponding to the selected viewport. To input a point at the command line area, you use either the construction plane coordinate system or the world coordinate system. To help you construct objects in the graphics area, you use various drawing aids. Rhino objects are organized into layers. Objects you construct are placed on the current layer.

Review Questions

1. Using simple sketches, illustrate the three types of computer models.
2. What types of objects can you construct using Rhino?
3. Give a brief account of the file formats supported by Rhino.
4. Describe the Rhino user interface.
5. Explain the concepts of the construction plane and the two types of coordinate systems.
6. Outline the content of the Rhino Help system.

Chapter 2

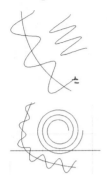

Wireframe Modeling and Curves: Part 1

■■ Objectives

The goals of this chapter are to introduce the key concepts of wireframe modeling, to explain the importance of 3D curves in solid and surface modeling, and to illustrate various methods of constructing basic Rhino curves. After studying this chapter, you should be able to:

❐ State the characteristics of a 3D wireframe model

❐ Appreciate the importance of 3D curves in making 3D solids and surfaces

❐ Construct basic 3D curves using Rhino

■■ Overview

As you learned in Chapter 1, there are three basic ways of representing an object in the computer: wireframe modeling, surface modeling, and solid modeling. Chapters 2 through 4 are written to let you gain a deep understanding of wireframe modeling. In this chapter you will examine the characteristics of wireframe modeling, explore the significance of curves in surface and solid modeling, and learn how to use Rhino as a tool to construct various types of basic curves.

In the next two chapters, you will learn more about Rhino curve tools and work on various curve construction projects. After equipping yourself with basic skills in 3D curve manipulation, in chapters 5 through 8 you will learn how to construct surfaces and solids, and in Chapter 9 you will consolidate your learning on curves, surfaces, and solids.

■■ Wireframe Modeling Concepts

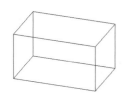

Fig. 2-1. Wires representing a rectangular block.

A 3D wireframe model represents an object by using curves to define the object's edges. Figure 2-1 shows the wireframe model of a rectangular block. In essence, this model has eight separate, unrelated curves. Between the curves, there is no information.

Representation of 3D Object

To illustrate curved surfaces in a wireframe model, such as a cylinder or a sphere, you construct additional curves to depict the contour of the surface. Figure 2-2 shows the wireframe model of a cylinder.

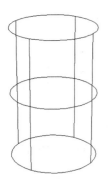

Fig. 2-2. Contour wires representing the curved surface of a cylinder.

Because all curves in a 3D wireframe model are independent entities and there is no information between them, a wireframe model represents an object only implicitly. To perceive a 3D object by viewing a wireframe model, you combine your visual perception of the curves with added meaning from your imagination on how the curves relate to one another. Hence, the wireframe model shown in figure 2-1 has several meanings. It can be simply a set of curves, just a framework. It can also be a box with an opening in one of its sides. (See figure 2-3.) Furthermore, it also implies a solid block. (See figure 2-4.)

Fig. 2-3. A box with an opening.   Fig. 2-4. A solid box.

Construction of Wireframe Models

Prior to constructing the wireframe model of an object, you "deconstruct" the object into discrete curves by thinking about the outlines and silhouettes of the object and the edges of the object where two faces meet. To make the model, you construct the curves in accordance with your perception of the object's appearance. Because making wireframe models requires you to determine the locations of the vertices of the curves, the task of wireframe construction is tedious. Despite the laborious work required, objects are not completely represented by wireframe

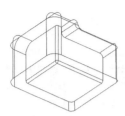

Fig. 2-5. Wireframe model of a complex object.

models. For example, construction of the fillet edges of the model shown in figure 2-5 is very time consuming.

Limitation of Wireframe Models

Because a wireframe model is simply a set of unrelated curves, there is no surface and volume information stored in the computer. As a result, a wireframe model has limited application in computerized downstream operations such as analysis, CNC machining, and rapid prototyping.

Curves for Surface and Solid Modeling

Although the use of curves alone in computer modeling is diminishing, you still need to learn how to construct 3D curves because they are required in making surfaces and solids; in particular, in free-form surface modeling.

Curves for Surface Modeling

There are many ways to construct surfaces in the computer. A fundamental way to make free-form surfaces is to use a set of curves to define the silhouette and contour of the surface and let the computer construct a surface patch on the curves. (You will learn surface modeling in chapters 5 through 7.) Figure 2-6 shows four 3D curves and a surface constructed from the curves.

Fig. 2-6. Curves and surface constructed.

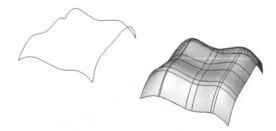

Curves for Solid Modeling

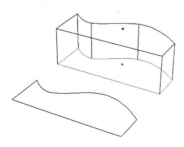

Among the many ways to represent a solid in the computer, one basic way is to construct a curve in a closed loop and sweep the curve in 3D space. (You will learn solid modeling in Chapter 8.) Figure 2-7 shows a closed-loop curve being extruded to form a solid.

Fig. 2-7. Extruding a closed-loop curve to construct a solid.

■■ Curves

To prepare yourself for making 3D solids and surfaces, you need to understand the characteristics of the curves and ways to construct them. Among the many types of curves you can use to create 3D surfaces and solids, the spline is the most important because it enables you to construct complex free-form shapes. Spline curves are discussed in the section that follows.

Spline Curve

A spline is a way of defining a free-form curve by specifying two end points and two or more tangent vectors that control the profile of the curve. There are many mathematical ways to define a spline. Some of the most popular ways are outlined in the sections that follow.

Polynomial Spline

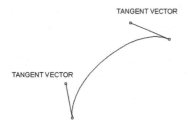

Fig. 2-8. Polynomial spline segment and its tangent vectors.

A polynomial spline is a set of spline segments. At the end points of each spline segment there is a tangent vector having a direction and magnitude. (See figure 2-8.) The effect of the tangent defines the curvature of the segment. Because the tangent vector is described by a polynomial equation, the spline is called a polynomial spline. How the tangent vector affects the shape of the spline segment is determined by the degree of the polynomial equation. The overall shape of a polynomial spline is the combined effect of all of its segments.

B-spline

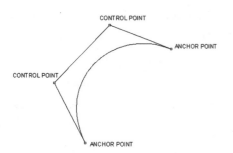

Fig. 2-9. B-spline segment and its control points.

A B-spline is also a multi-segment spline curve. A connection point between two contiguous spline segments is called a knot. Being an extension of the polynomial spline, each spline segment is formed from the weighted sum of four local polynomial basis functions. Hence, the spline is called a basis spline (or B-spline). The shape of a B-spline segment is controlled by four control points lying outside the spline. Movement of a control point affects only four segments of the curve. (See figure 2-9.)

Curves

Non-uniform Rational B-spline (NURBS)

A non-uniform rational B-spline curve is a derivative of the B-spline curve. It has two additional characteristics: non-uniform and rational. Unlike the uniform B-spline, in which each spline segment is defined by a uniform parameter domain, the parameter domain of a non-uniform B-spline need not be uniform.

Because of the non-uniform characteristic, different levels of continuity between the spline segments can be attained, unequal spacing between the knot points is allowed, and the spline can more accurately interpolate among a set of given points. With a rational form, the curve can better represent conic shapes. Using NURBS mathematics, you can trim a curve at any point and the curve will retain its original shape. NURBS curves and surfaces are used in most surface modeling tools.

Rhino Curves

Rhino uses NURBS mathematics to define curves and surfaces. To reiterate, a spline curve is a set of connected spline segments, and the joint between two contiguous spline segments is a knot. The degree of the polynomial basis equation, the control point location, and the weight of the control points determine the shape of each spline segment.

Polynomial Degree

The degree of a polynomial equation has a direct impact on the complexity of the curve's shape. For example, a line is a degree 1 NURBS curve, a circle is a degree 2 NURBS curve, and a free-form curve is a NURBS curve of degree 3 or above. You can raise or reduce the polynomial degree of a curve. Raising the degree of a polynomial spline curve does not change the curve's shape but does increase the number of control points.

In turn, more control points enable you to modify the curve to create a more complex shape. On the other hand, reducing the degree decreases the number of control points and hence simplifies the shape of the curve. Figure 2-10 shows curves with increasing degrees of polynomial equation.

Fig. 2-10. Degree 2, 3, and 4 curves.

Control Point Location

The location of control points directly affects the contour of a curve. Moving control points changes the curve's shape. (See figure 2-11.) Normally, control points lie outside the curve. In an open-loop curve, only the first and last control points coincide with the end points of the curve. In a closed-loop curve, all control points lie outside the curve. (See figure 2-12.)

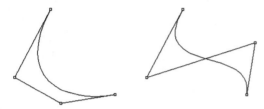

Fig. 2-11. Control point locations affecting the shape of the curve.

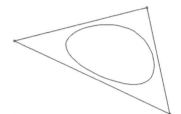

Fig. 2-12. Control points in a closed-loop curve.

Control Point Weight

The weight of control points also has a significant effect on the shape of a spline segment. You can regard the weight of a control point as a pulling force that pulls the spline curve toward the control point. The higher the weight, the closer the curve will be pulled to the control point. (See figure 2-13.)

Knot

A spline curve is a set of spline segments, and a knot defines the junction between two contiguous spline segments. Adding knots to a curve increases the number of spline segments without changing the shape of the curve. However, having more segments means that the curve has more control points. Subsequently, you can modify the curve to create a more complex shape. Figure 2-14 shows a knot added to a curve.

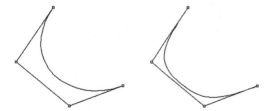

Fig. 2-13. Control points with different weights (from left to right, weight 1 and weight 7).

Fig. 2-14. More control points with additional knot (right).

Curves

Kink Point

A kink point is a special type of knot on a curve in which the tangent direction of the contiguous spline segments is not the same. A kink occurs when you join two curves with different tangent directions, or when you explicitly add a kink to a curve. Figure 2-15 shows the effect of moving a kink point of a curve.

Fig. 2-15. Result of kink point moved.

Periodic Curve

A closed curve with no kink point is called a periodic curve. Figure 2-16 shows a closed-loop curve with a kink and a periodic curve.

Edit Point

Because control points are normally not lying along the curve, you may find it difficult to make a curve pass through designated locations by manipulating the control point. To add flexibility in modifying a NURBS curve, Rhino enables you to use edit points along the curve. Edit points are independent of degree, control points, and knots of a curve. You move the edit points to change the shape of the curve. (See figure 2-17.)

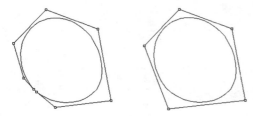

Fig. 2-16. Closed-loop curve with a kink and a periodic curve.

Fig. 2-17. Curve modified by moving edit points.

Handlebar Editor

Fig. 2-18. Use of the handlebar.

Edit points enable you to modify a curve only in a limited scope. To change the tangent direction and the location of any point along a curve, you use the handlebar, which is a special type of editing tool. It consists of a point on the curve and two tangent lines. Moving the central point of the handlebar changes the location of a point along the curve. Selecting and dragging the end points of the handlebar changes the weight and tangent direction of the curve at a selected location. (See figure 2-18.)

■■ Rhino Curve Construction Tools

Using Rhino, you construct point objects, basic curves, and derived curves, and edit and transform curves. To examine existing curves you use analysis tools.

Command Menu

As mentioned in Chapter 1, you use Rhino commands in three ways: by selecting an item from the pull-down menu, by selecting a button on a toolbar, and by typing a command name at the command line area. You will find the tools for making point objects and NURBS curves in the Curve pull-down menu. To edit and transform curves, you use the Edit and Transform pull-down menus. To examine and evaluate curves, you use the Analyze pull-down menu. These pull-down menu items are shown in figure 2-19.

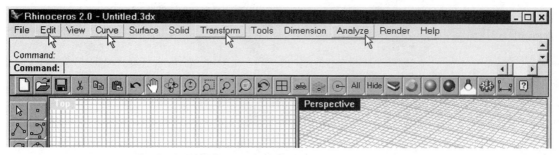

Fig. 2-19. Pull-down menu items.

Rhino Curve Construction Tools

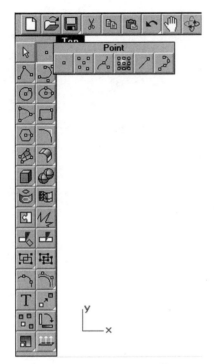

Fig. 2-20. Dragging the Point fly-out toolbar.

Toolbars concerning curve construction, editing, transforming, and analysis are integrated in the Main toolbar, which is displayed by default. Selecting a button from the toolbar that has a small white triangle at the lower right corner and holding down the left mouse button will bring out a fly-out toolbar. By selecting and dragging the header bar of a fly-out toolbar, you put the fly-out toolbar on your desktop as a floating toolbar. Figure 2-20 shows the Main toolbar and the Point fly-out toolbar.

Input for Point and Basic Curve Construction

To construct point objects and basic curves, you use a command and specify a location. To specify a location, you use your pointing device to select a point in one of the viewports, or input a set of coordinates at the command line area.

As explained in Chapter 1, there is a construction plane in each of the viewports. Selecting a point in one of the viewports specifies a location on the corresponding construction plane. To key in a set of coordinates, you use either the construction plane coordinate system or the world coordinate system.

The X and Y directions of the construction plane coordinates correspond to the X and Y axes of the construction plane. Therefore, typing the same construction plane coordinates (other than the origin) at the command line area results in different locations in 3D space depending on where the cursor is placed; for example, on the Top viewport as opposed to the Front viewport. In the following tutorials, you will use the pointing device to select points in viewports.

Constructing Point Objects and Basic Curves

A point is a node. Basic curves follow basic geometric patterns. There are eleven types of basic curves: line, free-form curve, circle, arc, ellipse, parabola, polygon, rectangle, conic, helix, and spiral. (See figures 2-21 through 2-23.)

Fig. 2-21. Line, circle, arc, polygon, and rectangle (left to right).

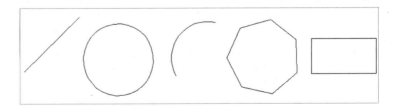

Fig. 2-22. Free-form curve.

Fig. 2-23. Ellipse, parabola, conic, helix, and spiral (left to right).

Points

Figure 2-24 shows the Point toolbar. Using point objects, you specify locations in 3D space, define vertices and definition points of curves, and portray locations on a surface. Point construction commands are discussed in the sections that follow.

Fig. 2-24. Point toolbar.

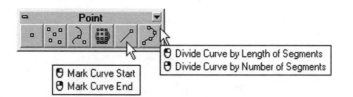

Point Construction Commands. Table 2-1 outlines the functions of point construction commands and their location in the pull-down menu.

Table 2-1: Point Construction Commands and Their Functions

Command Name	Pull-down Menu	Function
NOTE: Command name entry is not case sensitive. You can use any combination of small and capital letters.		
Point	Curve > Point Object > Single Point	For constructing a single point object
Points	Curve > Point Object > Multiple Points	For constructing a series of point objects
ClosestPt	Curve > Point Object > Closest Point	For constructing a point on a curve that is closest to a specified location
DrapePt	Curve > Point Object > Drape Point	For constructing a matrix of points on the surfaces described by a rectangle
CrvStart	Curve > Point Object > Mark Curve Start	For constructing a point at the start point of a curve

Rhino Curve Construction Tools

Command Name	Pull-down Menu	Function
CrvEnd	Curve > Point Object > Mark Curve End	For constructing a point at the end point of a curve
DivideByLength	Curve > Point Object > Divide Curve by > Length of Segments	For constructing a series of points along a curve by specifying the distances between consecutive points
Divide	Curve > Point Object > Divide Curve by > Number of Segments	For constructing a series of points along a curve by specifying the number of segments

Multiple Points. To construct a set of point objects for making curves, perform the following steps.

1. Start a new file. Use the metric (millimeter) template file.
2. Double click on the Top viewport title to maximize the viewport.
3. Select Curve > Point Object > Multiple Points, or the Multiple Points button from the Point toolbar.
   Command: Points
4. Select four locations in accordance with figure 2-25. (The exact location of points in this tutorial is unimportant.)

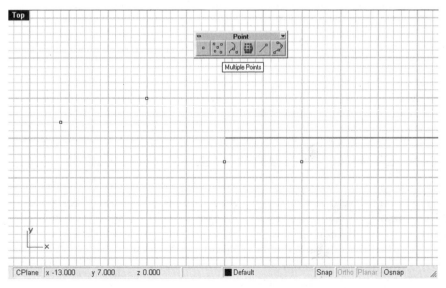

Fig. 2-25. Four point objects constructed.

5. Press the Enter key to terminate the command.

Four point objects are constructed. Save your file as *Point1.3dm*. You will use these points in line construction later.

Line

A line segment is a degree 1 curve. There are 15 ways to construct a line. Figure 2-26 shows the Lines toolbar.

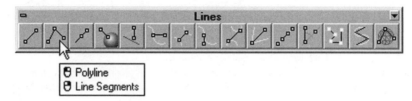

Fig. 2-26. Lines toolbar.

Table 2-2 outlines the functions of line construction commands and their location in the pull-down menu.

Table 2-2: Line Construction Commands and Their Functions

Command Name	Pull-down Menu	Function
Line	Curve > Line > Single Line	For constructing a single line segment by specifying two end points or a midpoint and an end point
Lines	Curve > Line > Line Segments	For constructing a string of line segments
Polyline	Curve > Line > Polyline	For constructing a string of connected line segments
LinePerp	Curve > Line > Perpendicular from Curve	For constructing a line perpendicular to a selected curve
LinePP	Curve > Line > Perpendicular to 2 Curves	For constructing a line perpendicular to two selected curves
LineTan	Curve > Line > Tangent from Curve	For constructing a line tangent to a selected curve
LineTT	Curve > Line > Tangent to 2 Curves	For constructing a line tangent to two selected curves
LineAngle	Curve > Line > Angled	For constructing a line at a specified angle to a reference line
Bisector	Curve > Line > Bisector	For constructing a line bisecting two selected reference lines
Line4Pt	Curve > Line > From 4 Points	For constructing a line by using two reference points to specify a direction and selecting two end points on the direction line

Rhino Curve Construction Tools

Command Name	Pull-down Menu	Function
Normal	Curve > Line > Normal to Surface	For constructing a line normal to a selected surface
LineV	Curve > Line > Vertical to CPlane	For constructing a line normal to the current construction plane
PolylineThroughPt	Curve > Line > Polyline Through Points	For fitting a polyline along a set of point objects
PolylineOnMesh	Curve > Line > Polyline on Mesh	For constructing a polyline on the face of a polygon mesh
ConvertToPolyline	Curve > Line > Convert Curve to Polyline	For constructing a polyline along a curve

Line Segments. To construct a single line segment by specifying two points, perform the following steps.

1 Start a new file. Use the metric (millimeter) template.

2 Select Curve > Line > Single Line, or the Line button from the Lines toolbar.

Command: Line

3 Select a point in the Top viewport and a point in the Front viewport. (See figure 2-27.)

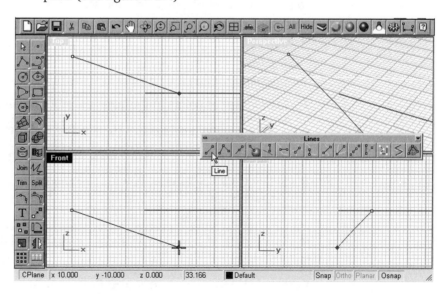

Fig. 2-27. Single line segment constructed.

To construct a line perpendicular to the construction plane of the Top viewport, continue with the following steps.

4 Select Curve > Line > Vertical to CPlane, or the Vertical Line button from the Lines toolbar.

Command: LineV

5 Select a point in the Top viewport to specify a location, and select a point in the Front viewport to specify the height. (See figure 2-28.)

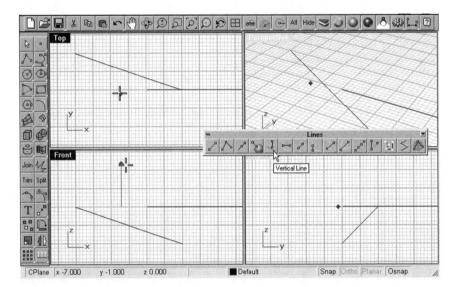

Fig. 2-28. Line perpendicular to the construction plane.

To construct a set of connected line segments, continue with the following steps.

6 Select Curve > Line > Line Segments, or the Line Segments button from the Lines toolbar.

Command: Lines

7 Select a set of points in the Top viewport and press the Enter key.

A set of connected line segments is constructed. (See figure 2-29.)

To construct a polyline, continue with the following steps.

8 Select Curve > Line > Polyline, or the Polyline button from the Lines toolbar.

Command: Polyline

9 Select a set of points in the Right viewport and press the Enter key. (See figure 2-30.)

Connected Line Segments and Polyline. At a glance, the connected line segments and the polyline seem to be similar. In fact, the polyline is a single object and the connected line segments are separate line segments with their end points coincident with each other. To separate the

Rhino Curve Construction Tools

Fig. 2-29. Set of connected line segments constructed.

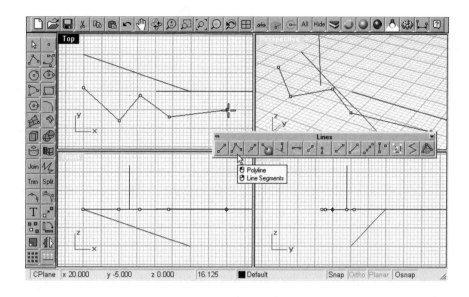

Fig. 2-30. Polyline constructed.

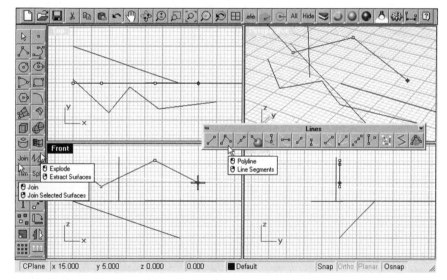

polyline into individual line segments, you explode it. To convert a set of connected line segments to a single polyline, you join the line segments. Perform the following steps.

1   Select Edit > Join, or the Join button from the Main toolbar.
Command: Join

2   Select the connected line segments and press the Enter key.

The line segments are joined to become a polyline.

3 Select Edit > Explode, or the Explode button from the Main toolbar.
   Command: Explode

4 Select the polyline and press the Enter key.

The polyline is exploded to show individual line segments. Save your file as *Lines.3dm*.

4-Point Line. To use the four point objects to draw a line, perform the following steps.

1 Open the file *Point1.3dm*.

2 Check the Osnap pane on the status bar to display the Osnap dialog box.

3 In the Osnap dialog box, check the Point box to set object snap mode to Point. This way, the cursor will snap to point objects.

4 Select Curve > Line > From 4 Points, or the Line by 4 Points button from the Lines toolbar.
   Command: Line4Pt

5 Select the two points (from left to right) indicated in figure 2-31 to indicate a direction.

Fig. 2-31. Points selected to indicate a direction.

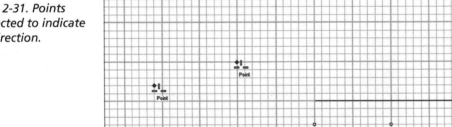

6 Select the point highlighted in figure 2-32 to indicate the start point of the line, and select the point highlighted in figure 2-33 to indicate the end point.

Fig. 2-32. Point indicating the start point of the line.

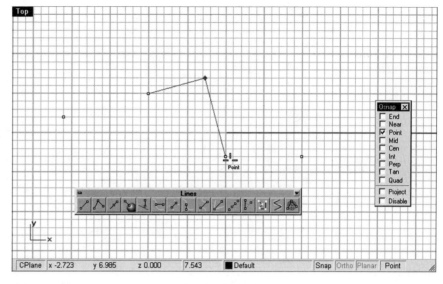

Fig. 2-33. Point indicating the end point of the line.

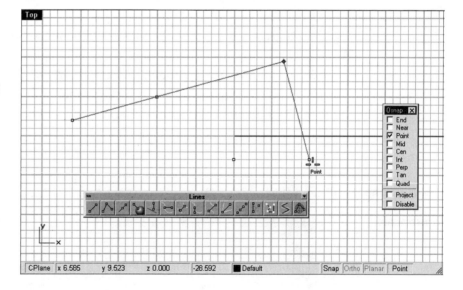

A line is constructed. (See figure 2-34.) Save your file.

Fig. 2-34. Line constructed.

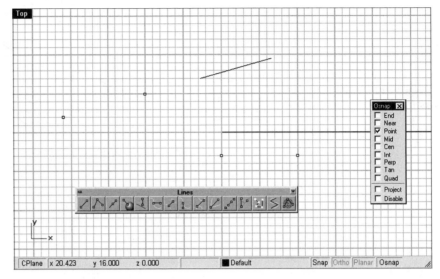

Line at an Angle. To construct a line at an angle to a reference line, perform the following steps.

1. Start a new file. Use the metric (millimeter) template.
2. Maximize the Top viewport.
3. Check the Osnap pane in the status bar.
4. In the Osnap dialog box, check the End box.
5. With reference to figure 2-35, construct a line segment.

Fig. 2-35. Line segment constructed.

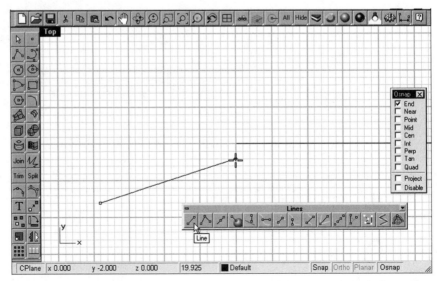

6. Select Curve > Line > Angled, or the Angled Line button from the Lines toolbar.

   Command: LineAngle

7 Select the end points of the line segment. (See figure 2-36.)
8 Type *25* to specify an angle.
9 Type *10* to specify the length of the line.

A line at 25 degrees from the selected line is constructed.

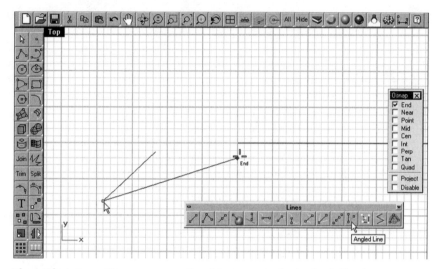

Fig. 2-36. Line at angle being constructed.

Line Bisector. To construct a line bisector, perform the following steps.

1 Select Curve > Line > Bisector, or the Bisector Line button from the Lines toolbar.
Command: Bisector

2 Select the end points indicated in figure 2-37 to indicate the start of the bisector line and the start of the angle to be bisected.

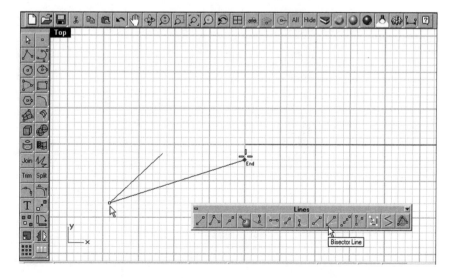

Fig. 2-37. Start of bisector and start of angle to be bisected specified.

46 CHAPTER 2: Wireframe Modeling and Curves: Part 1

3 Select the end point indicated in figure 2-38 to indicate the end of the angle to be bisected.

4 Type *20* to specify the length of the line.

A line bisector is constructed.

Fig. 2-38. End of angle to be bisected specified.

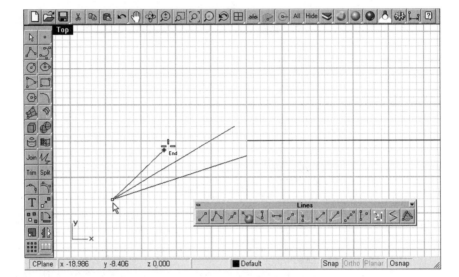

Polyline Through a Set of Points. To construct a polyline passing through a set of points, perform the following steps.

1 With reference to figure 2-39, construct a set of point objects.

Fig. 2-39. Points constructed.

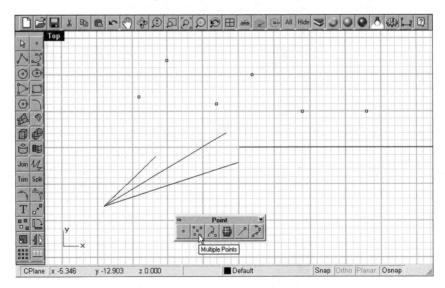

Rhino Curve Construction Tools

2 Select Curve > Line > Polyline Through Points, or the Polyline Through Points button from the Lines toolbar.

Command: PolylineThroughPt

3 Select the points and press the Enter key.

A polyline is constructed. (See figure 2-40.) Save your file as *LinePt.3dm*.

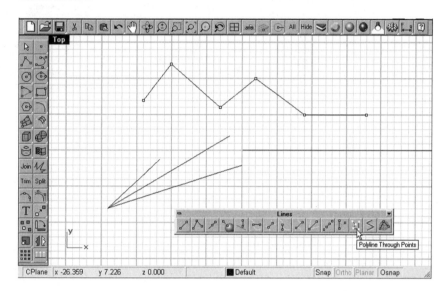

Fig. 2-40. Polyline constructed.

Free-form Curve

A flexible way to define a curve for making free-form surfaces is to use a free-form curve. A free-form curve is a degree 3 NURBS curve. There are seven ways to define a free-form curve. Figure 2-41 shows the Curve toolbar, which contains buttons for seven types of free-form curves, as well as conic, parabola, helix, and spiral options.

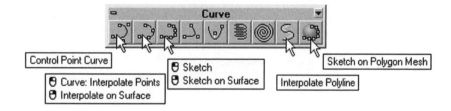

Fig. 2-41. Curve toolbar.

Table 2-3 outlines the functions of free-form curve commands and their location in the pull-down menu.

Table 2-3: Free-form Curve Commands and Their Functions

Command Name	Pull-down Menu	Function
Curve	Curve > Free-form > Control Points	For constructing a spline curve by specifying control points
InterpCrv	Curve > Free-form > Interpolate Points	For constructing a spline curve passing through specified points
Sketch	Curve > Free-form > Sketch	For using the pointing device (mouse) as a drawing pen to sketch a spline curve
InterpPolyline	Curve > Free-form > Interpolate Polyline	For constructing a spline curve that interpolates along the end points of the line segments of a polyline
InterpCrvOnSrf	Curve > Free-form > Interpolate on Surface	For constructing a spline curve on a selected surface
SketchOnSrf	Curve > Free-form > Sketch on Surface	For using the pointing device (mouse) as a drawing pen to sketch a spline curve on a selected surface
SketchOnMesh	Curve > Free-form > Sketch on Polygon Mesh	For using the pointing device (mouse) as a drawing pen to sketch a spline curve on a selected polygon mesh

In the following, you will construct a series of point objects. You will then construct two free-form curves by specifying control points and points through which the curve has to interpolate. To create the series of point objects, perform the following steps.

1 Start a new file. Use the metric (millimeter) template file.
2 Double click on the Top viewport to maximize the viewport.
3 Select Curve > Point Object > Multiple Points, or the Multiple Points button from the Point toolbar.
   Command: Points
4 Select four locations to construct four points in accordance with figure 2-42 and press the Enter key.

Four points are constructed.

To copy the points using the Copy command, perform the following steps.

5 Select Transform > Copy, or the Copy button from the Transform toolbar.
   Command: Copy

Rhino Curve Construction Tools

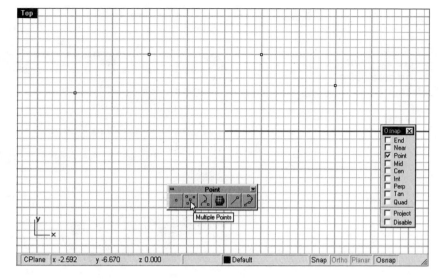

Fig. 2-42. Four points constructed.

6 Select all points and press the Enter key.

7 Select a point at any position as the point to copy from. This is the base point of copy. (See figure 2-43.)

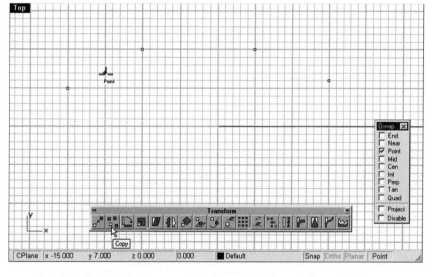

Fig. 2-43. Base point of copy selected.

8 Select a point as the point to copy to. Distance and direction between this point and the base point indicate the location of the copied objects in relation to the original objects. (See figure 2-44.)

9 Press the Enter key to terminate the command.

The points are copied.

Fig. 2-44. Reference point to the base point selected.

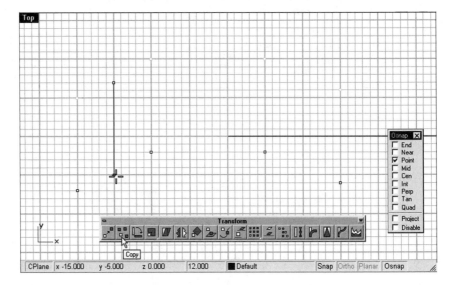

Control Point Curve. To construct a free-form curve by specifying control points, perform the following steps.

1. Check the Osnap button on the status bar.
2. In the Osnap dialog box, check the Point box.
3. Select Curve > Free-form > Control Points, or the Control Point Curve button from the Curve toolbar.

   Command: Curve

4. Select the points (from left to right) indicated in figure 2-45.

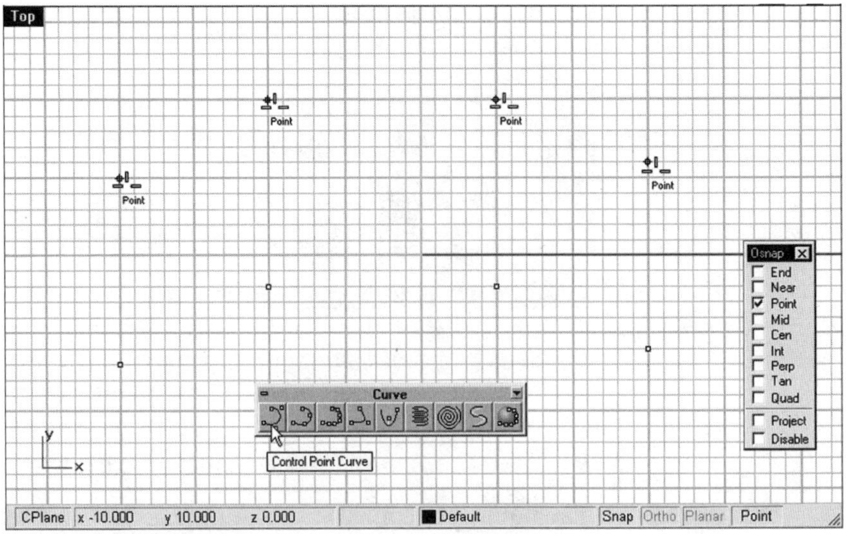

Fig. 2-45. Points selected (from left to right).

5. Press the Enter key to terminate the command.

Rhino Curve Construction Tools

A free-form curve is constructed.

Interpolate Curve. To construct a free-form curve to interpolate selected points, perform the following steps.

1 Select Curve > Free-form > Interpolate Points, or the Curve/Interpolate Points button from the Curve toolbar.
   Command: InterpCrv

2 Select the points (from left to right) indicated in figure 2-46.

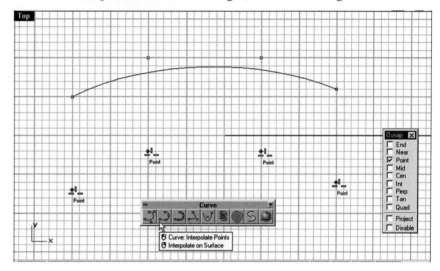

Fig. 2-46. Interpolation points being selected.

3 Press the Enter key to terminate the command.

Two free-form curves are constructed. (See figure 2-47.) Note the difference in shape.

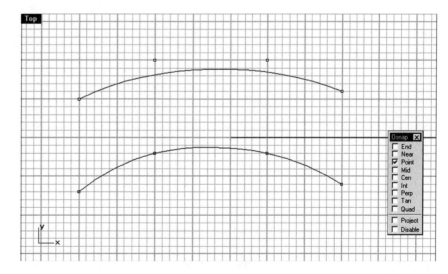

Fig. 2-47. Two free-form curves constructed.

Chapter 2: Wireframe Modeling and Curves: Part 1

Change Object's Layer. Now the point objects are not required. To keep these objects and hide them in the display, change the free-form curves to *Layer 01*, set the current layer to *Layer 01*, and turn off the default layer.

1 Select Edit > Layers > Change Object Layer, or the Change Layer button from the Standard toolbar.

   Command: ChangeLayer

2 Select the curves and press the Enter key.

3 In the Layer for Object dialog box, select *Layer 01* and the OK button. (See figure 2-48.)

The curves are moved to *Layer 01*. Now set the current layer to *Layer 01* and turn off the default layer where the point objects reside.

Fig. 2-48. Layer for Object dialog box.

4 Select Edit > Layers > Edit Layers, or the Edit Layers button from the Standard toolbar. (See figure 2-49.)

   Command: Layer

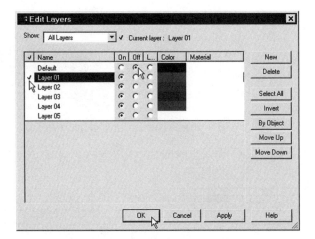

Fig. 2-49. Edit Layers dialog box.

5 In the Edit Layers dialog box, select the Off button of the *Default* layer to turn it off, select the check mark of *Layer 01* to set it as the current layer, and select the OK button to exit.

To appreciate the difference between the two free-form curves, you will turn on their control points.

6 Select Edit > Point Editing > Control Points On, or the Control Points On button from the Point Editing toolbar.

   Command: PtOn

7 Select the curves and press the Enter key.

Control Point. The control points are turned on. (See figure 2-50.) The upper curve (curve constructed by specifying four control points) has the control point located at the specified location. The lower curve (curve constructed by specifying the interpolated points) has six control points, with the curve showing more segments. Now turn off the control points.

Fig. 2-50. Control points turned on.

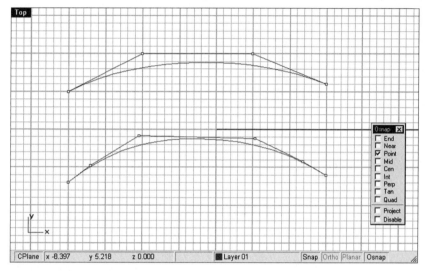

8 Select Edit > Point Editing > Points Off, or the Points Off button from the Point Editing toolbar.
Command: PtOff

9 Save the file as *Freeform1.3dm*.

Polyline and Free-form Curve. A polyline is a series of joined line segments. If you already constructed a polyline, you can construct a free-form curve to interpolate the end points of the polyline. On the other hand, if you have a free-form curve, you can convert it to a polyline.

1 Start a new file. Use the metric (millimeter) template.
2 Maximize the Top viewport.

Refer to figure 2-51 for constructing a polyline.

3 Select Curve > Free-form > Interpolate Polyline, or the Interpolate Polyline button from the Curve toolbar.
Command: InterpPolyline

4 Select the polyline and press the Enter key.

A free-form curve is interpolated along the end points of the polyline.

Fig. 2-51. Free-form curve being interpolated.

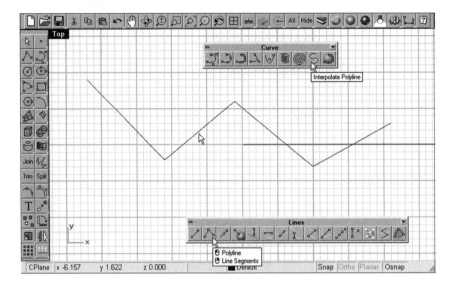

5   Select Curve > Line > Convert Curve to Polyline, or the Convert Curve to Polyline button from the Lines toolbar.

6   Select the free-form curve and press the Enter key.

7   Type *50* at the command line area to specify the angle tolerance.

8   A polyline is constructed from the free-form curve. Save your file as *PolylineCrv.3dm*. (See figures 2-52 and 2-53.)

Fig. 2-52. Polyline being constructed.

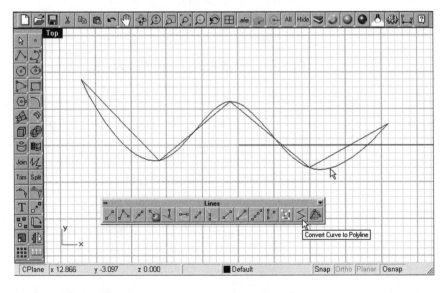

Points along Curve. To construct two free-form curves and points along those curves, perform the following steps.

1   Start a new file. Use the metric (millimeter) template.

Rhino Curve Construction Tools 55

Fig. 2-53. Polyline constructed.

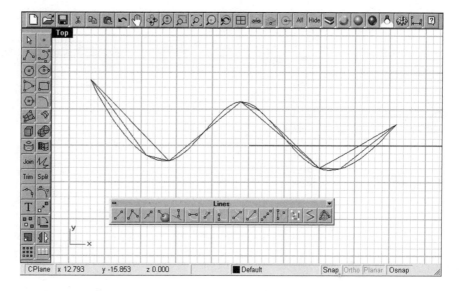

2 Maximize the Top viewport and construct two free-form curves. (See figure 2-54.)

Fig. 2-54. Curves constructed.

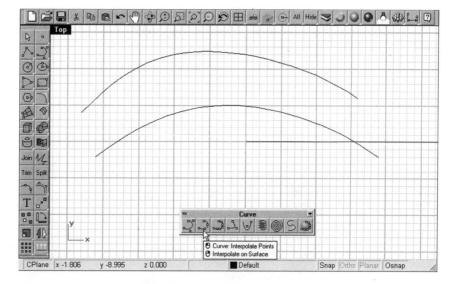

3 Select Curve > Point Object > Mark Curve Start, or the Mark Curve Start button from the Point toolbar.

Command: CrvStart

4 Select the curve indicated in figure 2-55.

5 Select Curve > Point Object > Mark Curve End, or the Mark Curve End button from the Point toolbar.

Command: CrvEnd

6 Select the curve indicated in figure 2-55.

Fig. 2-55. Point constructed at the start and end points of a curve.

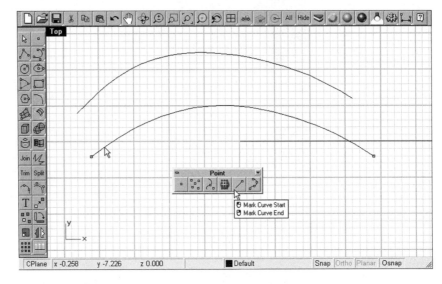

Points are constructed at the start and end points of the curve.

7 Select Curve > Point Object > Divide Curve by > Length of Segments, or the Divide Curve by Length of Segments button from the Point toolbar.

Command: DivideByLength

8 Select the curve indicated in figure 2-56 and press the Enter key.

9 Select two points to indicate the distance.

Fig. 2-56. Points constructed at specified spacing.

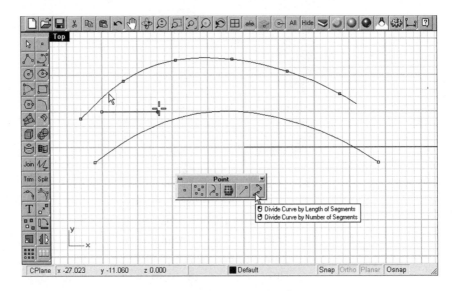

Rhino Curve Construction Tools 57

Points are constructed along the selected curve at a specified interval.

10 Select Curve > Point Object > Divide Curve by > Number of Segments, or the Divide Curve by Number of Segments button from the Point toolbar.
Command: Divide

11 Select the curve indicated in figure 2-57 and press the Enter key.

12 Type 3 at the command line area to specify the number of segments.

Two points are constructed at equal spacing along the curve, dividing the curve into three equal parts. Save your file as *PointCrv.3dm*.

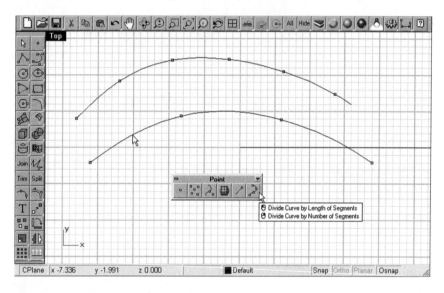

Fig. 2-57. Points constructed at equal spacing along the curve.

Line Perpendicular to a Curve. To construct a line perpendicular to a curve, perform the following steps.

1 Start a new file. Use the metric (millimeter) template.
2 Maximize the Top viewport.
3 With reference to figure 2-58, construct two free-form curves and two points in one of the curves at equal space.
4 Select Curve > Line > Perpendicular from Curve, or the Line Perpendicular from Curve button from the Lines toolbar.
Command: LinePerp
5 Select a point on the curve and a point to indicate the end point. (See figure 2-59.)

A line perpendicular to a curve is constructed.

58   Chapter 2: Wireframe Modeling and Curves: Part 1

Fig. 2-58. Curves and points constructed.

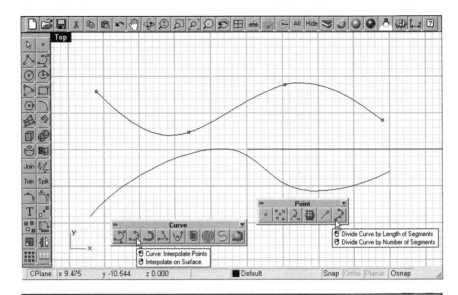

Fig. 2-59. Line being constructed perpendicular to a curve.

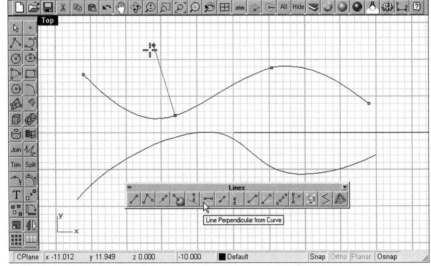

Line Perpendicular to Two Curves. To construct a line perpendicular to two curves, perform the following steps.

1   Select Curve > Line > Perpendicular to 2 Curves, or the Line Perpendicular to Two Curves button from the Lines toolbar.
Command: LinePP

2   Select the curves.

A line perpendicular to two curves is constructed. (See figure 2-60.)

Rhino Curve Construction Tools 59

Fig. 2-60. Line perpendicular to two curves constructed.

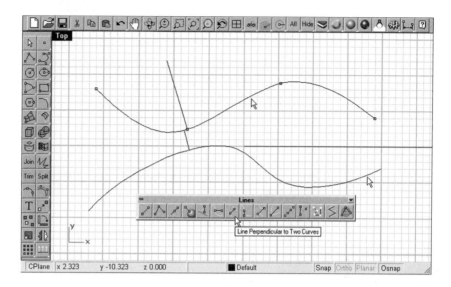

Line Tangent to a Curve. To construct a line tangent to a curve, perform the following steps.

1   Select Curve > Line > Tangent from Curve, or the Line Tangent from Curve button from the Lines toolbar.
    Command: LineTan

2   Select a point on the curve and a point to indicate the end point. (See figure 2-61.)

A line tangent to a curve at a point is constructed.

Fig. 2-61. Line tangent to a curve constructed.

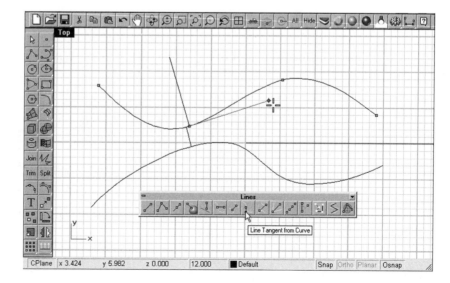

Line Tangent to Two Curves. To construct a line tangent to two curves, perform the following steps.

1 Select Curve > Line > Tangent to 2 Curves, or the Line Tangent to Two Curves button from the Lines toolbar.
 Command: LineTT

2 Select the curves indicated in figure 2-62.

A line tangent to two curves is constructed. Save your file as *LineCrv.3dm*.

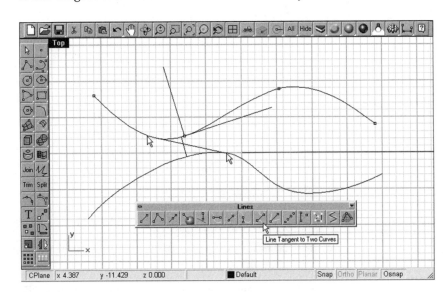

Fig. 2-62. Line tangent to two curves constructed.

Circle

A circle is a degree 2 curve. There are eight ways to construct a circle. You do this from the Circle toolbar, shown in figure 2-63.

Fig. 2-63 Circle toolbar.

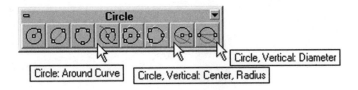

Table 2-4 outlines the functions of circle commands and their location in the pull-down menu.

Rhino Curve Construction Tools

Table 2-4: Circle Commands and Their Functions

Command Name	Pull-down Menu	Function
Circle	Curve > Circle > Center, Radius	For constructing a circle by specifying the center point and the radius. (Using the Vertical option, it enables you to construct a circle perpendicular to the construction plane. Using the AroundCurve option, it enables you to construct a circle with the center on a curve and perpendicular to the curve.)
CircleD	Curve > Circle > Diameter	For constructing a circle by specifying two diametric points. (Using the Vertical option, it enables you to construct a circle perpendicular to the construction plane.)
Circle3Pt	Curve > Circle > 3 Points	For constructing a circle by specifying three circumferential points.
CircleTTR	Curve > Circle > Tangent, Tangent, Radius	For constructing a circle by specifying two tangential curves and the radius.
CircleTTT	Curve > Circle > Tangent to 3 Curves	For constructing a circle by specifying three tangential curves.

Circle. To construct circles, perform the following steps.

1 Start a new file. Use the metric (millimeter) template.

2 Maximize the Top viewport.

3 Select Curve > Circle > Center, Radius, or the Circle/Center, Radius button from the Circle toolbar.
   Command: Circle

4 Select two points to indicate the center and radius. (See figure 2-64.)

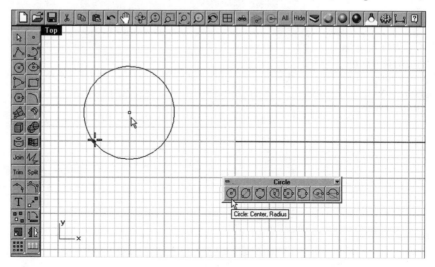

Fig. 2-64. Circle constructed by specifying the center and radius.

5   Select Curve > Circle > Diameter, or the Circle/Diameter button from the Circle toolbar.
    Command: CircleD

6   Select two points to indicate the diameter of the circle. (See figure 2-65.)

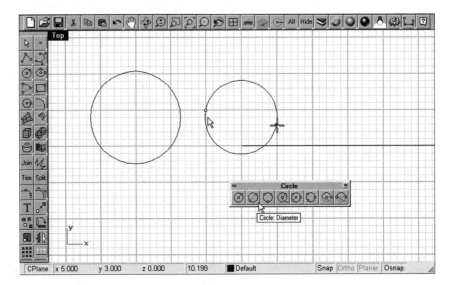

Fig. 2-65. Circle constructed by specifying the diameter.

7   Select Curve > Circle > 3 Points, or the Circle/3 Points button from the Circle toolbar.
    Command: Circle3Pt

8   Select three points to indicate three points on the circumference of the circle. (See figure 2-66.)

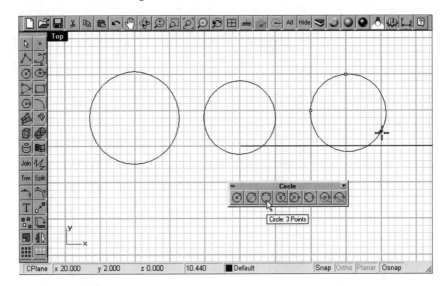

Fig. 2-66. Circle constructed by specifying three points on the circumference.

Rhino Curve Construction Tools

Three circles are constructed. Save your file as *Circle.3dm*.

Circle Around Curve. To construct a circle along and perpendicular to one of the free-form curves, perform the following steps.

1. Open the file *Freeform1.3dm*.
2. Select File > Save As and specify the file name as *Circle1.3dm*.
3. Select Curve > Circle > Center, Radius and type *A* (A stands for AroundCurve) at the command line area, or select the Circle/Around Curve button from the Circle toolbar.

   Command: Circle

   Center of circle (Vertical AroundCurve): A

4. Select a point along a free-form curve (see figure 2-67) to indicate the center location.
5. To specify a radius, you may type a value at the command line area or select a point to indicate the radius. (See figure 2-68.)

Fig. 2-67. Point selected along the curve.

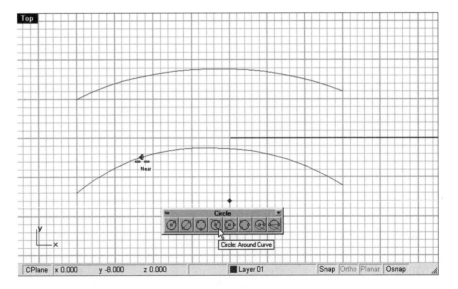

6. Double click on the Top viewport title to return to a four-viewport display. (See figure 2-69.)

A circle is constructed around a curve. Save your file.

Fig. 2-68. Radius specified.

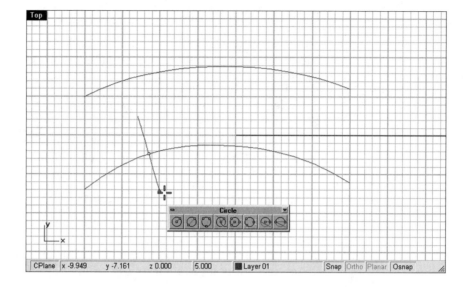

Fig. 2-69. Circle constructed perpendicular to the selected curve.

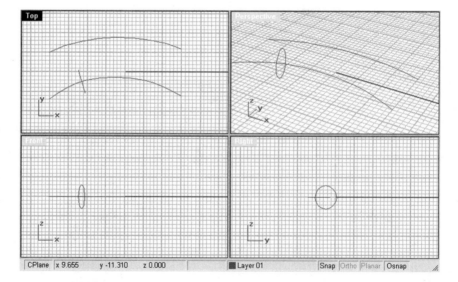

Tangent Circles. To construct tangent circles, perform the following steps.

1 Start a new file. Use the metric (millimeter) template.
2 Maximize the Top viewport.
3 With reference to figure 2-70, construct three free-form curves.
4 Select Curve > Circle > Tangent, Tangent, Radius, or the Circle/Tangent, Tangent, Radius button from the Circle toolbar.
Command: CircleTTR

Rhino Curve Construction Tools 65

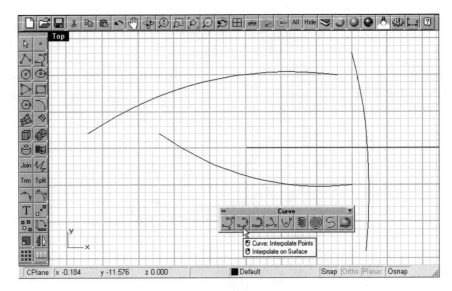

Fig. 2-70. Curves to be constructed.

5 Select the curves indicated in figure 2-71.

6 Select two points to indicate the radius.

A circle tangent to two curves is constructed.

7 Select Curve > Circle > Tangent to 3 Curves, or the Circle/Tangent, Tangent, Tangent button from the Circle toolbar.
Command: CircleTTT

8 Select the curves indicated in figure 2-72.

A circle tangent to three curves is constructed. Save your file as *Circle2.3dm*.

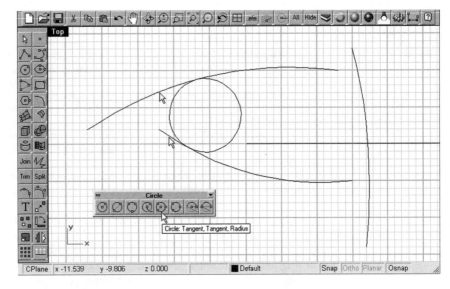

Fig. 2-71. Circle tangent to two curves constructed.

Fig. 2-72. Circle constructed tangent to three curves.

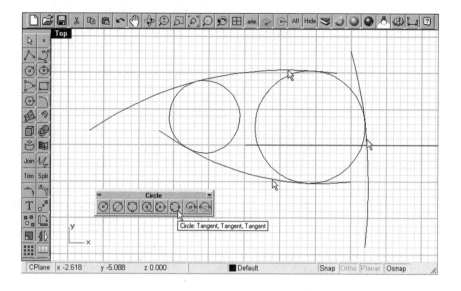

Arc

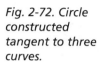

Fig. 2-73. Arc toolbar.

An arc is a degree 2 curve. There are five ways to construct an arc. You do this from the Arc toolbar, shown in figure 2-73.

Table 2-5 outlines the functions of arc commands and their location in the pull-down menu.

Table 2-5: Arc Commands and Their Functions

Command Name	Pull-down Menu	Function
Arc	Curve > Arc > Center, Start, Angle	For constructing an arc by specifying the center, starting point, and chord angle
Arc3Pt	Curve > Arc > 3 Points	For constructing an arc by specifying two end points and a point on the arc
ArcDir	Curve > Arc > Start, End, Direction	For constructing an arc by specifying two end points and a tangent direction
ArcSER	Curve > Arc > Start, End, Radius	For constructing an arc by specifying two end points and the radius
ArcTTR	Curve > Arc > Tangent, Tangent, Radius	For constructing an arc tangent to two curves and the radius

In the following, you will construct arcs in two other ways.

1. Start a new file. Use the metric (millimeter) template.
2. Maximize the Top viewport.

Rhino Curve Construction Tools

3 Select Curve > Arc > Center, Start, Angle, or the Arc/Center, Start, Angle button from the Arc toolbar.

Command: Arc

4 Select a point to indicate the radius, a point to indicate the start point, and a point to indicate the angle. (See figure 2-74.)

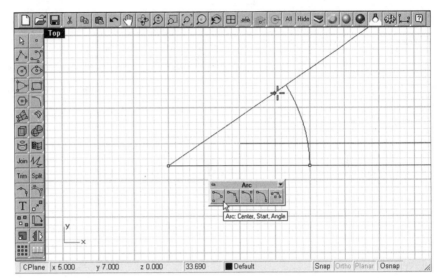

Fig. 2-74. Arc being constructed by specifying the center, start, and angle.

5 Select Curve > Arc > Start, End, Direction, or the Arc/Start, End, Direction button from the Arc toolbar.

Command: ArcDir

6 Select a point to indicate the start point, a point to indicate the end point, and a point to indicate the direction. (See figure 2-75.)

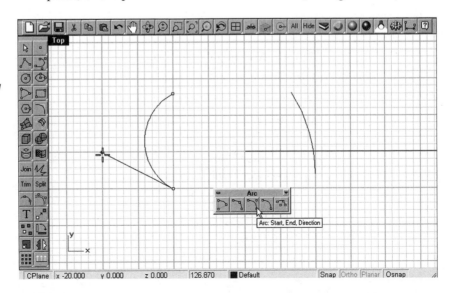

Fig. 2-75. Arc being constructed by specifying the start, end, and direction.

Two arcs are constructed. Save your file as *Arc.3dm*.

Ellipse

An ellipse is also a degree 2 curve. There are five ways to construct an ellipse. You do this from the Ellipse toolbar, shown in figure 2-76.

Fig. 2-76. Ellipse toolbar.

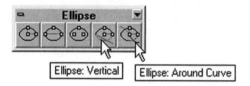

Ellipse Commands. Table 2-6 outlines the functions of ellipse commands and their location in the pull-down menu.

Table 2-6: Ellipse Commands and Their Functions

Command Name	Pull-down Menu	Function
Ellipse	Curve > Ellipse > From Center	For constructing an ellipse by specifying the center, major axis, and minor axis. (Using the Vertical option, it enables you to construct an ellipse perpendicular to the construction plane. Using the AroundCurve option, it enables you to construct an ellipse with the center along a curve and perpendicular to the curve.)
EllipseD	Curve > Ellipse > Diameter	For constructing an ellipse by specifying the major axis end points and the minor axis.
Ellipse fromfoci	Curve > Ellipse > From Foci	For constructing an ellipse by specifying the focus points and a point on the ellipse.

Ellipse Around Curve. To construct an ellipse around the lower curve, perform the following steps.

1 Open the file *Circle1.3dm*.
2 Select File > Save As and specify the file name as *Ellipse1.3dm*.
3 Select Curve > Ellipse > From Center, Radius and type *A* at the command line area, or select the Ellipse/Around Curve button from the Ellipse toolbar.
   Command: Ellipse
   Center of ellipse (Vertical AroundCurve FromFoci): A
4 Move the cursor over the Top viewport and select a point along a curve indicated in figure 2-77 to specify the center of the ellipse.

Rhino Curve Construction Tools

Fig. 2-77. Center point of the ellipse selected on the curve.

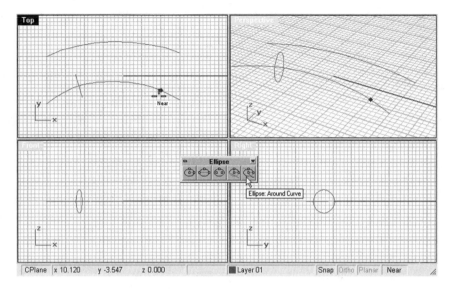

5   With reference to figure 2-78, select a point in the Top viewport to specify the first axis and a point in the Perspective viewport to specify the second axis.

Fig. 2-78. First and second axes specified.

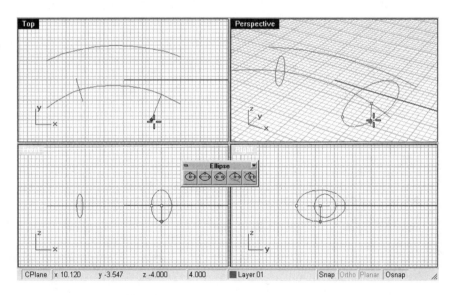

An ellipse is constructed around a curve. (See figure 2-79.) Save your file.

Fig. 2-79. Ellipse constructed perpendicular to the selected curve.

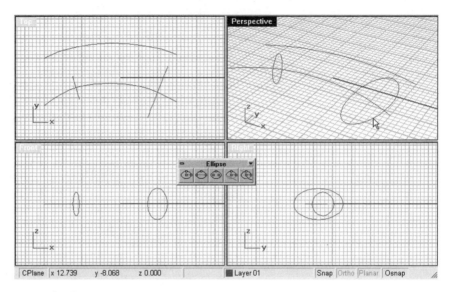

Parabola

A parabola is degree 2 curve. There are two ways to construct a parabola. You do this from the Parabola button of the Curve toolbar, shown in figure 2-80.

Fig. 2-80. Parabola button of the Curve toolbar.

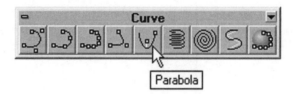

Table 2-7 outlines the functions of parabola commands and their location in the pull-down menu.

Table 2-7: Parabola Commands and Their Functions

Command Name	Pull-down Menu	Function
Parabola	Curve > Parabola > Focus, Direction	For constructing a parabola by specifying the focus point, direction, and an end point of the parabola
Parabola Vertex	Curve > Parabola > Vertex, Focus	For constructing a parabola by specifying the vertex, focus point, and end point of the parabola

To start a new drawing and construct a set of curves (including parabola, conic, helix, spiral, and polygon), perform the following steps.

Rhino Curve Construction Tools

1. Select File > New to start a new file. Use the metric (millimeter) template. Then double-click on the Top viewport to set the display to the selected viewport.
2. Select Curve > Parabola > Vertex, Focus (or select the Parabola button from the Curve toolbar), type *V*, and press the Enter key to use the Vertex option.

   Command: Parabola

   Parabola focus (Vertex MarkFocus = No Half = No): Vertex
3. Select a point indicated by the arrow in figure 2-81 to specify the vertex, and select another point to specify the focus point.
4. Select a point to indicate the end point. (See figure 2-82.)

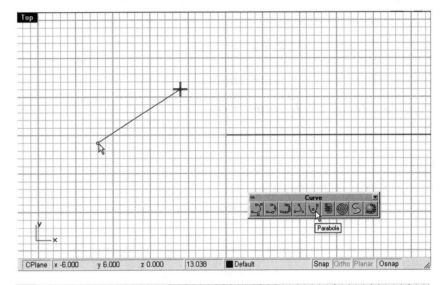

Fig. 2-81. Vertex and focus point selected.

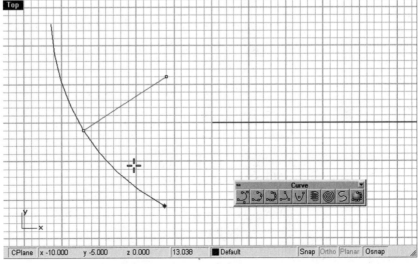

Fig. 2-82. End point indicated.

A parabola is constructed. Save your file as *Parabola.3dm*.

Conic

A conic is also a degree 2 curve. You construct conics from the Conic button of the Curve toolbar, shown in figure 2-83.

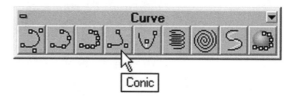

Fig. 2-83. Conic button of the Curve toolbar.

Table 2-8 outlines the function of the Conic command and its location in the pull-down menu.

Table 2-8: Conic Command and Its Function

Command Name	Pull-down Menu	Function
Conic	Curve > Conic	For constructing a conic curve by specifying the start point, reference vertex, end point, and a curvature point

To construct a conic curve, perform the following steps. Open the file *Parabola.3dm* if you already closed it.

1 Select File > Save As and specify the file name as *Conic.3dm*.

2 Select Curve > Conic, or the Conic button from the Curve toolbar.
Command: Conic

3 Select a point indicated by the arrow in figure 2-84 to specify the start point of the conic, and another point to indicate the vertex.

4 Select a point indicated by the arrow in figure 2-85 to specify the end point, and another point to indicate a point on the conic.

A conic curve is constructed. Save your file.

Helix

A helix is a degree 3 curve. You construct a helix from the Helix button of the Curve toolbar, shown in figure 2-86.

Table 2-9 outlines the function of the Helix command and its location in the pull-down menu.

Rhino Curve Construction Tools

Fig. 2-84. Start point and reference vertex selected.

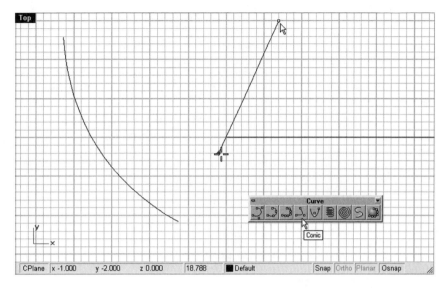

Fig. 2-85. End point and point on the conic selected.

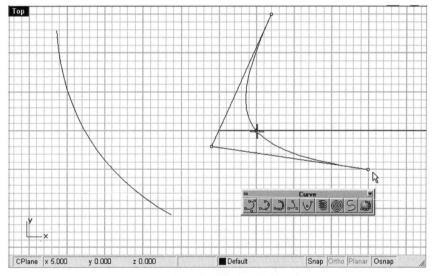

Fig. 2-86. Helix button of the Curve toolbar.

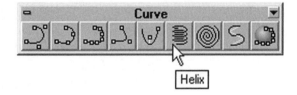

CHAPTER 2: Wireframe Modeling and Curves: Part 1

Table 2-9: Helix Command and Its Function

Command Name	Pull-down Menu	Function
Helix	Curve > Helix	For constructing a helix by specifying the axis, radius, and number of turns (pitch)

In the following you will construct two helix curves. Open the file *Conic.3dm* if you already closed it.

1 Select File > Save As and save the file as *Helix.3dm*.

2 Select Curve > Helix, or the Helix button from the Curve toolbar.
Command: Helix

3 Select a point indicated by the arrow in figure 2-87 to specify the start point of the axis of the helix, and another point to indicate the axis end point.

Fig. 2-87. Start and end points of the helix axis specified.

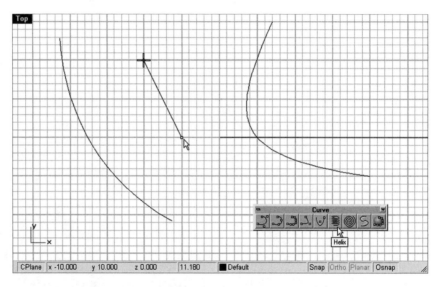

4 Specify the radius by selecting the point indicated in figure 2-88.

5 In the Helix/Spiral dialog box, specify the number of turns (or pitch) and click on the OK button.

6 Press the Enter key to repeat the Helix command. Type *A* and press the Enter key to use the AroundCurve option.
Command: Helix
Start of axis (Vertical AroundCurve): A

7 Select the parabola curve indicated in figure 2-89.

Fig. 2-88. Radius indicated and number of turns specified.

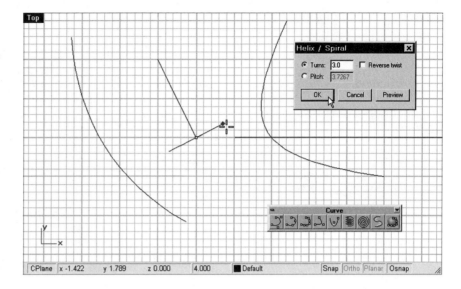

8 Specify the radius by selecting the point indicated in figure 2-90.

9 In the Helix/Spiral dialog box, specify three (3) turns and click on the OK button.

A helix around the parabola curve is constructed. Save your file.

Fig. 2-89. Parabola curve selected.

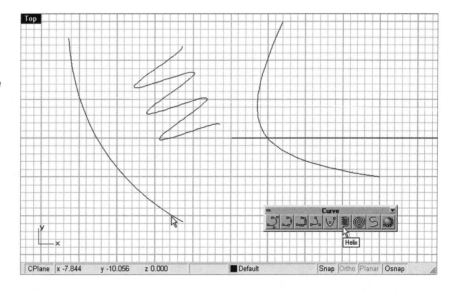

76   Chapter 2: Wireframe Modeling and Curves: Part 1

Fig. 2-90. Radius indicated and number of turns specified.

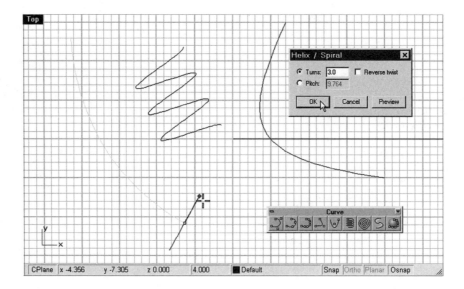

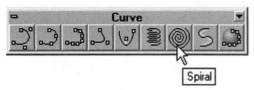

Fig. 2-91. Spiral button of the Curve toolbar.

Spiral

A spiral is a degree 3 curve. You construct spirals from the Spiral button of the Curve toolbar, shown in figure 2-91.

Table 2-10 outlines the function of the Spiral command and its location in the pull-down menu.

Table 2-10: Spiral Command and Its Function

Command Name	Pull-down Menu	Function
Spiral	Curve > Spiral	For constructing a spiral by specifying the axis, the radii at the axis end points, and the number of turns (pitch)

In the following you will construct two spiral curves: a flat spiral and a spiral around a curve. Open the file *Helix.3dm* if you already closed it.

1. Select File > Save As and specify the file name as *Spiral.3dm*.
2. Select Curve > Spiral, or the Spiral button from the Curve toolbar.
   Command: Spiral
3. Type *F* to use the Flat option.
4. Select the point indicated by the arrow in figure 2-92 to specify the center of the flat spiral, and another point to indicate a radius at one end of the spiral.

Rhino Curve Construction Tools 77

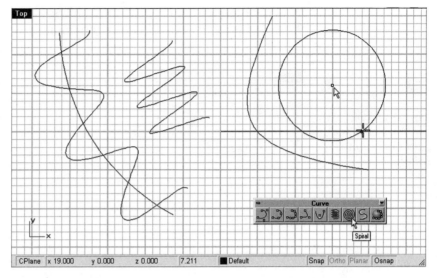

Fig. 2-92. Center and radius at one end of the spiral selected.

5 Select a point to indicate a radius at the other end of the spiral. (See figure 2-93.)

6 In the Helix/Spiral dialog box, specify three (3) turns and click on the OK button.

A flat spiral is constructed.

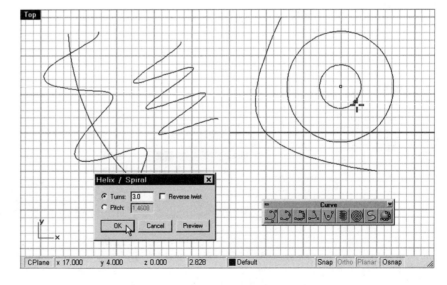

Fig. 2-93. Radius at the second end point of the spiral selected and number of turns specified.

To construct a spiral around a curve, continue with the following steps.

7 Press the Enter key to repeat the command.
Command: Spiral

8 Type *A* to use the AroundCurve option.

9 Select the conic curve indicated in figure 2-94 and a point to specify the start radius.

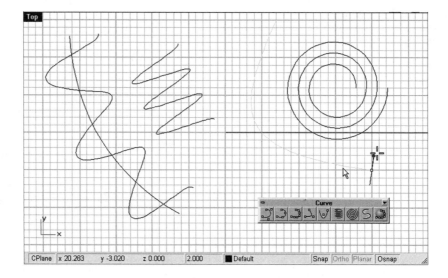

Fig. 2-94. Conic curve selected and start radius of the spiral specified.

10 Select a point to indicate the end radius. (See figure 2-95.)

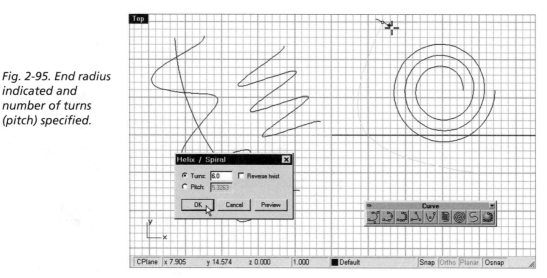

Fig. 2-95. End radius indicated and number of turns (pitch) specified.

11 In the Helix/Spiral dialog box, specify six (6) turns and click on the OK button.

A spiral around the conic curve is constructed. Save your file.

Polygon

Fig. 2-96. Polygon toolbar.

A regular polygon is a set of connected lines. A polygon is a degree 1 curve. To construct a regular polygon, you specify the inscribing circle, the circumscribing circle, or the length of a side. You do this from the Polygon toolbar, shown in figure 2-96.

Polygon Commands. Table 2-11 outlines the functions of polygon commands and their location in the pull-down menu.

Table 2-11: Polygon Commands and Their Functions

Command Name	Pull-down Menu	Function
Polygon	Curve > Polygon > Center, Radius	For constructing a regular polygon by specifying the number of sides and the center and radius of the inscribed or circumscribed circle
PolygonEdge	Curve > Polygon > By Edge	For constructing a regular polygon by specifying the number of sides and two end points of a side

Polygon Around Curve. To construct a regular polygon around a helix curve, perform the following steps. Open the file *Spiral.3dm* if you already closed it.

1 Select File > Save As and specify the file name as *Polygon.3dm*.
2 Select Curve > Polygon > Center, Radius, or the Polygon/Center, Radius button from the Polygon toolbar.
   Command: Polygon
3 Type *A* to use the AroundCurve option.
4 Type *N* to use the Number option. Then type 6 to specify a six-sided polygon.
5 Select the helix curve indicated in figure 2-97.
6 Specify the corner of the polygon by selecting the point indicated in figure 2-98.

A six-sided polygon (hexagon) is constructed around the helix.

80 CHAPTER 2: Wireframe Modeling and Curves: Part 1

Fig. 2-97. Helix curve selected.

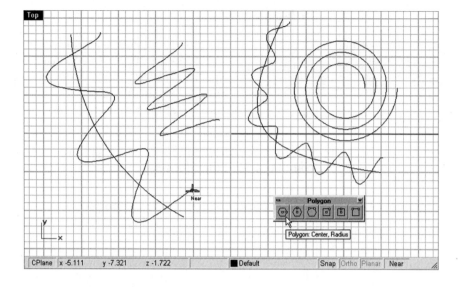

Fig. 2-98. Corner of polygon specified.

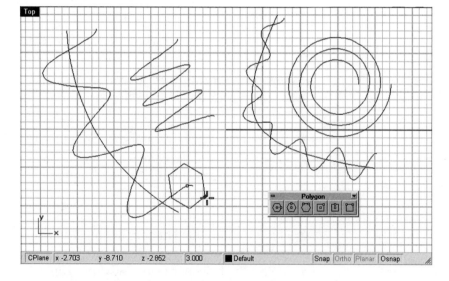

7 Double click on the Top viewport title to set the display to a four-viewport display. (See figure 2-99.)

Save your file.

Fig. 2-99. Polygon constructed.

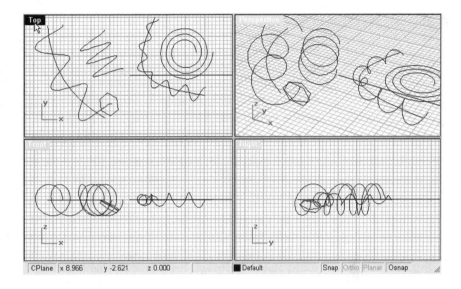

Rectangle

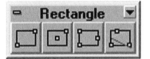

Fig. 2-100. Rectangle toolbar.

A rectangle is a set of connected lines and is a degree 1 curve. There are four ways to construct a rectangle. You do this from the Rectangle toolbar, shown in figure 2-100.

Table 2-12 outlines the functions of rectangle commands and their location in the pull-down menu.

Table 2-12: Rectangle Commands and Their Functions

Command Name	Pull-down Menu	Function
Rectangle	Curve > Rectangle > Corner to Corner	For constructing a rectangle by specifying its diagonal corners
RectangleCen	Curve > Rectangle > Center, Corner	For constructing a rectangle by specifying the center of the rectangular area and a corner of the rectangle
Rectangle3Pt	Curve > Rectangle > 3 Points	For constructing a rectangle by specifying two end points of an edge and the width
RectangleV	Curve > Rectangle > Vertical	For constructing a rectangle perpendicular to the current construction plane by specifying two end points of an edge and the height of the rectangle

■■ Summary

There are three major types of computer models: wireframe, surface, and solid. Among these, the wireframe model has very limited application in design and downstream computerized manufacturing systems. Therefore, it is not sensible to exhaust your effort to create complex 3D curves just to represent the edges and silhouettes of a 3D object.

However, curves are extremely useful and important in making free-form surfaces and solids because you use them as frameworks on which surfaces are constructed. Therefore, you need to have a good understanding of the characteristics of curves and methods of constructing 3D curves.

To construct free-form surfaces and solids, NURBS (non-uniform rational basis spline) curves are commonly used in modeling applications. A NURBS curve is a set of polynomial spline segments. The shape of each spline segment in a curve is influenced by the degree of polynomial equation used to define the curve. A line is degree 1, a circle is degree 2, and a free-form curve is degree 3 or above. The higher the degree, the more control points exist in a spline segment. For a curve of degree 3, each spline segment has four control points. Between two contiguous segments of a spline curve is a knot.

Adding or removing knots increases or reduces the number of segments in a curve. A special type of knot on a curve is a kink. It is a junction between two contiguous segments where the tangent directions are not congruent. A closed curve with no kink is called a periodic curve. The overall shape of a NURBS curve is affected by the number of segments in the curve, the degree of polynomial equation, the location of control points, and the weight of control points.

There are many ways to construct NURBS curves. In this chapter you learned the Rhino user interface, and construction of point objects and basic curves. In the next chapter you will learn how to trace a curve from a background image, derive curves from existing objects, edit curves, and transform curves.

■■ Review Questions

1. Outline the key concepts of wireframe modeling and explain the limitations of the wireframe model.
2. Explain the significance of curves in computer modeling.
3. With the aid of sketches, explain the characteristics and key features of a NURBS curve.
4. Give a brief account of the types of basic geometric point objects and curves you can construct using Rhino.

Chapter 3

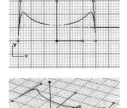

Wireframe Modeling and Curves: Part 2

■■ Objectives

The goals of this chapter are to demonstrate how to use Rhino as a tool to trace background images, construct derived curves, edit curves, and transform curves. After studying this chapter, you should be able to:

- ❏ Trace a curve from background images
- ❏ Construct derived curves
- ❏ Edit curves
- ❏ Transform curves

■■ Overview

This chapter is a continuation of Chapter 2. Here you will learn how to trace a curve from a background image embedded in the display viewport, derive curves from existing objects, edit curves, and transform curves. In the next chapter you will learn how to analyze curves and further enhance your knowledge by working on several curve construction projects.

■■ More on Curves

The sections that follow pick up where Chapter 2 left off. These sections cover tracing background images, constructing derived curves, editing curves, and transforming curves.

Tracing Background Images

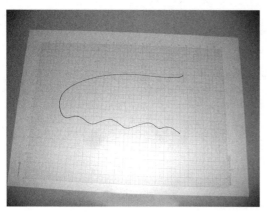

Fig. 3-1. Sketch drawn on a piece of graph paper.

Knowing that the fundamental issue of point and basic curve construction is to specify a location, you need to be able to use the pointing device to select a point or input the coordinates at the command line area. To help determine the locations, you might sketch your design idea on graph paper. From the graph paper, you extract point locations. If you already have a physical object, you can use a digitizer to obtain the required coordinates. (You will learn how to use a digitizer in Chapter 5.) Figure 3-1 shows a set of curves drawn on graph paper.

Another way of using the sketch is to scan it as a digital image and insert it in the viewport as a background image. If the image is properly scaled, you may directly sketch in the viewport. To try this way of sketching, perform the following steps.

1. Obtain a piece of graph paper.

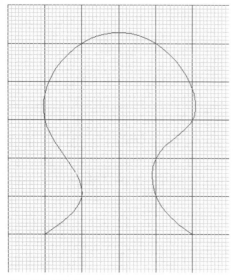

Fig. 3-2. Bitmap image.

2. Use a pencil to construct a free-hand sketch depicting the curve you want to construct in the computer.
3. Scan the sketch as a digital image.
4. After scanning, use an image processing application to trim away the unwanted portion of the image and log the size of the image per grid-line measure. For example, the bitmap image shown in figure 3-2 is 30 mm wide and 35 mm tall.
5. Incorporate the digital image as a background image in the Front viewport.
6. Select File > New and use the metric (millimeter) template to start a new file.
7. Double click the Top viewport title to maximize the viewport.
8. Check the Snap button on the status bar to turn on snap mode.

More on Curves 85

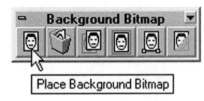

Fig. 3-3. Background Bitmap toolbar.

9 Select View > Background Bitmap > Place, or the Place Background Bitmap button from the Background Bitmap toolbar. (Figure 3-3 shows the Background Bitmap toolbar.)

Command: PlaceBackgroundBitmap

10 In the Open Bitmap File dialog box (figure 3-4), select the bitmap you prepared. (You can see the types of image files supported in the *Files of type* pull-down box.)

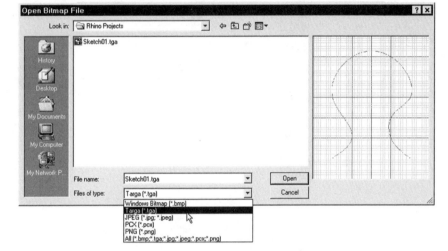

Fig. 3-4. Open Bitmap File dialog box and image file types.

11 Select a point on the Front viewport. Then select a point 30 mm to the right of the first selected point, because the width of the bitmap used in this example is 30 mm wide. (See figure 3-5.)

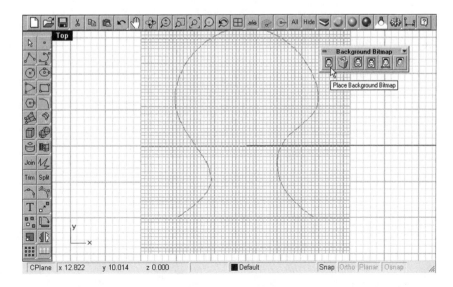

Fig. 3-5. Image placed on the background of Front viewport.

12 Select Curve > Free-form > Sketch, or the Sketch button from the Curve toolbar. (Alternatively, you can use the Curve from Control Points option or the Interpolated Curve option to trace images.)

Command: Sketch

13 Hold down the left mouse button to sketch a free-form curve. (See figure 3-6.)

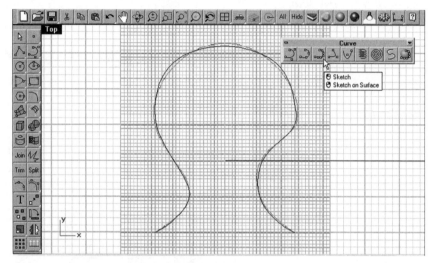

Fig. 3-6. Free-form sketch constructed.

14 Select View > Background Bitmap > Hide, or the Hide Background Bitmap button from the Background Bitmap toolbar.

Command: HideBackgroundBitmap

A sketch is constructed and the bitmap is hidden. (See figure 3-7.) Save your file as *Sketch.3dm*.

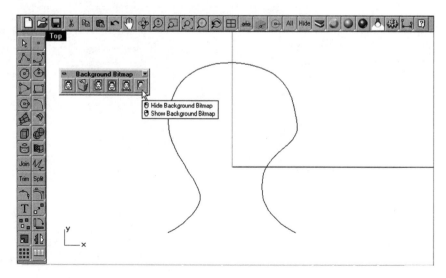

Fig. 3-7. Image hidden and grid mesh turned off.

More on Curves

Constructing Derived Curves

As well as constructing basic geometric curves, you can construct curves from existing objects, such as curves, surfaces, and solids. From existing curves, you can derive seven types of curves. You can extend a curve, construct a fillet or a chamfer at the intersection of two curves, offset a curve, blend two curves, construct a curve from two curves residing on two viewports, or construct a set of cross-section curves across a set of longitudinal profile curves. In this chapter you will learn curves derived from curves. In Chapter 9 you will learn other types of derived curves.

Extend

Extending derives a new curve by increasing the curve's length. You extend a curve by extending it to a boundary curve, dragging an end of the curve to a new position, adding a straight line segment to the curve, adding an arc segment to the curve, or extending it to meet a surface. You perform these actions from the Extend toolbar in the Curve Tools toolbar, shown in figure 3-8.

Fig. 3-8. Extend toolbar from the Curve Tools toolbar.

Table 3-1 outlines the functions of extend commands and their location in the pull-down menu.

Table 3-1: Extend Commands and Their Functions

Command Name	Pull-down Menu	Function
Extend	Curve > Extend > Extend Curve	For extending a curve by specifying the boundary of extension or extension length. (Using the Dynamic option, it enables you to drag a curve dynamically to extend it.)
ExtendByLine	Curve > Extend > By Line	For adding a line segment tangent to an end point of a curve.
ExtendByArc	Curve > Extend > By Arc	For adding an arc tangent to an end point of a curve by specifying the radius and the angle of the arc.
ExtendByArcToPt	Curve > Extend > By Arc to Point	For adding an arc tangent to an end point of a curve by specifying the end point of the arc.
ExtendCrvOnSrf	Curve > Extend > Curve on Surface	For extending a curve to a surface.

Now start a new file. In the following, you will construct two free-form curves using the methods you learned previously. Based on these curves, you will construct various types of derived curves.

1. Select File > New to start a new file. Use the metric (millimeter) template file.
2. With reference to figure 3-9, construct two free-form curves.

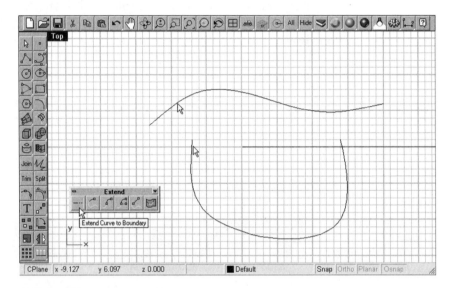

Fig. 3-9. Boundary curve (upper) and curve to be extended selected.

Extend to Boundary Curve. To extend a curve to a boundary curve, perform the following steps.

1. Select Curve > Extend > Extend Curve, or the Extend Curve to Boundary button from the Extend toolbar.
   Command: Extend
2. Select the upper curve to be used as the boundary curve.
3. Press the Enter key to terminate selection of the boundary curve.
4. Select the lower curve as the curve to be extended.
5. Press the Enter key to terminate the command.

The lower curve is extended to meet the upper curve. (See figure 3-10.)

Dynamic Extend. To extend the other end of the curve in a dynamic way, perform the following steps.

1. Press the Enter key to repeat the command.
   Command: Extend
2. Press the Enter key to use dynamic extension.
3. Select the curve indicated in figure 3-10.

More on Curves

Fig. 3-10. Curve extended to a boundary curve.

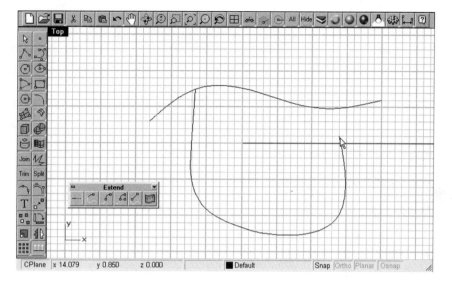

4 Move the cursor to a new position and left-click. (See figure 3-11.)
5 Press the Enter key.

The curve is extended. Save your file as *Extend.3dm*.

Fig. 3-11. Curve extended by dragging.

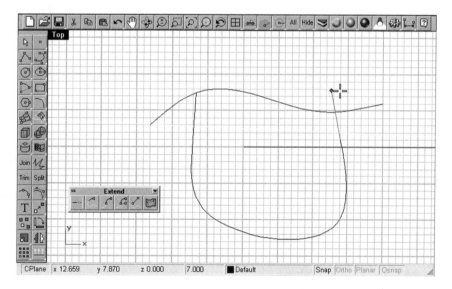

Fillet

A fillet curve derives from two nonparallel curves. It is a circular curve tangent to two selected curves. You construct fillet curves from the Fillet button of the Curve Tools toolbar, shown in figure 3-12.

Fig. 3-12. Fillet button of the Curve Tools toolbar.

Table 3-2 outlines the function of the Fillet command and its location in the pull-down menu.

Table 3-2: Fillet Command and Its Function

Command Name	Pull-down Menu	Function
Fillet	Curve > Fillet	For constructing a tangent arc at the intersection of two curves

In the following, you will derive a fillet curve. Open the file *Extend.3dm* if you already closed it.

1. Select File > Save As and specify the file name as *Fillet.3dm*.
2. Double click on the Top viewport to maximize it.
3. Select Curve > Fillet, or the Fillet button from the Curve Tools toolbar.

   Command: Fillet
4. Type *R* at the command line area to use the Radius option.
5. Type *3* to specify a radius of 3 mm.
6. Select the curves indicated in figure 3-13.

Fig. 3-13. Curves selected to derive a fillet curve.

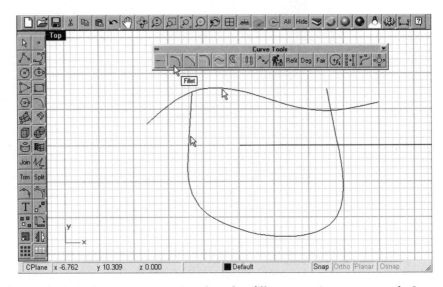

The selected curves are trimmed and a fillet curve is constructed. Save your file.

More on Curves 91

Chamfer

A chamfer curve also derives from two nonparallel curves. It joins the selected curves with a beveled line. You construct chamfer curves from the Chamfer button of the Curve toolbar, shown in figure 3-14. Table 3-3 outlines the function of the Chamfer command and its location in the pull-down menu.

Fig. 3-14. Chamfer button of the Curve Tools toolbar.

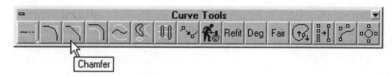

Table 3-3: Chamfer Command and Its Function

Command Name	Pull-down Menu	Function
Chamfer	Curve > Chamfer	For constructing a bevel edge at the intersection of two curves

In the following, you will construct a chamfer curve. Open the file *Fillet.3dm* if you already closed it.

1. Select File > Save As and specify the file name as *Chamfer.3dm*.
2. Select Curve > Chamfer, or the Chamfer button from the Curve Tools toolbar.
   Command: Chamfer
3. Type *D* at the command line area to use the Distance option.
4. Set the first chamfer distance and second chamfer distance to 3.
5. Select the curves indicated in figure 3-15.

Fig. 3-15. Curves selected to derive a chamfer curve.

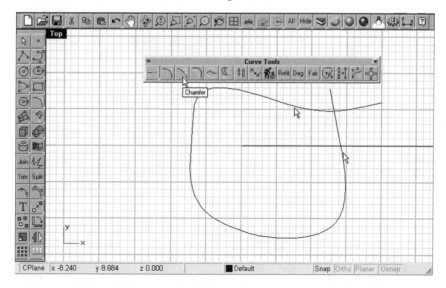

A chamfer curve is constructed. Save your file.

Offset

Offsetting derives a new curve at a specified distance from an existing curve. You construct offset curves from the Offset button of the Curve Tools toolbar, shown in figure 3-16.

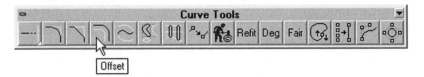

Fig. 3-16. Offset button of the Curve Tools toolbar.

Table 3-4 outlines the function of Offset command and its location in the pull-down menu.

Table 3-4: Offset Command and Its Function

Command Name	Pull-down Menu	Function
Offset	Curve > Offset	For constructing a curve that offsets at a distance from a selected curve

In the following, you will construct two offset curves. Open the file *Chamfer.3dm* if you already closed it.

1. Select File > Save As and specify the file name as *Offset.3dm*.
2. Select Curve > Offset, or the Offset button from the Curve Tools toolbar.
   Command: Offset
3. Type *D* at the command line area to use the Offset Distance option.
4. Type *4* to indicate an offset distance of 4 mm.
5. With reference to figure 3-17, select the curve indicated by the arrow and select a location above the selected curve to indicate the offset direction.

An offset curve is constructed.

6. Press the Enter key to repeat the command.
7. With reference to figure 3-18, select the curve and indicate the offset direction.

The second offset curve is constructed. Save your file.

More on Curves

Fig. 3-17. Curve selected and offset direction indicated.

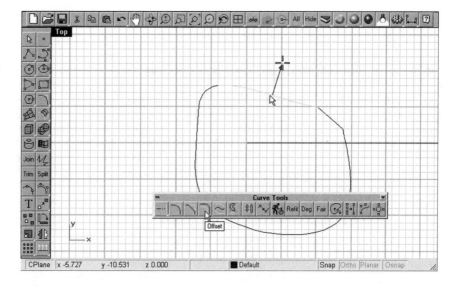

Fig. 3-18. Second curve selected and offset direction indicated.

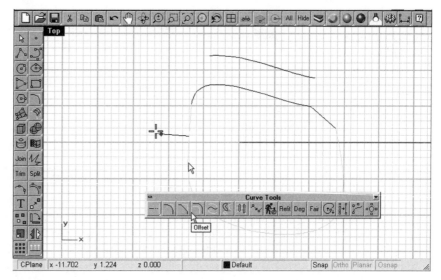

Blend

Blending derives a curve that fits smoothly between two selected curves. You construct blend curves from the Blend button of the Curve Tools toolbar, shown in figure 3-19.

Fig. 3-19. Blend button of the Curve Tools toolbar.

94   CHAPTER 3: Wireframe Modeling and Curves: Part 2

Table 3-5 outlines the function of the Blend command and its location in the pull-down menu.

Table 3-5: Blend Command and Its Function

Command Name	Pull-down Menu	Function
Blend	Curve > Blend	For constructing a blend curve between two curves

In the following, you will construct a blend curve. Open the file *Blend.3dm* if you already closed it.

1   Select File > Save As and specify the file name as *Blend.3dm*.

2   Select Curve > Blend, or the Blend button from the Curve Tools toolbar.

Command: Blend

3   Select the curves at locations indicated in figure 3-20.

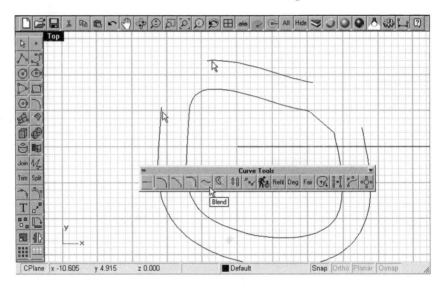

Fig. 3-20. Curves selected.

A blend curve is constructed. (See figure 3-21.) Save your file.

3D Curve from Two Planar Curves

If you already have an idea of the shape of a 3D curve in two orthographic drawing views, you can first construct two planar curves residing on two adjacent viewports and derive the 3D curve from the planar curves. Figure 3-22 shows the command button for making a 3D curve from two planar curves.

More on Curves

Fig. 3-21. Blend curve constructed.

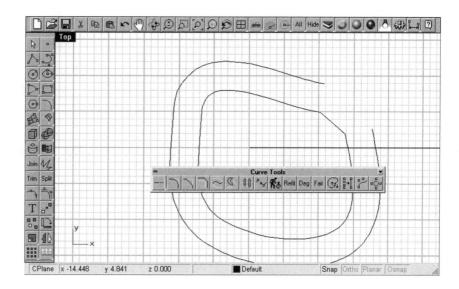

Fig. 3-22. Curve From 2 Views button of the Curve Tools toolbar.

Table 3-6 outlines the function of Curve From 2 Views command and its location in the pull-down menu.

Table 3-6: Curve From 2 Views Command and Its Function

Command Name	Pull-down Menu	Function
Crv2View	Curve > From 2 Views	For constructing a 3D curve from two planar curves residing on two different construction planes

In the following, you will start a new file, construct two planar curves that correspond to the front and right view of the 3D curve, and let the computer derive a 3D curve from the planar curves.

1 Start a new file. Use the metric (millimeter) template.

2 Construct a free-form curve on the Front viewport and another curve on the Right viewport. These curves represent the front and right views of a 3D curve. (See figure 3-23.)

Fig. 3-23. Curves constructed on the Front and Right viewports.

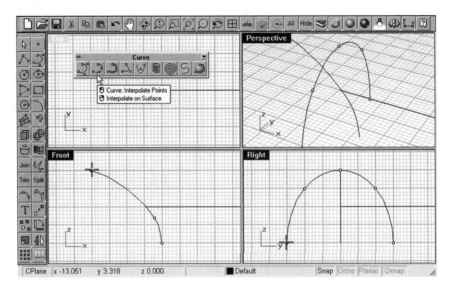

3 Select Curve > From 2 Views, or the From 2 Views button from the Curve Tools toolbar.

Command: Crv2View

4 Select the curves on the Front and Right viewports.

A 3D curve is derived. (See figure 3-24.) Save your file as *2Curves.3dm*.

Fig. 3-24. 3D curve derived from two planar curves residing on two views.

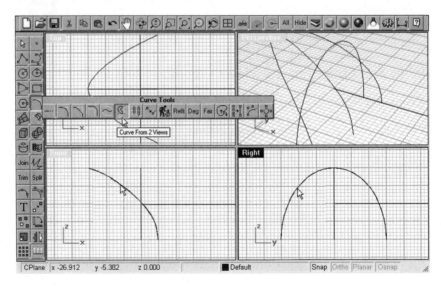

Cross-section Profiles

A cross-section profile is a closed-loop free-form curve interpolating the intersection points between a set of longitudinal curves and a specified

More on Curves

section plane across the curves. You work with cross-section profiles from the CSec Profiles button of the Curve Tools toolbar, shown in figure 3-25.

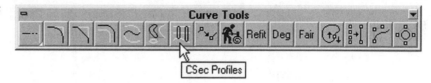

Fig. 3-25. CSec Profiles button of the Curve Tools toolbar.

Table 3-7 outlines the function of the Csec command and its location in the pull-down menu.

Table 3-7: Csec Command and Its Function

Command Name	Pull-down Menu	Function
Csec	Curve > CSect Profiles	For constructing cross-section curves across a set of longitudinal profile curves

To construct four profile curves and derive a series of cross-section profiles from them, perform the following steps.

1 Start a new file. Use the metric (millimeter) template.

2 Construct two free-form curves in the Top viewport in accordance with figure 3-26.

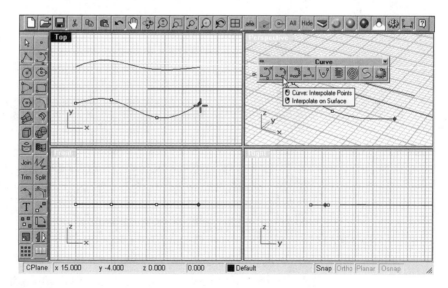

Fig. 3-26. Two free-form curves constructed in the Top viewport.

3 Construct two free-form curves in the Front viewport in accordance with figure 3-27.

Fig. 3-27. Two free-form curves constructed on the Front viewport.

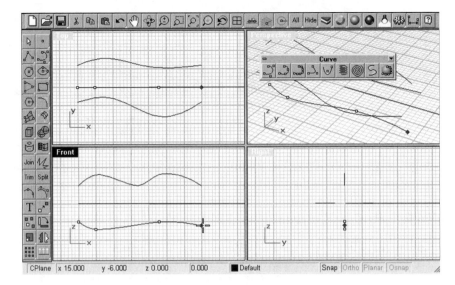

4   Select Curve > Csec Profiles, or the CSec Profiles button from the Curve Tools toolbar.

Command: CSec

5   Select the profile curves indicated in figure 3-28 (in either clockwise or counterclockwise fashion).

Fig. 3-28. Profile curves selected.

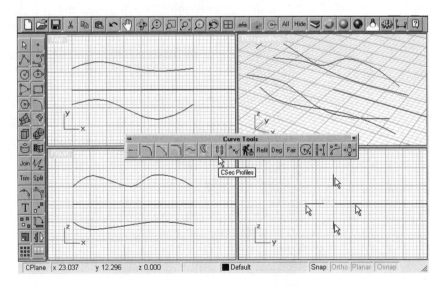

6   Press the Enter key.

7   With reference to figure 3-29, pick two points in the Front viewport to define a section plane. A section profile is constructed.

More on Curves

Fig. 3-29. Section plane indicated.

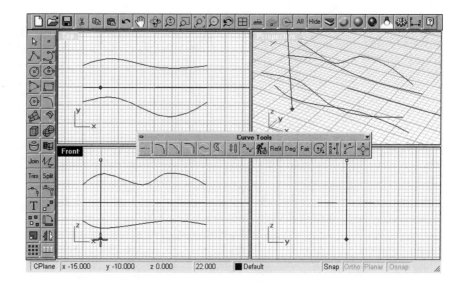

8 Pick two more points to define another section plane. (See figure 3-30.)

Fig. 3-30. Second section plane indicated.

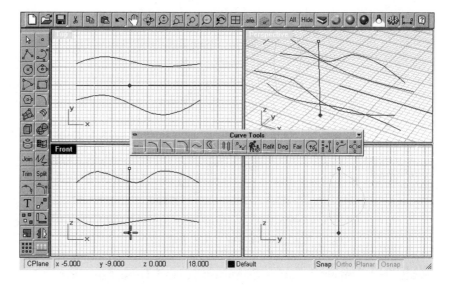

9 Press the Enter key.

Two section profiles are constructed. (See figure 3-31.) Save your file as *CSec.3dm*.

Fig. 3-31. Two section profiles derived.

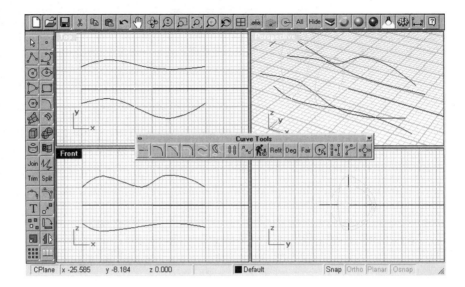

Curves from Objects

There are 13 ways to derive curves from existing surfaces. (You will learn them in Chapter 9.) You construct such curves from the Curve From Object toolbar, shown in figure 3-32.

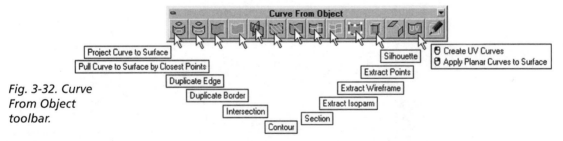

Fig. 3-32. Curve From Object toolbar.

Table 3-8 outlines the functions of the derived-curve commands and their locations in the pull-down menu.

Table 3-8: Derived-Curve Commands and Their Functions

Command Name	Pull-down Menu	Function
Project	Curve > From Objects > Project	For constructing curves and points on surfaces by projecting selected curves and points in a direction perpendicular to the current construction plane
Pull	Curve > From Objects > Pullback	For constructing curves and points on surfaces by projecting selected curves and points toward the selected surface, regardless of the orientation of the construction plane

Command Name	Pull-down Menu	Function
DupEdge	Curve > From Objects > Duplicate Edge	For constructing curves by duplicating selected edges of surfaces
DupBorder	Curve > From Objects > Duplicate Border	For constructing curves by duplicating all edges of selected surfaces
Intersect	Curve > From Objects > Intersection	For constructing curves or points at the intersection of selected surfaces or of surfaces and curves
Contour	Curve > From Objects > Contour	For constructing a set of contour curves of selected surfaces
Section	Curve > From Objects > Section	For constructing section curves of selected surfaces
Silhouette	Curve > From Objects > Silhouette	For constructing silhouette curves of selected surfaces
ExtractIsoparm	Curve > From Objects > Extract Isoparm	For constructing isoparm curves at specified locations of a selected surface
ExtractPt	Curve > From Objects > Extract Points	For constructing point objects at the control points of selected surfaces
ExtractWireframe	Curve > From Objects > Extract Wireframe	For constructing a set of isoparm and edge curves of selected surfaces
CreateUVCrv	Curve > From Objects > Create UV Curves	For mapping the edges and untrimmed edges of a selected surface on the current construction plane
ApplyCrv	Curve > From Objects > Apply UV Curves	For mapping selected planar curves on a selected surface

Because these curves derive from existing objects, you will learn how to construct them after learning how to construct surfaces and solids in chapters 5 through 8.

Editing Curves

You have learned how to construct point objects, basic curves, and derived curves. In this section you will learn how to edit existing curves. Methods of editing curves can be grouped into the following three main categories.

- ❒ Edit by joining, exploding, trimming, and splitting
- ❒ Edit by manipulating control points, edit points, knots, and points along the curve
- ❒ More advanced editing concerning polynomial degree and fit tolerance

Joining, Exploding, Trimming, and Splitting

To meet various design needs, you may have to join two or more curves into one, explode a joined curve into individual curve components, trim a curve to remove unwanted portions, or split a curve into two curves. You perform these operations from the Join, Explode, Trim, and Split buttons of the Main toolbar, shown in figure 3-33.

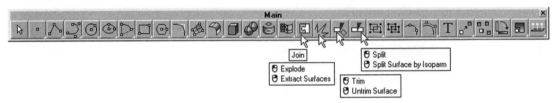

Fig. 3-33. Edit tools on the Main toolbar.

Table 3-9 outlines the Join, Explode, Trim, and Split commands and their locations in the pull-down menu.

Table 3-9: Join, Explode, Trim, and Split Commands and Their Functions

Command Name	Pull-down Menu	Function
Join	Edit > Join	For joining two or more curves into a single curve
Explode	Edit > Explode	For exploding a joined curve into individual curve components
Trim	Edit > Trim	For trimming a curve
Split	Edit > Split	For splitting a curve into two curves

Joining Curves. In the following, you will join several curves to form a single curve. Open the file *Blend.3dm*.

1 Select File > Save As and specify the new file name *Edit1.3dm*.
2 Select Edit > Join, or the Join button from the Main toolbar.
   Command: Join
3 Select the curves indicated in figure 3-34.
4 Press the Enter key to terminate the command.

The selected curves are joined. Now repeat the Join command.

5 Press the Enter key.
6 Select the curves indicated in figure 3-35.

More on Curves

Fig. 3-34. Curves selected.

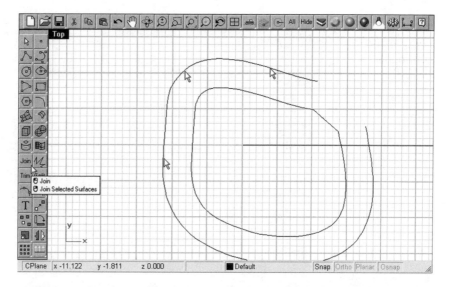

Fig. 3-35. Curves selected.

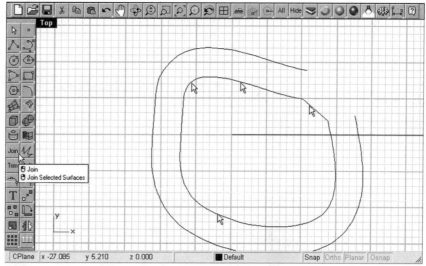

Because the selected curves form a closed loop, the command terminates automatically after you select the last curve. Two curves are derived by joining existing curves. Save your file.

Continuity of Contiguous Curves. Continuity refers to the smoothness at the junction between two contiguous curves or surfaces. There are three types of continuity: positional (G0 continuity), tangent (G1 continuity), and curvature (G2 continuity).

In a G0 continuity joint, the control points at the end points of contiguous curves coincide. The chamfer joints indicated in figure 3-36 have G0 continuity.

Fig. 3-36. G0 continuity.

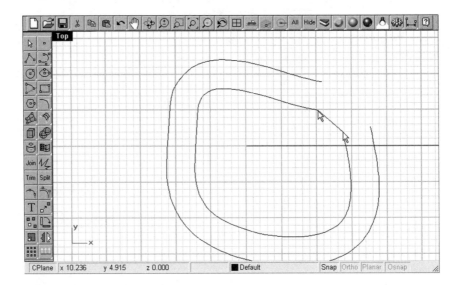

In a G1 continuity joint, the tangent direction of the control points at the end points of the curves is the same. The fillet joints indicated in figure 3-37 have G1 continuity.

Fig. 3-37. G1 continuity.

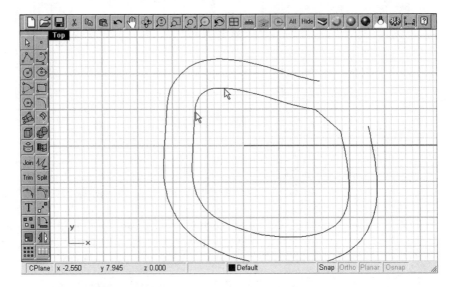

In a G2 continuity joint, the curvature and tangent direction of the control points at the end points of the curves are the same. The blend-curve joints indicated in figure 3-38 have G2 continuity.

More on Curves

Fig. 3-38. G2 continuity.

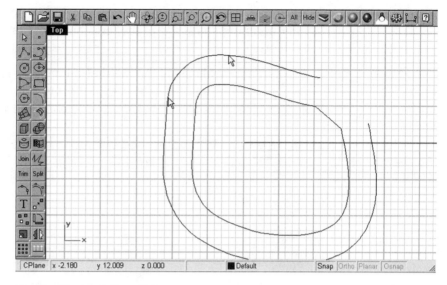

Exploding a Joined Curve. Contrary to joining, you can explode a joined curve into separate, individual curves, as follows.

1 Select Edit > Explode, or the Explode button from the Main toolbar.
Command: Explode

2 Select the curve indicated in figure 3-39. The curve is exploded.

3 Because the joined curves are needed in tutorials later in this chapter, revert the Explode command by selecting the Undo command.
Command: Undo

Save your file.

Fig. 3-39. Curve exploded and operation undone.

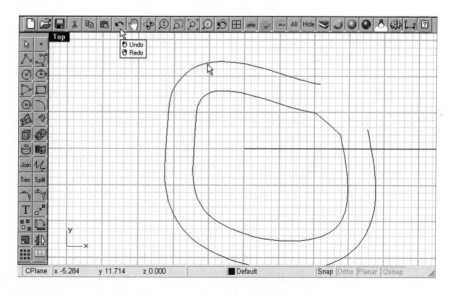

CHAPTER 3: Wireframe Modeling and Curves: Part 2

Trimming Curves. Unwanted portions of a curve can be trimmed. After trimming, the shape of the remaining portion of the curve remains unchanged.

1. Start a new file using the metric (millimeter) template. Double click on the Top viewport title to maximize the viewport.
2. Construct two free-form curves in accordance with figure 3-40.

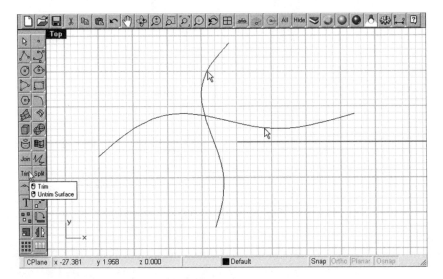

Fig. 3-40. Curve being trimmed.

To trim away the upper portion of the vertical curve at the intersection of the curves, continue with the following steps.

3. Select Edit > Trim, or the Trim button from the Main toolbar.
   Command: Trim
4. Select the horizontal curve as the trimming curve and press the Enter key.
5. Select the upper part of the vertical curve and press the Enter key.

The upper part of the vertical curve is trimmed. (See figure 3-41.)

Splitting a Curve into Two Curves. To split the horizontal curve into two curves at the intersection of the vertical curve, perform the following steps.

1. Select Edit > Split, or the Split button from the main toolbar. (See figure 3-42.)
   Command: Split
2. Select the horizontal curve. This is the curve to be split.
3. Select one of the vertical curves. This is the cutting object.
4. Press the Enter key.

More on Curves

Fig. 3-41. Vertical curve trimmed.

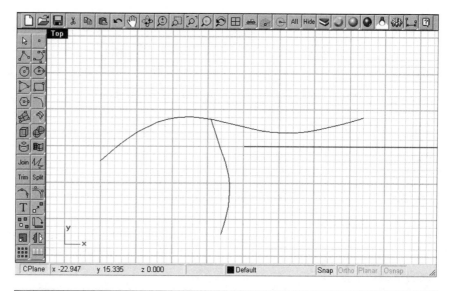

Fig. 3-42. Curve being split.

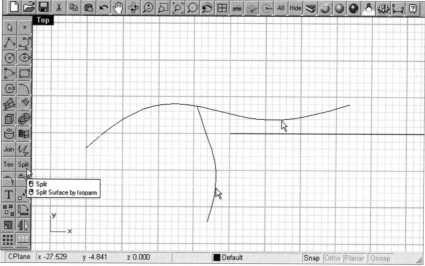

The horizontal curve is split into two. Save your file as *Split.3dm*.

Point Editing

The basic ways of modifying the shape of a curve are to manipulate its control points, edit points, knots, kinks, and selected locations along the curve. You perform these operations from the Point Editing toolbar, shown in figure 3-43.

Table 3-10 outlines the point editing commands and their locations in the pull-down menu.

Fig. 3-43. Point Editing toolbar.

Table 3-10: Point Editing Commands and Their Functions

Command Name	Pull-down Menu	Function
PtOn	Edit > Point Editing > Control Points On	For turning on the control points of selected curves and surfaces
EditPtOn	Edit > Point Editing > Edit Points On	For turning on the edit points of selected curves
PtOff	Edit > Point Editing > Points Off	For turning off the control points and edit points of curves, and the control points of surfaces
PtOffSelected	Edit > Point Editing > Points Off Selected	For turning off the control points and edit points of selected curves, and the control points of surfaces
Weight	Edit > Point Editing > Edit Weight	For adjusting the weight of selected control points
InsertKnot	Edit > Point Editing > Insert Knot	For inserting knot points to a curve or surface
RemoveKnot	Edit > Point Editing > Remove Knot	For removing knot points from a curve or surface
InsertKink	Edit > Point Editing > Insert Kink	For inserting kink points in a curve
InsertEditPoint	Edit > Point Editing > Insert Edit Point	For inserting edit points in a curve
Hbar	Edit > Point Editing > Handlebar Editor	For using the handlebar to modify a curve or surface
MoveUVN	None	For moving selected control points of a surface in the U and V directions, as well as along the normal direction of a surface (see Chapter 6)

To change the shape of a curve by point editing, perform the following.

1. Start a new file using the metric (millimeter) template.
2. Double click on the Top viewport title to maximize the viewport.
3. Turn off the grid display.
   Command: Grid
4. Construct a free-form curve in accordance with figure 3-44.

More on Curves

Fig. 3-44. Free-form curve constructed.

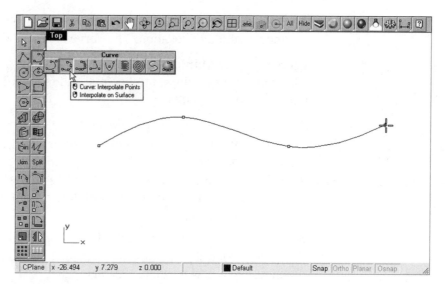

Edit Point Manipulation. To turn on the edit points of the curve and manipulate the points to modify the curve, perform the following steps.

1 Select Edit > Point Editing > Edit Points On, or the Edit Points On button from the Point Editing toolbar.

Command: EditPtOn

2 Select the curve and press the Enter key.

The edit points are now turned on. Note that the number and location of the edit points of your curve may not be the same as that indicated in figure 3-45.

Fig. 3-45. Edit point being manipulated.

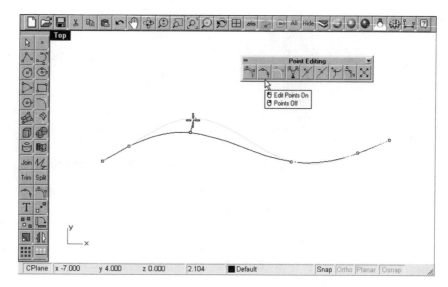

3 Select an edit point, hold down the left mouse button, and drag the mouse to a new location. The curve is modified.

Control Point Manipulation. In the following, you will turn off the edit points, turn on the control points, and modify the curve by manipulating the control points.

1 Select Edit > Point Editing > Points Off, or right-click the Points Off button on the Point Editing toolbar.

Command: PtOff

2 Select Edit > Point Editing > Control Points On, or the Control Points On button from the Point Editing toolbar.

Command: PtOn

3 Select the curve and press the Enter key.

4 Select a control point, hold down the left mouse button, and drag the mouse to a new location. (See figure 3-46.) The curve is modified.

5 Compare the results of manipulating the edit points and the control points.

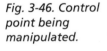
Fig. 3-46. Control point being manipulated.

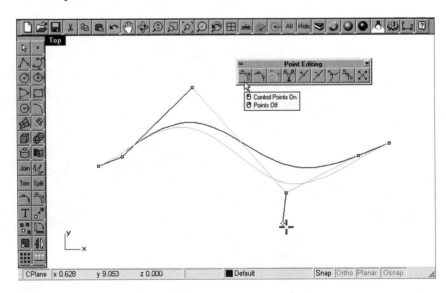

To edit the weight of a control point, continue with the following steps.

6 Select Edit > Point Editing > Edit Weight, or the Edit Control Point Weight button from the Point Editing toolbar.

Command: Weight

7 Select a control point (you can select multiple points) and press the Enter key.

More on Curves

8 In the Set Control Point Weight dialog box, set the weight of the selected point(s) to 2 and click on the OK button. (See figure 3-47.)

Note that the curve is being pulled more closely to the selected control point.

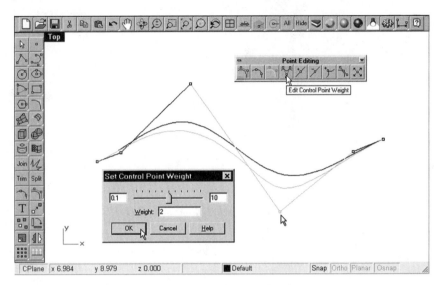

Fig. 3-47. Control point weight being changed.

Insertion of Knot Point. In the following, you will insert a knot to the curve. Adding knots to a curve increases the number of spline segments. As a result, there will be more control points. However, the shape of the curve will not change until you manipulate the control points.

1 Select Edit > Point Editing > Insert Knot, or the Insert Knot button from the Point Editing toolbar.

Command: InsertKnot

2 Select the curve.

3 Select a point(s) along the curve where you want to insert a knot and press the Enter key. (See figure 3-48.)

4 Select Edit > Point Editing > Control Points On, or the Control Points On button from the Point Editing toolbar.

Command: PtOn

Compare figure 3-47 with figure 3-49 to find out the change in the number of control points.

Insertion of Kink Point. In the following, you will insert a kink to the curve. A kink point is a special type of knot on a curve where the tangent direction of contiguous spline segments is not the same.

Fig. 3-48. Knot being inserted in the curve.

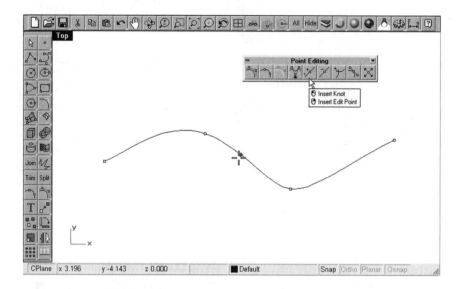

Fig. 3-49. Number of control points increased.

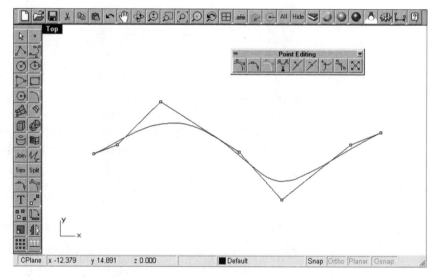

1 Select Edit > Point Editing > Insert Kink, or the Insert Kink button from the Point Editing toolbar.

Command: InsertKink

2 Select the curve.

3 Select a point(s) along the curve where you want to insert a kink and press the Enter key.

A kink is inserted. (See figure 3-50.)

4 Select the kink point, hold down the left mouse button, and drag the mouse to a new location to see the effect of a kink. (See figure 3-51.)

More on Curves

Fig. 3-50. Kink inserted in the curve.

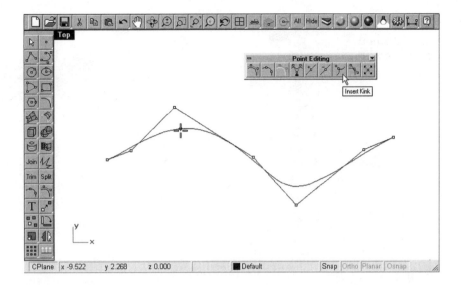

Fig. 3-51. Control point at the kink being manipulated.

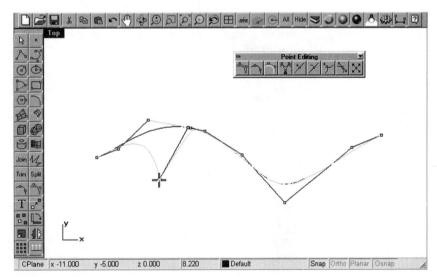

Removal of Knot and Kink Points. To remove the kink knot and other knot(s) of the curve, perform the following steps.

1 Select Edit > Point Editing > Remove Knot, or the Remove Knot button from the Point Editing toolbar.

Command: RemoveKnot

2 Select the knots indicated in figure 3-52 and press the Enter key.

Fig. 3-52. Knots being removed.

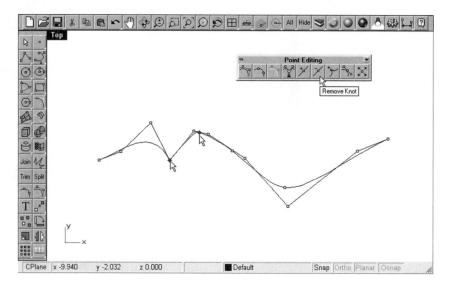

Some knots of the curve are removed. (See figure 3-53.) After you remove a knot from a curve, the number of control points decreases. As a result, the shape of the curve is simplified and changed.

Fig. 3-53. Knots removed.

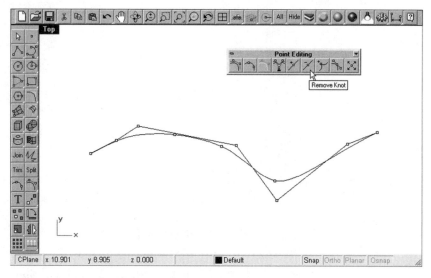

Handlebar Editor. Apart from manipulating the edit points and the control points, you can use a handlebar to edit a curve, as follows.

1 Select Edit > Point Editing > Handlebar Editor, or the Handlebar Editor button from the Point Editing toolbar.

Command: HBar

2 Select the curve. (See figure 3-54.)

More on Curves

Fig. 3-54. Handlebar being manipulated.

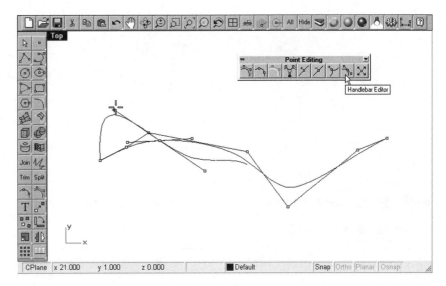

3 Click on an end point of the handlebar and drag.

The curve is modified. (See figure 3-55.) Using a handlebar to edit the shape of a curve, the curve profile changes but the location of the selected point at the handlebar does not change. Save your file as *Edit2.3dm*.

Fig. 3-55. Curve modified.

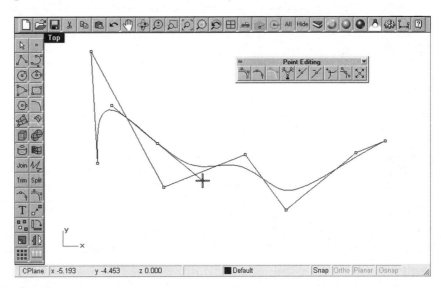

Advanced Editing

Now you will learn more advanced ways to edit a curve. The advanced editing tools are located in the Curve Tools toolbar, shown in figure 3-56.

Chapter 3: Wireframe Modeling and Curves: Part 2

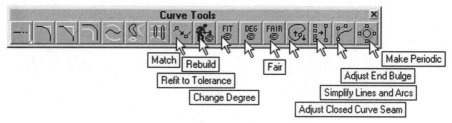

Fig. 3-56. Advanced editing tools on the Curve Tools toolbar.

Table 3-11 outlines the advanced curve editing commands and their locations in the pull-down menu.

Table 3-11: Advanced Curve Editing Commands and Their Functions

Command Name	Pull-down Menu	Function
Match	Curve > Edit Tools > Match	For matching an end point of a curve with the end point of another curve by modifying the position, tangency, and/or curvature of the selected curves
Rebuild	Curve > Edit Tools > Rebuild	For reconstructing the selected curve with a new polynomial degree setting and new number of control points
FitCrv	Curve > Edit Tools > Refit to Tolerance	For constructing a curve from another curve with fewer or more control points by setting a new tolerance value to the curve
ChangeDegree	Curve > Edit Tools > Change Degree	For modifying a curve by changing the polynomial degree setting
Fair	Curve > Edit Tools > Fair	For removing large curvature variations of a curve by limiting the geometry change to specified tolerance
CrvSeam	Curve > Edit Tools > Adjust Closed Curve Seam	For moving and flipping the direction of the seam point of a closed curve
SimplifyCrv	Curve > Edit Tools > Simplify Lines and Arcs	For changing line and arc segments of a curve to NURBS line and arc segments
EndBulge	Curve > Edit Tools > Adjust End Bulge	For adjusting the end bulge of a curve without changing the tangent direction and curvature of the curve
MakeCrvPeriodic	Curve > Edit Tools > Make Periodic	For constructing a closed-loop curve without a kink from a curve

In the following, you will work with various curve-editing tools.

1 Start a new file using the metric (millimeter) template.
2 Double click on the Top viewport title to maximize the viewport.
3 Turn off grid display.
Command: Grid

More on Curves

4 Select Curve > Free-form > Sketch, or the Sketch button from the Curve Tools toolbar to construct two free-form curves in accordance with figure 3-57.

Command: Sketch

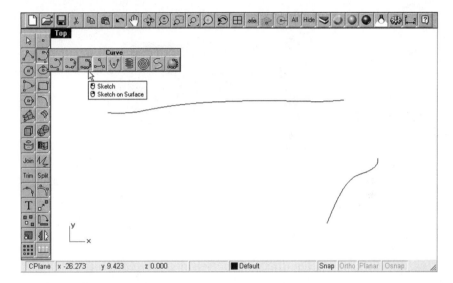

Fig. 3-57. Free-form sketch constructed by sketching.

Matching Two Curves. To modify the shape of the upper curve for one of its end points to match the other curve, perform the following steps.

1 Select Curve > Edit Tools > Match, or the Match button from the Curve Tools toolbar.

Command: Match

2 Select the upper curve. This curve will be modified. (See figure 3-58.)

Fig. 3-58. Curve selected.

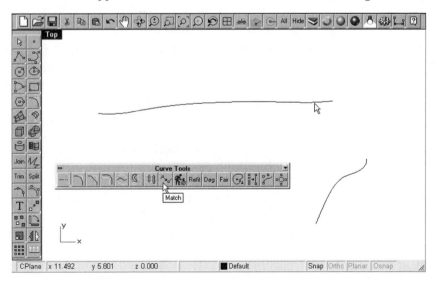

Fig. 3-59. Match Curve Options dialog box.

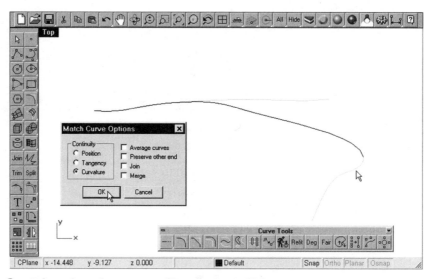

3 Select the other curve. (See figure 3-59.)

4 Try out various options in the Match Curve Options dialog box. Then click on the OK button.

The shape of the selected curve is modified to match the other curve.

Rebuilding. In the following, you will join the curves, and then rebuild the curve with a different number of control points.

1 Select Edit > Join, or the Join button from the Main toolbar.
Command: Join

2 Select the curves and press the Enter key.

3 Select Curve > Edit Tools > Rebuild, or the Rebuild button from the Curve Tools toolbar.
Command: Rebuild

4 Select the curve and press the Enter key. (See figure 3-60.)

5 In the Rebuild Curve dialog box, set the number of control points to 5. (The default value shown may not be the same as yours.) Leave the degree unchanged.

6 Click on the OK button. (See figure 3-61.)

The selected curve is rebuilt with a curve consisting of different number of control points. The Rebuild command always removes kinks. This is a technique commonly used by designers to both simplify curves and avoid problems with kinks. (If you reduce the control point number, the curve simplifies and the shape changes. If you increase the control point number, the shape will not change until you manipulate the control points.)

More on Curves

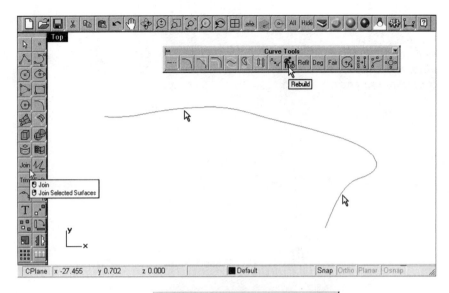

Fig. 3-60. Curves joined.

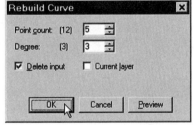

Fig. 3-61. Rebuild Curve dialog box.

Fitting Polyline. To construct a polyline and refit a free-form curve from it, perform the following steps.

1. Select Curve > Line > Polyline, or the Polyline button from the Lines toolbar.

   Command: Polyline

2. Construct a polyline with seven segments in accordance with figure 3-62.

3. Select Curve > Edit Tools > Refit to Tolerance, or the Refit to Tolerance button from the Curve Tools toolbar. (See figure 3-63.)

   Command: FitCrv

4. Select the polyline and press the Enter key.

You may keep the original curve or delete it by using the Delete Input option. Here the input curve is deleted.

5. Accept the default degree of polynomial by pressing the Enter key.

A free-form curve is refit from the selected polyline.

Fig. 3-62. Curve rebuilt with a different number of control points and a polyline constructed.

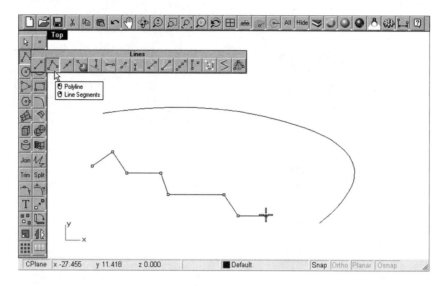

Fig. 3-63. Curve being refit from a polyline.

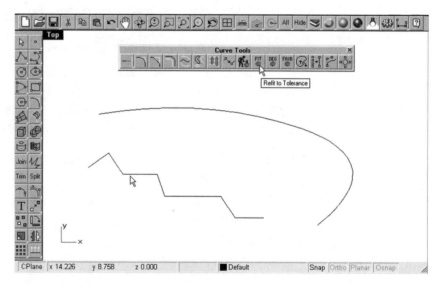

Changing Polynomial Degree. In the following, you will change the polynomial degree of a curve. To see the change, you will turn on the control points.

1 Select Edit > Point Editing > Control Points On, or the Control Points On button from the Point Editing toolbar.

Command: PtOn

2 Select the curve and press the Enter key.

Now change the polynomial degree and watch the change in the number of control points in the curve.

More on Curves

3. Select Curve > Edit Tools > Change Degree, or the Change Degree button from the Curve Tools toolbar.

   Command: ChangeDegree

4. Select the curve and press the Enter key.
5. Type *4* at the command line area to specify a degree 4 curve.

The degree of polynomial of the selected curve is changed. To reiterate, you do not need to turn on the control points to change the polynomial degree. (See figure 3-64.)

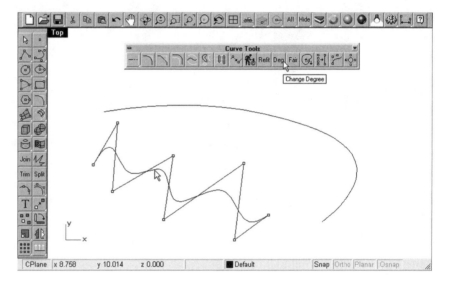

Fig. 3-64. Degree of polynomial of the selected curve being changed.

Fairing. In the following, to remove large curvature variation, you will fair the curve.

1. Select Curve > Edit Tools > Fair, or the Fair button from the Curve Tools toolbar.

   Command: Fair

2. Select the curve and press the Enter key.
3. Type *4* at the command line area to specify a tolerance of 4 mm.
4. Turn off the control point.

The curve is faired. (See figure 3-65.)

Simplifying. To replace arc segments in a curve with a true NURBS curve, you simplify the curve. In the following, you will add an arc segment to a curve and then simplify it.

1. Select Curve > Extend > By Arc, or the Extend by Arc button from the Extend toolbar.

   Command: ExtendByArc

Fig. 3-65. Curve being faired.

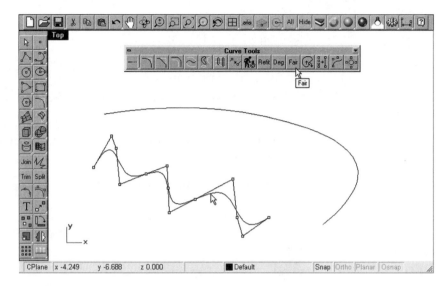

2 Select the curve indicated in figure 3-66.
3 Specify a point to indicate the radius of the arc.
4 Specify a point to indicate the end point of the arc.
5 Select Curve > Edit Tools > Simply Lines and Arcs, or the Simply Lines and Arcs button from the Curve Tools toolbar.
Command: SimplifyCrv
6 Select the curve indicated in figure 3-67 and press the Enter key.

The curve is extended with an arc segment and the arc segment of the curve is replaced with a true NURBS segment. You will not find much change on the shape of the curve.

Fig. 3-66. Curve being extended.

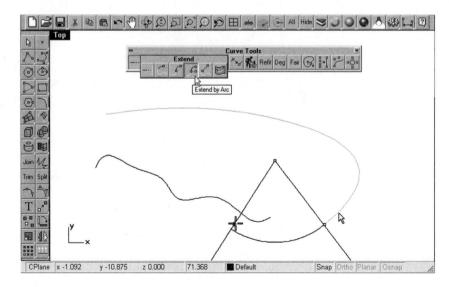

Fig. 3-67. Curve extended.

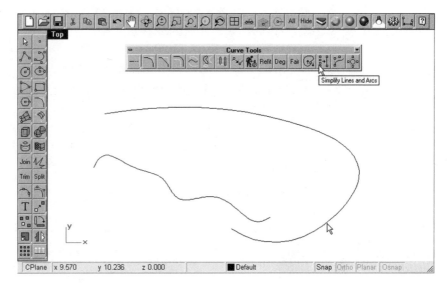

Editing End Bulge. To modify the shape of a curve without changing the tangent direction at the end points, you edit the end bulge. As a result, the continuity at the junction of the curves remains unchanged.

1 Select Curve > Edit Tools > Adjust End Bulge, or the Adjust End Bulge button from the Curve Tools toolbar.
   Command: EndBulge
2 Select the curve indicated in figure 3-68.
3 Select and drag the control points.
4 Press the Enter key.

The end bulge is modified.

Fig. 3-68. End bulge being modified.

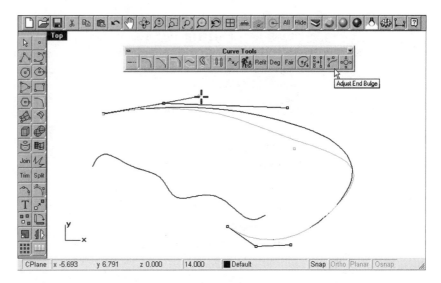

Making a Curve Periodic. In the following, you will make the curve periodic. A periodic curve is a closed curve with no kink point. If you make an open-loop curve periodic, the curve becomes a closed-loop curve.

1 Select Curve > Edit Tools > Make Periodic, or the Make Periodic button from the Curve Tools toolbar.

Command: MakeCrvPeriodic

2 Select the curve indicated in figure 3-69.

3 Press the Enter key.

The curve becomes a periodic curve.

Fig. 3-69. Curve being made periodic.

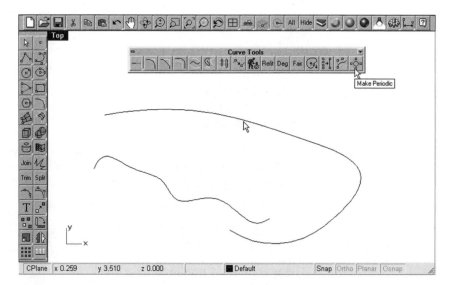

Adjusting Seam Location. A closed-loop curve, like an open-loop curve, has a start point and an end point. The difference between a closed-loop curve and an open-loop curve is that the start point and end point coincide. The coincident point is called the seam point. In the following, you will adjust the location of the seam point of a closed curve.

1 Select Curve > Edit Tools > Adjust Closed Curve Seam, or the Adjust Closed Curve Seam button from the Curve Tools toolbar.

Command: CrvSeam

2 Select the curve indicated in figure 3-70 and press the Enter key.

3 Select the seam point

4 Select a point along the curve to indicate a new position for the seam point.

5 Press the Enter key.

More on Curves

The seam is relocated. Save your file as *Edit3.3dm*.

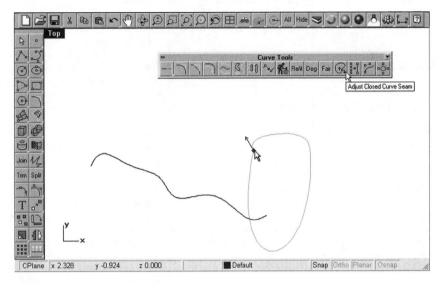

Fig. 3-70. Seam point being adjusted.

Transforming Curves

Transforming concerns translation and deformation. Some of the transform tools described here will apply not only to curves but to surfaces and solids. You perform transform operations from the Transform toolbar, shown in figure 3-71.

Fig. 3-71. Transform toolbar.

Table 3-12 outlines the transform commands and their locations in the pull-down menu.

Table 3-12: Transform Commands and Their Functions

Command Name	Pull-down Menu	Function
Move	Transform > Move	For moving selected objects.
Copy	Transform > Copy	For copying selected objects.
Rotate	Transform > Rotate	For rotating selected objects about a point on the current construction plane.
Rotate3D	Transform > Rotate 3-D	For rotating selected objects about a specified axis.

Command Name	Pull-down Menu	Function
Scale	Transform > Scale > Scale 3-D	For uniformly scaling the X, Y, and Z dimensions of selected objects.
Scale2D	Transform > Scale > Scale 2-D	For uniformly scaling the X and Y dimensions (with Z dimension unchanged) of selected objects.
Scale1D	Transform > Scale > Scale 1-D	For scaling selected objects in a specified direction.
ScaleNU	Transform > Scale > Non-Uniform Scale	For scaling selected objects by specifying individual scale factors for the X, Y, and Z dimensions.
Shear	Transform > Shear	For deforming selected objects by shearing them about a shear plane. (The area of a sheared surface and the volume of a sheared box remain unchanged.)
Mirror	Transform > Mirror	For constructing a mirror copy of selected objects.
Orient	Transform > Orient > 2 Points	For transforming selected objects to a specified location and orientation and uniformly scaling the selected objects by specifying two reference points and two target points.
Orient3Pt	Transform > Orient > 3 Points	For translating selected objects to a specified location by defining target point, base direction, and orientation.
OrientOnSrf	Transform > Orient > On Surface	For translating selected objects to specified locations on a surface.
OrientPerpToCrv	Transform > Orient > Perpendicular to Curve	For translating selected objects perpendicular to a curve.
RemapCPlane	Transform > Orient > Remap to CPlane	For translating selected objects to a specified construction plane.
Array	Transform > Array > Rectangular	For constructing a rectangular pattern of selected objects.
ArrayPolar	Transform > Array > Polar	For constructing a circular pattern of selected objects.
ArrayCrv	Transform > Array > Along Curve	For constructing a pattern of selected objects along a curve.
ArraySrf	Transform > Array > Along Surface	For constructing a rectangular pattern of selected objects along the U and V directions of a surface.
SetPt	Transform > Set Points	For deforming selected objects by aligning selected control points.
ProjectToCPlane	Transform > Project to Cplane	For projecting selected objects to a specified construction plane.

More on Curves

Command Name	Pull-down Menu	Function
Twist	Transform > Twist	For deforming selected curves or surfaces by twisting them about a specified axis.
Bend	Transform > Bend	For deforming selected curves or surfaces by bending them about a bend axis.
Taper	Transform > Taper	For deforming selected curves or surfaces by tapering the objects.
Flow	Transform > Flow along Curve	For deforming selected curves or surfaces by specifying an original backbone curve and a new backbone curve.
Smooth	Transform > Smooth	For smoothing selected curves and surfaces by removing unwanted details and loops.

In the sections that follow, you will learn to use various transform tools on curves.

Move, Copy, Rotate, and Mirror

The simplest methods of transforming a curve are to move or copy it to a new location, rotate it about an axis, or make a mirror copy. After moving or copying, the orientation of the curve remains unchanged. To change the orientation of a curve, you rotate it about a point on the construction plane or about a 3D axis. To rotate a curve about a 3D axis, perform the following steps.

1 Start a new file. Use the metric (millimeter) template.
2 Construct a free-form curve in accordance with figure 3-72.

Fig. 3-72. Start and end of rotate axis specified.

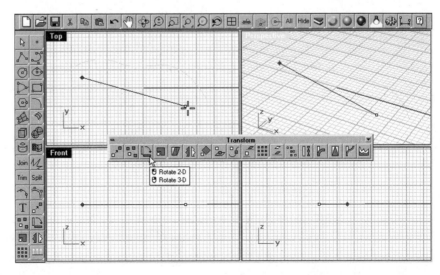

3 Select Transform > Rotate 3-D, or the Rotate 3-D button from the Transform toolbar.
   Command: Rotate3D
4 Select the curve and press the Enter key.
5 Indicate the start of the rotation axis and the end of the rotation axis. (See figure 3-72.)
6 Specify the rotation angle or indicate the first reference point and the second reference point.

The curve is rotated. (See figure 3-73.)

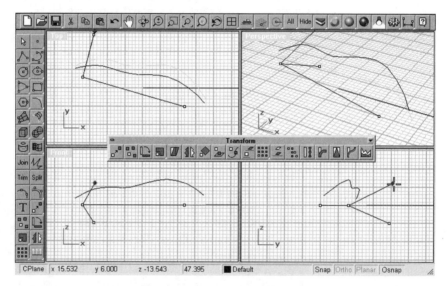

Fig. 3-73. First and second reference point indicated.

To rotate a control point of the curve, perform the following steps.

7 Turn on the control points of the curve.
8 Use the Rotate3D command again.
9 Select a control point and press the Enter key. (See figure 3-74.)
10 Indicate the start and end of the rotation axis.
11 Specify the rotation angle or indicate the first and second reference point.

The selected control point is rotated.

12 Turn off the control points.

Scale

There are four ways to scale a curve. You scale the entire curve or selected control points of the curve uniformly in X, Y, and Z dimensions, in two

More on Curves

Fig. 3-74. Selected control point being rotated about a 3D axis.

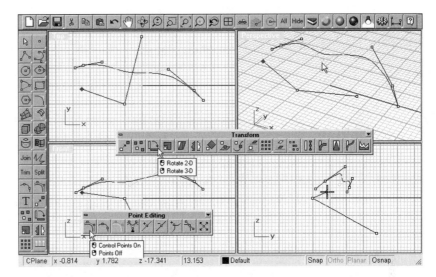

dimensions, in one dimension, or non-uniformly in three dimensions. When you scale the control points, you translate the locations of the selected control points in relation to a reference point. To scale a curve uniformly in the X, Y, and Z dimensions, perform the following steps.

1 Select Transform > Scale > Scale 3-D, or the Scale 3-D button from the Transform toolbar.

Command: Scale

2 Select the curve and press the Enter key.

3 Specify the scale factor at the command line area or indicate first and second reference point. (See figure 3-75.)

The entire curve is scaled.

Fig. 3-75. Curve being scaled.

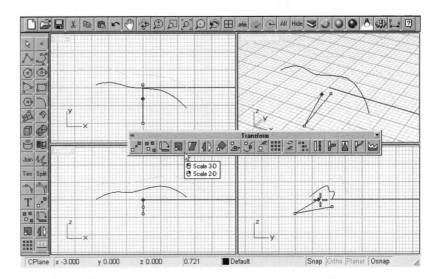

Shear

You can shear the entire curve or selected control points of the curve. Now you will deform the entire curve by shearing.

1. Select Transform > Shear, or the Shear button from the Transform toolbar.

   Command: Shear

2. Select the curve and press the Enter key.
3. Specify an origin point.
4. Specify a reference point.
5. Specify a shear angle or select a point to indicate the shear angle. (See figure 3-76.)

The entire curve is sheared. Save your file as *Transform.3dm*.

Fig. 3-76. Curve being sheared.

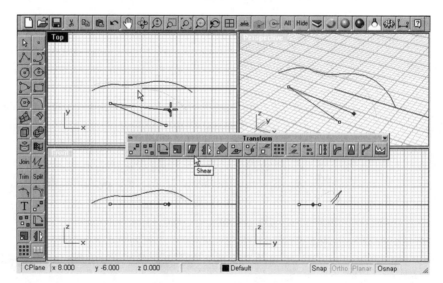

Orient

There are five ways to orient selected objects. These methods are discussed in the sections that follow.

Repositioning and Scaling. To construct a free-form curve and a circle and orient the end points of the curve to the quadrant points of the circle, perform the following steps.

1. Start a new file. Use the metric (millimeter) template. Double click on the Top viewport title to maximize the viewport.
2. Construct a free-form curve and a circle in accordance with figure 3-77.

More on Curves

Fig. 3-77. Curve and circle.

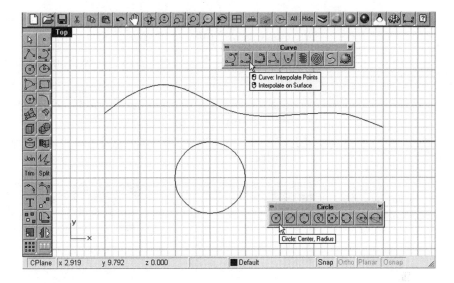

3 Select Transform > Orient > 2 Points, or the Orient/2 Points button from the Transform toolbar.

Command: Orient

4 Select the curve and press the Enter key.
5 Check the Osnap button on the status bar.
6 Check the End box and the Quad box in the Osnap dialog box.
7 Select the end points of the curve as the reference points.
8 Select the quadrant points of the circle as the target points. (See figure 3-78.)

Fig. 3-78. Curve oriented to the circle.

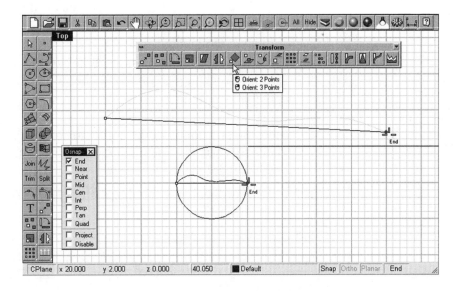

The curve is oriented to the circle. Note the change in the overall scale of the curve. Save your file as *Orient.3dm*.

Repositioning. To orient a polyline without scaling it, perform the following steps.

1 Start a new file. Use the metric (millimeter) template. Double click on the Top viewport to maximize it.

2 Construct a free-form curve and a triangle (polyline) in accordance with figure 3-79.

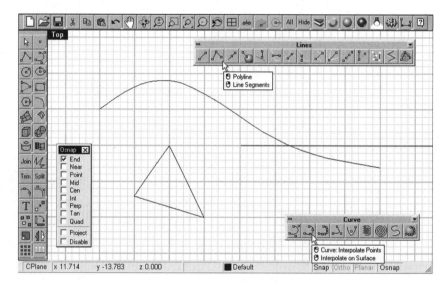

Fig. 3-79. Free-form curve and polyline.

3 Select Transform > Orient > 3 Points, or the Orient/3 Points button from the Transform toolbar.

Command: Orient3Pt

4 Select the triangle and press the Enter key.

5 Select two end points of the triangle (to form a reference line) and a point above (to indicate the reference direction) as the reference points.

6 Select the end points of the curve (to serve as a target line) and a point above (to indicate the target direction) as the target points. (See figure 3-80.)

The triangle is oriented. Note that the oriented object does not change shape. Save your file as *Orient3Pt.3dm*.

Relocating Perpendicular to a Curve. In the following, you will orient a curve perpendicular to another curve. Note that this command is construction plane dependent. Open the file *Orient3Pt.3dm* if you already closed it.

More on Curves 133

Fig. 3-80. Polyline being oriented.

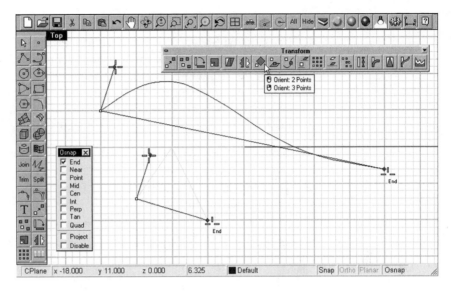

1. Select File > Save As and specify the file name as *OrientPerpToCrv.3dm*.
2. Select Transform > Orient > Perpendicular to Curve, or the Orient Perpendicular to Curve button from the Transform toolbar.
   Command: OrientPerpToCrv
3. Select the polyline and press the Enter key.
4. Check the Cen box of the Osnap dialog box.
5. Select the center of the polyline as the base point. (See figure 3-81.)
6. Select the free-form curve.
7. Select the end point of the curve.

Fig. 3-81. Polyline being oriented perpendicular to the curve.

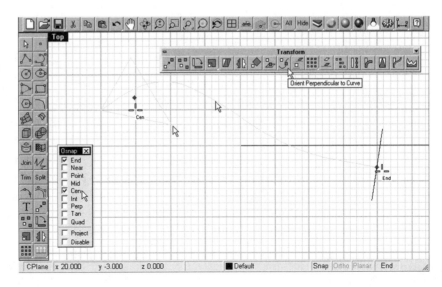

The polyline is oriented perpendicular to the curve. Save your file.

Relocating to Selected Construction Plane. The fourth way to orient a curve is to reposition it to a selected construction plane, which you will do in the following. This is especially useful for orienting 2D drawings to 3D construction planes. For example, you can map the front view to the Front viewport or the right view to the Right viewport. Open the file *OrientPerpToCrv.3dm* if you already closed it.

1 Select File > Save As and specify the file name as *RemapCPlane.3dm*.

2 Double click on the Top viewport title to return to a four-viewport display. (See figure 3-82.)

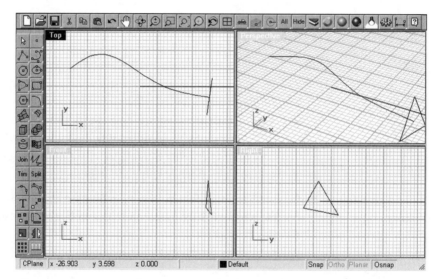

Fig. 3-82. Four-viewport display.

3 Select Transform > Remap to Cplane, or the Remap to CPlane button from the Transform toolbar.

Command: RemapCPlane

4 Select the polyline and press the Enter key.

5 Click on the Front viewport.

The polyline is oriented to the front viewport. (See figure 3-83.) Save your file.

Relocating to a Surface. The fifth way to orient an object is to orient it to a surface. In the following, you will construct a spherical surface and a circle and orient the circle to the surface. (You will learn surface and solid construction in later chapters.)

1 Start a new file. Use the metric (millimeter) template.

More on Curves

Fig. 3-83. Polyline oriented to the front viewport.

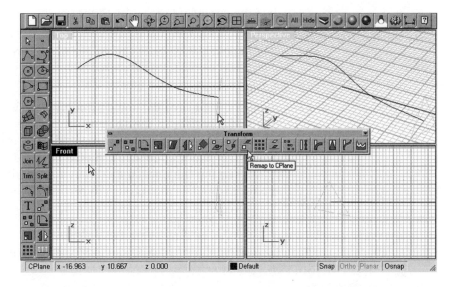

Fig. 3-84. A circle and a sphere.

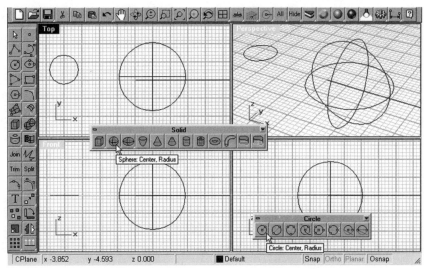

2. Construct a circle and sphere in accordance with figure 3-84.
3. To construct a sphere, select Solid > Sphere > Center, Radius, or the Sphere/Center, Radius button from the Solid toolbar.
   Command: Sphere
4. Select a point to indicate the center location.
5. Type a value or select a point to indicate the radius.
6. Select Transform > Orient > On Surface, or the Orient on Surface button from the Transform toolbar.
   Command: OrientOnSrf

7 Select the circle and press the Enter key.

8 Type *CEN* at the command line area to specify a center object snap.

9 Select the circle to specify the center of the circle as the point to orient from.

10 Select the sphere.

11 Move the cursor over the sphere and note how the circle transforms to the surface of the sphere. You will see two arrows: one white and one red. The white arrow indicates the normal direction of the circle, and the red arrow indicates the U direction of the circle. You can flip the normal direction and change the alignment of the object from U direction to V direction.

12 Select a location and press the Enter key. (You may make multiple selections before pressing the Enter key.)

The circle is oriented. (See figure 3-85.) Save your file as *OrientOn-Srf.3dm*.

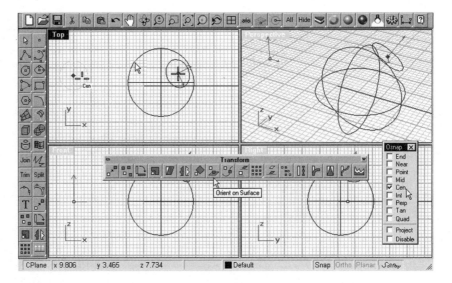

Fig. 3-85. Circle being oriented onto the surface of a sphere.

Arraying

An array is a set of repeated objects in a pattern. There are four ways to construct an array. These methods are discussed in the sections that follow.

Rectangular Array. In the following, you will construct a rectangular array of a polyline. In the array, there are two objects in the X direction at a spacing of 10 mm, three objects in the Y direction at a spacing of 12 mm, and one object in the Z direction.

More on Curves

1 Start a new file. Use the metric (millimeter) template. Double click on the Top viewport to maximize it.
2 Construct a polyline in accordance with figure 3-86.
3 Select Transform > Array > Rectangular, or the Rectangular Array button from the Array toolbar.
Command: Array
4 Select the polyline and press the Enter key.
5 Type *3* to specify the number of objects in the X direction.
6 Type *2* to specify the number of objects in the Y direction.
7 Type *1* to specify the number of objects in the Z direction.
8 Type *10* to specify X spacing.
9 Type *12* to specify Y spacing.

A rectangular array is constructed. Save your file as *Array.3dm*.

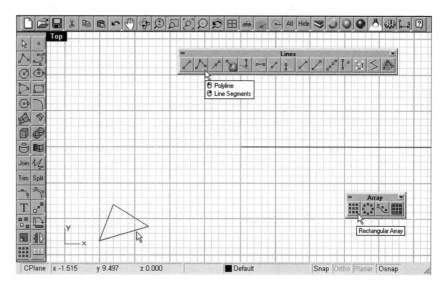

Fig. 3-86. Polyline constructed and being arrayed.

Polar Array. To construct a polar array, perform the following steps.

1 Select File > Save As and specify the file name as *ArrayPolar.3dm*.
2 Select Transform > Array > Polar, or the Polar Array button from the Array toolbar.
Command: ArrayPolar
3 Select the polyline indicated in figure 3-87 and press the Enter key.
4 Indicate the center of the polar array. (See figure 3-87.)
5 Type *3* to specify the number of elements in the array.
6 Type *360* (or press the Enter key if the default is 360) to specify the angle to fill.

A polar array is constructed. Save your file.

Fig. 3-87 Polar array being constructed.

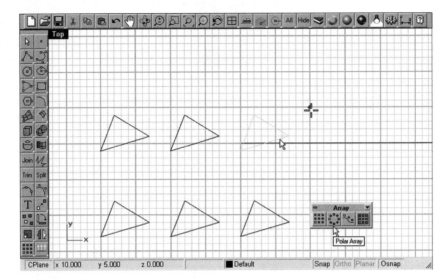

Array Along a Curve. To construct an array along a curve, perform the following steps.

1 Start a new file. Use the metric (millimeter) template. Double click on the Top viewport to maximize it.
2 Construct a triangle (polyline) and a free-form curve in accordance with figure 3-88.
3 Select Transform > Array > Along Curve, or the Array Along Curve button from the Array toolbar.
   Command: ArrayCrv

Fig. 3-88. Polyline being arrayed along the curve.

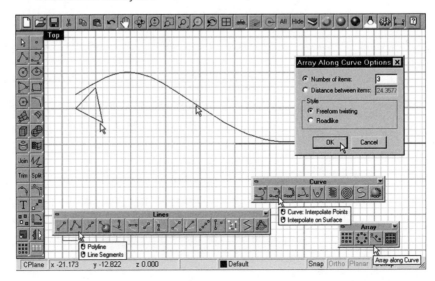

More on Curves

4 Select the polyline and press the Enter key.

5 Select the free-form curve to use it as the path curve.

6 In the Array Along Curve Options dialog box, specify 3 items in the *Number of items* box, check the *Freeform twisting* option, and click on the OK button.

The polyline is arrayed along the curve. (See figure 3-89.) Note that the triangle rotates about its own axis as it arrays along the curve. Save your file as *ArrayCrv.3dm*.

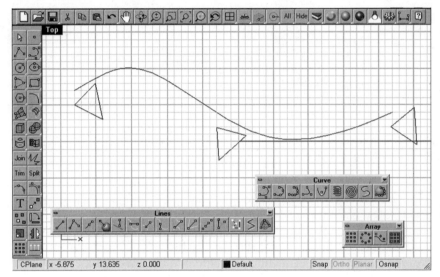

Fig. 3-89. Polyline arrayed along the curve.

Array on a Surface. To array a curve on a cylindrical surface, perform the following steps.

1 Start a new file. Use the metric (millimeter) template.

2 Construct a circle and a cylinder.

3 To construct a cylinder, select Solid > Cylinder, or the Cylinder button from the Solid toolbar.

Command: Cylinder

4 Select a point to indicate the center location.

5 Select a point to indicate the radius.

6 Select a point to indicate the location of the end of the cylinder.

7 Select Transform > Array > Along Surface, or the Array on Surface button from the Array toolbar.

Command: ArraySrf

8 Select the circle and press the Enter key.

9 Select the center of the circle to indicate the base point. (See figure 3-90.)

Fig. 3-90. Circle being arrayed along the cylindrical face of the cylinder.

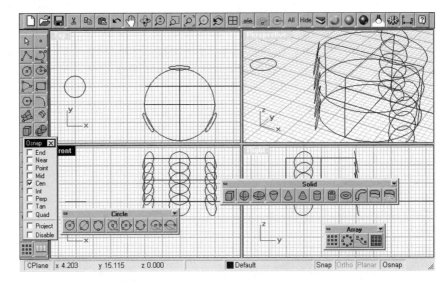

10 In the Front viewport, select a point to indicate the normal direction.
11 Select the cylindrical face of the cylinder to indicate a surface to array along.
12 Type *4* to specify the number of elements in the U direction.
13 Type *5* to specify the number of elements in the V direction.

An array is constructed. (See figure 3-91.) Because the cylindrical face is a closed surface, the first set of U elements overlaps with the last set of U elements. Therefore, you see what appears to be three U elements in the figure. Save your file as *ArraySrf.3dm*.

Fig. 3-91. Circle arrayed.

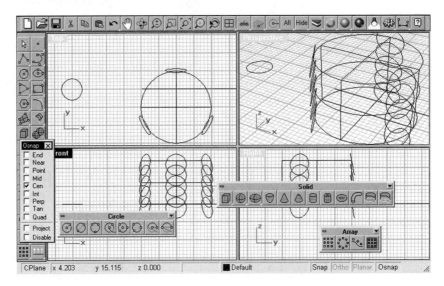

More on Curves

Setting Points

To transform a curve by setting its control points, perform the following.

1. Start a new file. Use the metric (millimeter) template. Double click on the Top viewport to maximize it.
2. Construct a free-form curve. (See figure 3-92.)
3. Turn on the control points.
4. Select Transform > Set Points, or the Set Points button from the Transform toolbar.
   Command: SetPt
5. Select the control points indicated in figure 3-93.

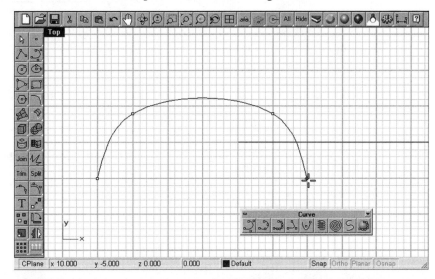

Fig. 3-92. Free-form curve being constructed.

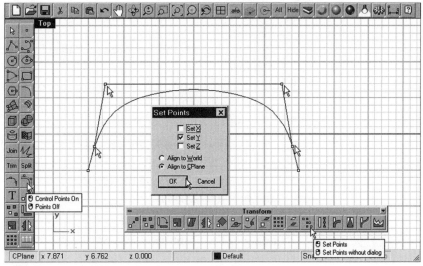

Fig. 3-93. Control points selected.

6 In the Set Points dialog box, check the Set Y box and clear the other two boxes, select Align to CPlane, and click on the OK button.

7 Select a point on the screen to indicate a Y coordinate. (See figure 3-94.)

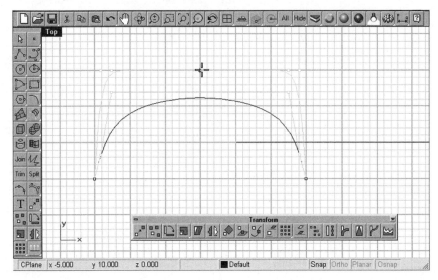

Fig. 3-94. Selected control points being set.

The Y coordinate of the selected control points is aligned. Save your file as *SetPt.3dm*.

Projection

You can transform a curve by projecting the entire curve or selected control points of the curve onto a selected construction plane. To project selected control points of a curve, perform the following steps.

1 Select File > Save As and specify a new file name of *ProjectToC-Plane.3dm*.

2 Double click on the Top viewport title to return to a four-viewport display.

3 Select Transform > Project to Cplane, or the Project to CPlane button from the Transform toolbar.
Command: ProjectToCPlane

4 Select the Front viewport, select the control points indicated in figure 3-95, and press the Enter key.

5 In the ProjectToCPlane dialog box, select the Yes button to delete the original objects.

The selected control points are projected to the construction plane corresponding to the Front viewport. Save your file.

More on Curves

Fig. 3-95. Points being projected.

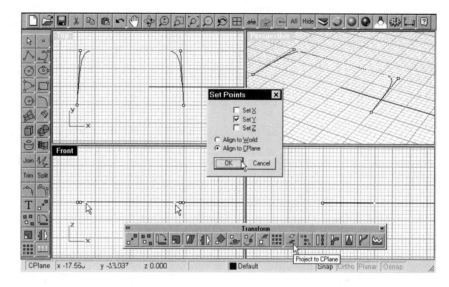

Twisting

You can transform a curve by twisting the entire curve or selected control points of the curve. To twist the selected control points of a curve, perform the following steps.

1 Select File > Save As and specify a new file name of *Twist.3dm*.

2 Select Transform > Twist, or the Twist button from the Transform toolbar.

Command: Twist

3 Select the control points indicated in figure 3-96 and press the Enter key.

Fig. 3-96. Control points selected and twist axis specified.

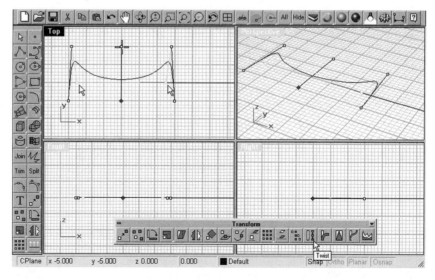

Chapter 3: Wireframe Modeling and Curves: Part 2

4 Specify two points in the Top viewport to indicate the twist axis.

5 In the Front viewport, specify two points to indicate the twist angle. (See figure 3-97.) (You may type a number at the command line area to specify the angle.)

The selected control points are twisted. Save your file.

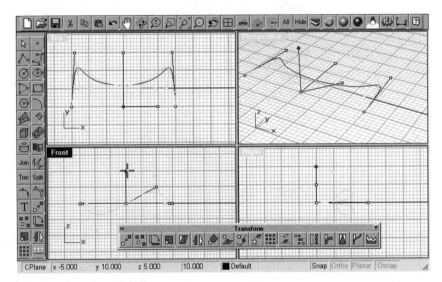

Fig. 3-97. Twist angle being specified.

Bending

You can transform a curve by bending the entire curve or selected control points of the curve. To bend the selected control points of a curve, perform the following steps.

1 Select File > Save As and specify a new file name of *Bend.3dm*.

2 Select Transform > Bend, or the Bend button from the Transform toolbar.

   Command: Bend

3 Select the control points indicated in figure 3-98 and press the Enter key.

4 Indicate two points to specify a reference spline.

5 Indicate a point to specify the amount of bend.

The selected control points are bent. Save your file.

Tapering

You can transform a curve by tapering the entire curve or selected control points of the curve, as follows.

More on Curves

Fig. 3-98. Selected control points being bent.

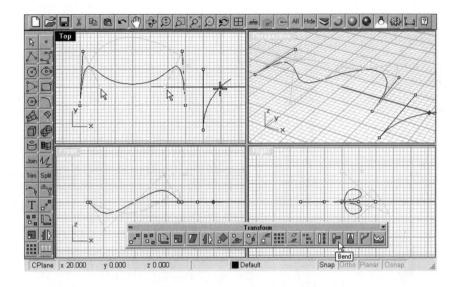

Fig. 3-99. Control points selected.

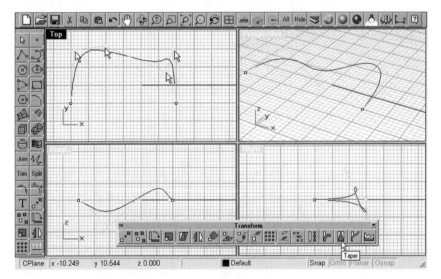

1 Select File > Save As and specify a new file name of *Taper.3dm*.

2 Select Transform > Taper, or the Taper button from the Transform toolbar.

Command: Taper

3 Select the control points indicated in figure 3-99 and press the Enter key.

4 Specify two points to indicate the taper axis.

5 Specify two points to indicate the start and end distances. (See figure 3-100.)

The selected control points are tapered. Save your file.

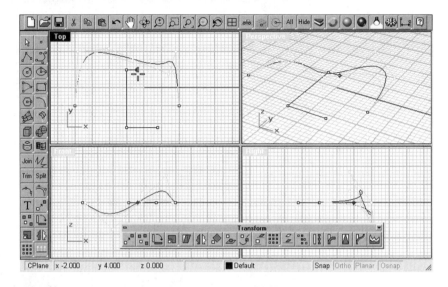

Fig. 3-100. Selected control points being tapered.

Smoothing

Smoothing transforms a curve by averaging its control points. To smooth a curve, perform the following steps.

1 Select File > Save As and specify a new file name of *Smooth.3dm*.
2 Turn off the control points.
3 Select Transform > Smooth, or the Smooth button from the Transform toolbar.

   Command: Smooth

4 Select the curve and press the Enter key.
5 In the Smooth dialog box, select Smooth X, Smooth Y, and Smooth Z, and the OK button. (See figure 3-101.)

The curve is smoothed. Save your file.

6 Turn on the control points again. (See figure 3-102.) Compare it with figure 3-100.

More on Curves

Fig. 3-101. Curve being smoothed.

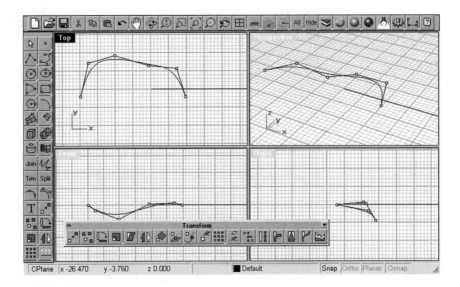

Fig. 3-102. Curve smoothed.

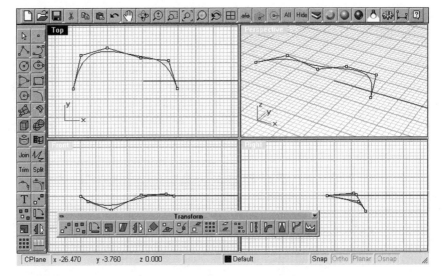

Flowing Along a Curve

You can transform a curve by setting the curve or its control points to flow along a selected curve, as follows.

1 Select File > Save As and specify a new file name of *Flow.3dm*.
2 Double click on the Top viewport title to maximize the viewport.
3 Construct a closed-loop free-form curve and a line in accordance with figure 3-103.

Fig. 3-103. Closed-loop curve and line constructed.

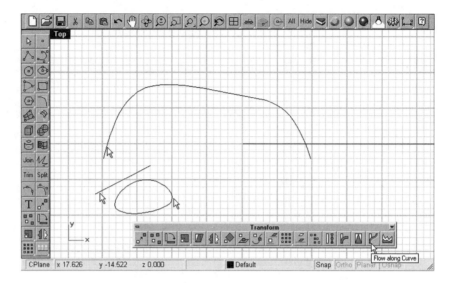

4 Select Transform > Flow, or the Flow button from the Transform toolbar.

Command: Flow

5 Select the closed-loop curve and press the Enter key.
6 Select the line to use as the original backbone curve.
7 Select the other curve to use as the new backbone curve.

The closed-loop curve is transformed to flow along the other curve. (See figure 3-104.)

Fig. 3-104. Curve transformed.

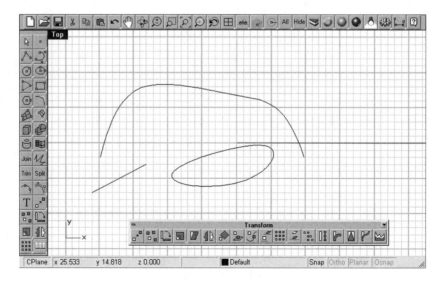

Summary

This chapter focused on four major issues: curve tracing, deriving curves, editing curves, and transforming curves. If you already have a design sketch and want to construct curves from your sketch, you scan the sketch as a digital image, place the digital image in the viewport's background, and trace a curve using Rhino.

By deriving a curve from existing objects, you get a curve that has some form or shape correlation with existing objects. The objects from which you derive a curve can be curves, surfaces, or solids. You learned how to derive curves from existing curves in seven ways: extending a curve, filleting two curves, chamfering two curves, offsetting a curve, blending two curves, deriving a 3D curve from two planar curves residing in two viewports, and deriving a set of cross-section curves across a set of longitudinal profile curves.

To meet design needs in surface and solid construction, you edit and transform curves. The following are the three domains of curve editing.

- ❑ Joining several contiguous curves into a single curve, exploding a set of joined curves into individual curves, trimming curves, and splitting curves
- ❑ Manipulation of the curve's control points, edit points, knot points, and selected locations along the curve
- ❑ Changing the curve's polynomial degree and fit tolerance

Curve transformation concerns translation and deformation. You transform curves by moving, copying, rotating, and arraying. You deform curves by scaling, shearing, orienting, setting points, projecting, twisting, bending, tapering, smoothing, and flowing.

Review Questions

1. Explain how to use free-hand sketches to help construct curves in the computer.
2. Depict the methods of deriving curves from existing curves.
3. Illustrate the methods of editing a curve by manipulating its control points, edit points, knots, and points along it.
4. Illustrate how a curve's polynomial degree and fit tolerance can be modified.
5. Explain methods of transforming a curve.

Chapter 4

Wireframe Modeling and Curves: Part 3

■■ Objectives

The goals of this chapter are to explain methods of analyzing 3D curves and to let you practice 3D curve construction using Rhino. After studying this chapter, you should be able to:

- ❒ Use various curve analysis tools
- ❒ Use Rhino as a tool in constructing 3D curves

■■ Overview

After realizing the importance and significance of 3D curves and learning how to construct, edit, and transform various types of curves in chapters 2 and 3, you will now learn how to analyze 3D curves and work on a number of projects. You will use knowledge gained here to construct 3D curves (for creating surfaces) in subsequent chapters.

■■ Analyzing Curves

You have learned how to construct point objects, basic curves, and derived curves; to edit curves; and to transform curves. In this chapter you will learn how to analyze point objects and curves. You perform analysis operations from the Analyze toolbar, shown in figure 4-1.

Fig. 4-1. Analyze toolbar.

Table 4-1 outlines the functions of the curve and point analysis commands and their location in the pull-down menu.

Table 4-1: Curve and Point Analysis Commands and Their Functions

Command Name	Pull-down Menu	Function
EvaluatePoint	Analyze > Point	For finding out the coordinates of a selected point
Length	Analyze > Length	For measuring the length of a selected curve or edge of a surface
Distance	Analyze > Distance	For measuring the distance between two selected points
Angle	Analyze > Angle	For measuring the angle between two lines by indicating the end points of the lines
Radius	Analyze > Radius	For measuring the radius of curvature at selected locations along a curve
BoundingBox	Analyze > Bounding Box	For constructing a bounding box to enclose a selected curve, surface, or polysurface
CurvatureGraphOn	Analyze > Curve > Curvature Graph On	For displaying a graph along a curve to illustrate the radius of curvature along the curve
CurvatureGraphOff	Analyze > Curve > Curvature Graph Off	For closing the curvature graph
Curvature	Analyze > Curve > Curvature Circle	For constructing a tangent circle along a curve to depict the radius of curvature at a selected location
Gcon	Analyze > Curve > Geometric Continuity	For discerning the geometric continuity at a join between two curves
CrvDeviation	Analyze > Curve > Deviation	For discerning the deviation between two curves
List	Analyze > Diagnostics > List	For obtaining the data of selected objects
Check	Analyze > Diagnostics > Check	For diagnosing a selected object for error
SelBadObjects	Analyze > Diagnostics > Select Bad Objects	For selecting any object that does not pass the check test
Dir	Analyze > Direction	For displaying and flipping the direction of selected objects

Evaluate Point

To discern the coordinates of selected points, perform the following steps.

Analyzing Curves

1 Open the file *Point1.3dm*.
2 Click on the Osnap button in the status bar.
3 In the Osnap dialog box, check the Point option.
4 Select Analyze > Point, or the Evaluate Point button from the Analyze toolbar.
Command: EvaluatePoint
5 Select a point object. (See figure 4-2.)

The coordinates of the selected point are displayed in the command line area.

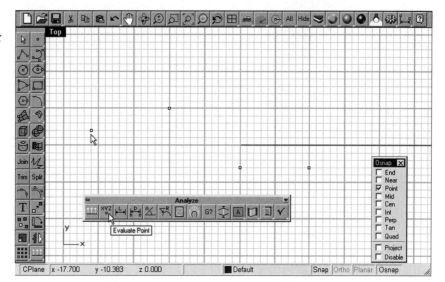

Fig. 4-2. Point object being evaluated.

6 Open the file *Edit1.3dm*.
7 Check the End option of the Osnap dialog box.
8 Select Analyze > Point, or the Evaluate Point button from the Analyze toolbar.
Command: EvaluatePoint
9 Select the end point indicated in figure 4-3.

The coordinates of the selected end point are displayed.

Length Measurement

To measure the length of a curve, perform the following steps.

1 Select Analyze > Length, or the Length button from the Analyze toolbar.
Command: Length

Fig. 4-3. End point of a curve selected.

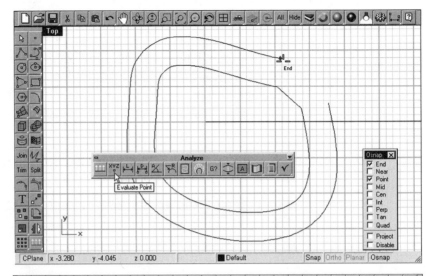

Fig. 4-4. Length of a curve being analyzed.

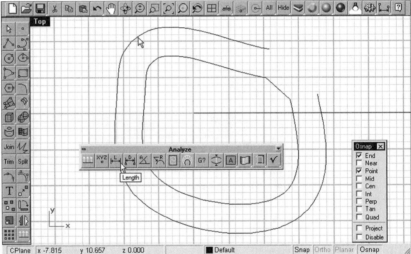

2 Select the curve indicated in figure 4-4.

The length of the selected curve is displayed.

Distance Measurement

To measure the distance between two selected points, perform the following steps.

1 Select Analyze > Distance, or the Distance button from the Analyze toolbar.

Command: Distance

2 Select the end points indicated in figure 4-5.

Analyzing Curves

The distance between the selected points is displayed.

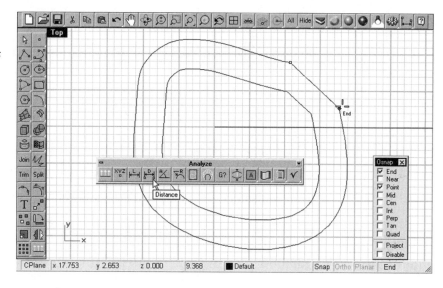

Fig. 4-5. Distance between two points being evaluated.

Angle Measurement

To measure the angle between two lines defined by four selected points, perform the following steps.

1 Select Analyze > Angle, or the Angle button from the Analyze toolbar.

Command: Angle

2 Select the points indicated in figure 4-6 to specify the first and second line.

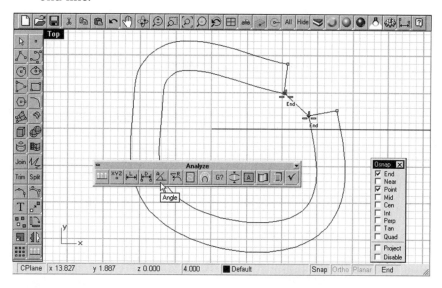

Fig. 4-6. Angle between the first and second line being evaluated.

The angle between the first and second line is displayed.

Curvature Radius

To discern the radius of curvature of a selected point along a curve, perform the following steps.

1. Check the Near option of the Osnap dialog box.
2. Select Analyze > Radius, or the Radius button from the Analyze toolbar.

   Command: Radius
3. Select a point along the curve indicated in figure 4-7.

The radius of curvature of the selected location is displayed.

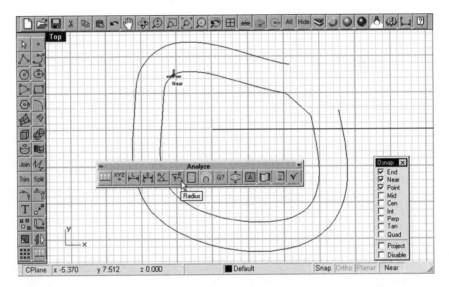

Fig. 4-7. Radius of curvature being evaluated.

Bounding Box Construction

To construct a bounding box, perform the following steps.

1. Select Analyze > Bounding Box, or the Curve Bounding Box button from the Analyze toolbar.

   Command: BoundingBox
2. Select the curve indicated in figure 4-8.

A bounding box is constructed.

Curvature Graph

To manipulate the curvature graph of a curve, perform the following steps.

Analyzing Curves

Fig. 4-8. Bounding box being constructed.

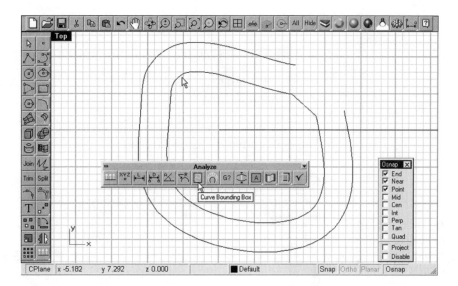

1 Select Analyze > Curve > Curvature Graph On, or the Curvature Graph On button from the Analyze toolbar.

Command: CurvatureGraphOn

2 Select the curve indicated in figure 4-9.

Fig. 4-9. Curvature graph displayed.

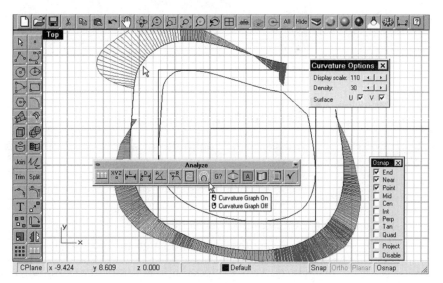

The curvature graph is displayed.

3 Select Analyze > Curve > Curvature Graph Off, or the Curvature Graph Off button from the Analyze toolbar.

Command: CurvatureGraphOff

The curvature graph is turned off.

Curvature Circle Construction

To construct a curvature circle, perform the following steps.

1 Select Analyze > Curve > Curvature Circle.
   Command: Curvature

2 Select the curve and point indicated in figure 4-10.

The curvature circle is constructed.

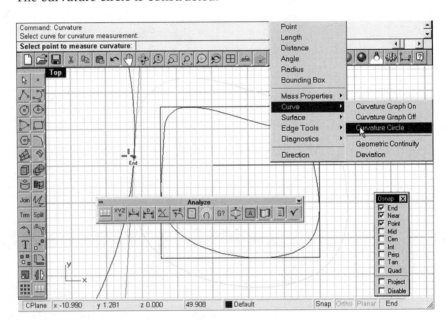

Fig. 4-10. Curvature of selected point along a curve displayed.

Geometric Continuity

To discern the continuity of a joint in a curve, perform the following steps.

1 Select Edit > Explode, or the Explode button from the Main toolbar.
   Command: Explode

2 Select the curve indicated in figure 4-11.

The curve is exploded into several curve components.

3 Select Analyze > Curve > Geometric Continuity, or the Geometric Continuity of 2 Curves button from the Analyze toolbar.
   Command: GCon

4 Select the curves indicated in figure 4-11.

The geometric continuity is displayed.

Analyzing Curves

Fig. 4-11. Geometric continuity evaluated.

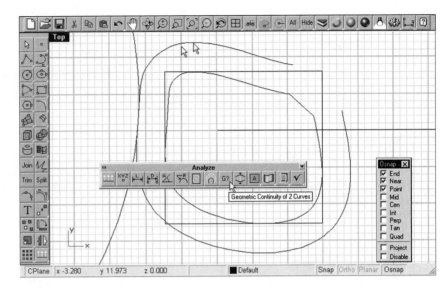

Deviation Between Curves

To analyze curves for deviation, perform the following steps.

1 Select Analyze > Curve > Deviation, or the Curve Deviation button from the Analyze toolbar.

Command: CrvDeviation

2 Select the curves indicated in figure 4-12.

A report is displayed.

Fig. 4-12. Curves selected.

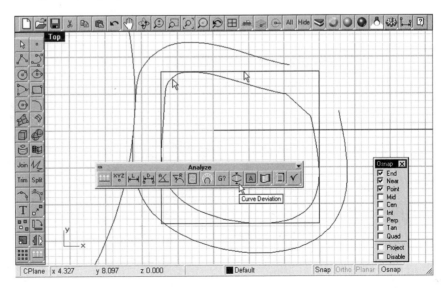

Database Listing

To display the database of a selected object in list form, perform the following steps.

1. Select Analyze > Diagnostics > List, or the List Object Database button from the Diagnostics toolbar.

   Command: List

2. Select the curve indicated in figure 4-13.

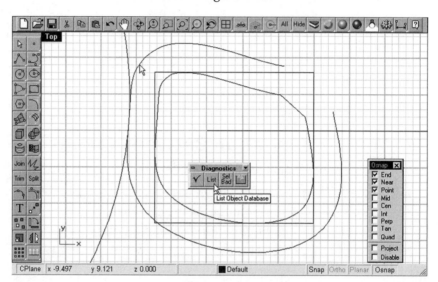

Fig. 4-13. Curve data listed.

The database of the selected objected is displayed as a list in the command line area.

Curve Direction

To display the direction of a curve, perform the following steps.

1. Select Analyze > Direction, or the Direction button from the Analyze toolbar.

   Command: Dir

2. Select the curve indicated in figure 4-14.

The direction of the curve is displayed.

Bad Object Report

To check a curve for errors, perform the following steps.

Analyzing Curves

Fig. 4-14. Direction of the curve displayed.

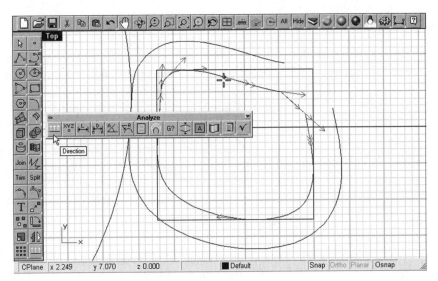

1 Select Analyze > Diagnostics > Check, or the Check Object button from the Analyze toolbar.

Command: Check

2 Select the curve indicated in figure 4-15.

The selected curve is checked. To check the entire file for bad objects, continue with the following.

3 Select Analyze > Diagnostics > Select Bad Objects, or the Bad Objects button from the Analyze toolbar.

Command: SelBadObjects

A report is displayed.

Fig. 4-15. Curve checked for errors.

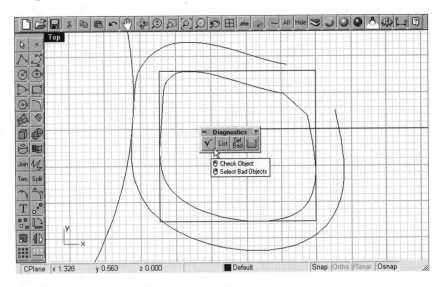

■■ Curve Construction Projects

Curve construction is a tedious job. However, curves are crucial in surface modeling. To enhance your skill on curve construction, you will work on a number of curve construction projects. With these curves, you will continue to construct the surface models in Chapter 7.

Because the location and shape of the curves have a direction impact on the shape of surfaces constructed from them, the particulars of most curves in these projects are given to you. While working on these projects, you should try to relate the 3D curves to the 3D surfaces. It is hoped that you reverse the process, seeing the curves when a surface is given.

JoyPad

Figure 4-16 shows the completed surface model of a joypad, and figure 4-17 shows the curves you will construct. To construct this model, perform the steps that follow.

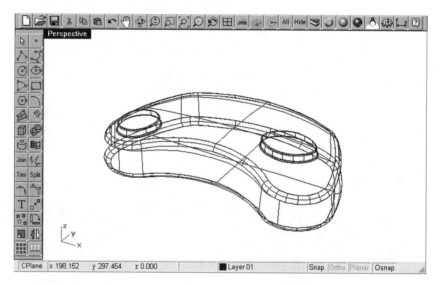

Fig. 4-16. Joypad model.

1. Start a new file. Use the metric (millimeter) template.
2. Turn off the grid and grid axes display in all viewports.
3. Maximize the Top viewport.
4. Construct a circle. Select Curve > Circle > Center, Radius, or the Circle/Center, Radius button from the Circle toolbar.
   Command: Circle
5. Select the Top viewport and type *50,50* at the command line area to specify the center.
6. Type *30* to specify the radius.

Curve Construction Projects   163

Fig. 4-17. Curves for making the joypad model.

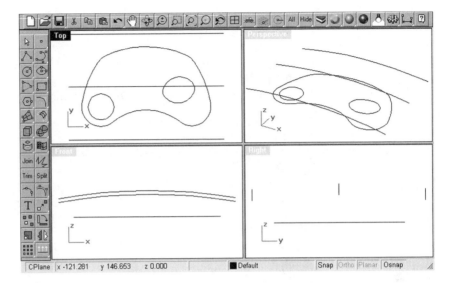

7 Repeat the Circle command to construct another circle of 30-mm radius at 70,100.

8 Select Zoom Extents from the Main toolbar.

Two circles are constructed on the Top viewport construction plane. (See figure 4-18.)

Fig. 4-18. Two circles constructed.

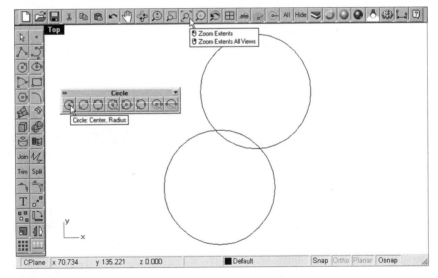

9 Mirror the circles. Select Transform > Mirror, or the Mirror button from the Transform toolbar.

Command: Mirror

10 Select the circles and press the Enter key.

164 CHAPTER 4: Wireframe Modeling and Curves: Part 3

11 Type *120,0* at the command line area to specify the start of the mirror plane.

12 Type *120,1* at the command line area to specify the end of the mirror plane.

13 Select Zoom Extents from the Main toolbar.

The circles are mirrored. (See figure 4-19.)

Fig. 4-19. Circles mirrored.

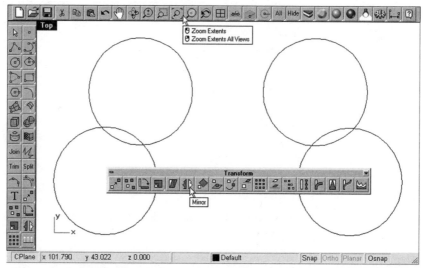

14 Construct a tangent circle. Select Curve > Circle > Tangent, Tangent, Radius, or the Circle/Tangent, Tangent, Radius button from the Circle toolbar.

Command: CircleTTR

15 Select the circles indicated in figure 4-20.

Fig. 4-20. Tangent circle constructed.

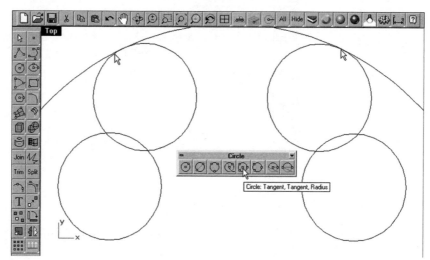

Curve Construction Projects

16 Type *150* at the command line area to specify the radius.

A tangent circle is constructed.

17 Repeat the CircleTTR command to construct another circle of radius 80 units, in accordance with figure 4-21.

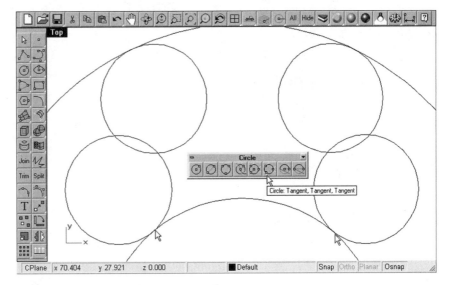

Fig. 4-21. Second tangent circle constructed.

Another tangent circle constructed.

18 Trim the tangent circles. Select Edit > Trim, or the Trim button from the Main toolbar.

Command: Trim

19 Select the four small circles and press the Enter key.

20 Select the two tangent circles indicated in figure 4-22.

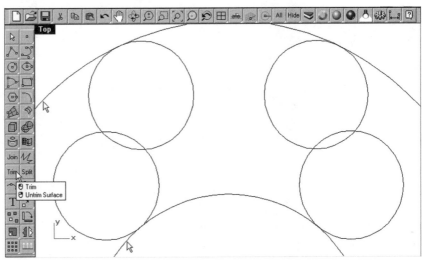

Fig. 4-22. Circles to be trimmed.

The tangent circles are trimmed.

21 With reference to figure 4-23, construct two tangent circles of radius 150 units.

Fig. 4-23. Two tangent circles constructed.

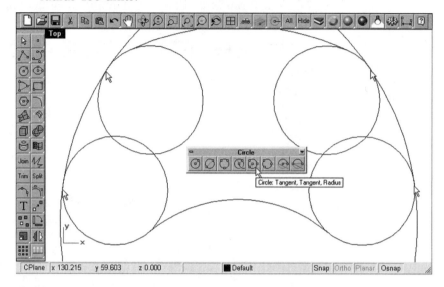

22 Trim the tangent circles in accordance with figure 4-24.

Fig. 4-24. Tangent circles trimmed.

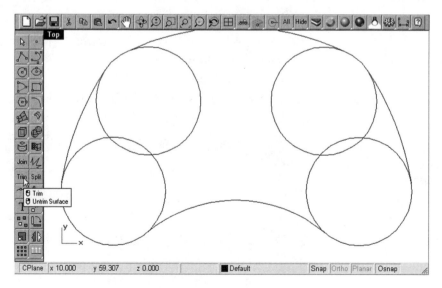

23 With reference to figure 4-25, trim the circles.

24 Join the arcs into a single curve. Select Edit > Join, or the Join button from the Main toolbar.

Command: Join

25 Select all arcs and press the Enter key.

Curve Construction Projects

Fig. 4-25. Circles trimmed.

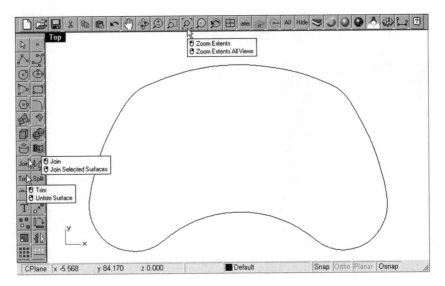

The circles are trimmed and joined.

26 Maximize the Front viewport. You will construct a free-form curve in this viewport.

27 Select Curve > Free-form > Interpolate Points, or the Curve/Interpolate Points button from the Curve Tools toolbar.
Command: InterpCrv

28 Type *0,20* at the command line area to specify the first point.

29 Type *120,30* at the command line area to specify the second point.

30 Type *240,20* at the command line area to specify the third point.

31 Press the Enter key to terminate the command.

A free-form curve is constructed. (See figure 4-26.)

32 Maximize the Perspective viewport.

33 Construct two copies of the free-form curve. Select Transform > Copy, or the Copy button from the Transform toolbar.
Command: Copy

34 Select the free-form curve and press the Enter key.

35 Select any point on the screen to specify a point to copy from.

36 Type *r0,80,5* at the command line area to specify a relative distance of a copy.

37 Type *r0,160* at the command line area to specify the distance of the second copy.

38 Press the Enter key.

Two copies of the free-form curve are constructed. (See figure 4-27.)

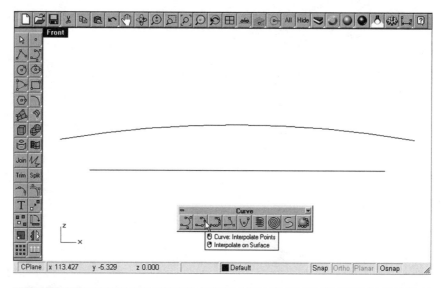

Fig. 4-26. Free-form curve constructed in the Front viewport.

Fig. 4-27. Free-form curves copied.

39 Construct a circle with center at 50,50 and a radius of 20 units.

40 Construct an ellipse. Select Curve > Ellipse > From Center, or the Ellipse/From Center button from the Ellipse toolbar.

Command: Ellipse

41 Type *170,75* at the command line area to specify the center.

42 Type *r25 < 10* at the command line area to specify the location of the end of the first axis. (r25 < 10 is a polar coordinate. It means a distance of 25 units within 10 degrees counterclockwise of the last point.)

43 Type *r20 < 110* at the command line area to specify the location of the end of the second axis.

Curve Construction Projects

A circle and an ellipse are constructed. Because an ellipse is a degree 2 curve and an ellipse has four kink points, extruding it may result in a polysurface of four surfaces instead of a single surface. Alternatively, you can construct a circle or ellipse without kink points by using the Deformable option. Hence, you will learn how to edit an ellipse with four kink points, change its polynomial degree to 3, and make it periodic. A periodic curve is a closed curve with no kink point. (See figure 4-28.)

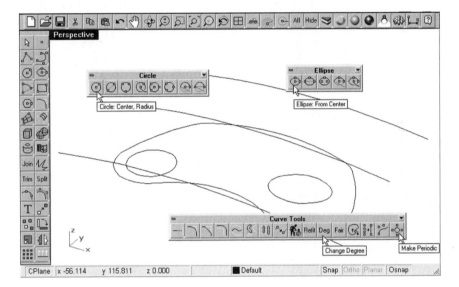

Fig. 4-28. Circle and ellipse constructed, and ellipse edited.

44 Select Curve > Edit Tools > Change Degree, or the Change Degree button from the Curve Tools toolbar.

Command: ChangeDegree

45 Select the ellipse and press the Enter key.

46 Type 3 at the command line area to change the polynomial degree to 3.

47 Select Curve > Edit Tools > Make Periodic, or the Make Periodic button from the Curve Tools toolbar.

Command: MakeCrvPeriodic

48 Select the ellipse and press the Enter key.

The curves for making the joypad model are complete. Save your file as *Joypad.3dm*.

JoyStick

In the following, you will construct the curves for making a joystick. Figure 4-29 shows the completed surface model, and figure 4-30 shows the curves.

170 CHAPTER 4: Wireframe Modeling and Curves: Part 3

Fig. 4-29. Surface model of the joystick.

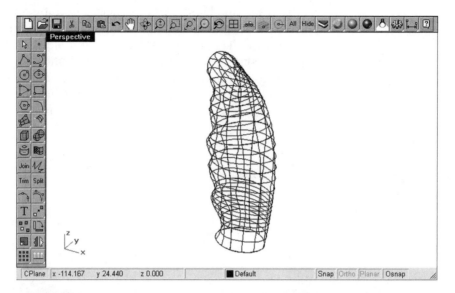

Fig. 4-30. Curves for making the joystick.

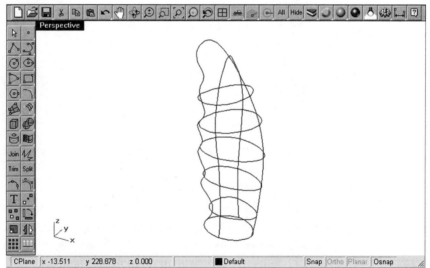

1. Start a new file. Use the metric (millimeter) template.
2. Obtain a piece of graph paper. Make a sketch of the front and side view of the joystick in accordance with figure 4-31.
3. If you have a digital scanner, scan the sketch to obtain two digital images (front and side view). (See figure 4-32.)
4. Maximize the Front viewport.
5. Check the Snap pane in the status bar.

Curve Construction Projects

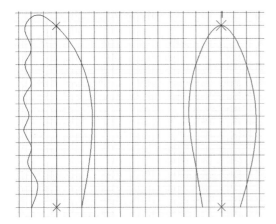

Fig. 4-31. Sketches for the joystick.

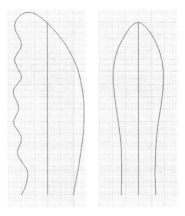

Fig. 4-32. Scanned bitmap images.

6 Select View > Background Bitmap > Place, or the Place Background Bitmap from the Background Bitmap toolbar.
Command: PlaceBackgroundBitmap

7 Select the bitmap file (*sketchfront.tga*) for the front view of the joystick and select the Open button.

Figure 4-32 shows the width of the images (30 units). Therefore, you will select two points in the Front viewport, 30 units apart.

8 Select the two points (-15,0 and 15,5) indicated in figure 4-33.

Fig. 4-33. Bitmap image being placed.

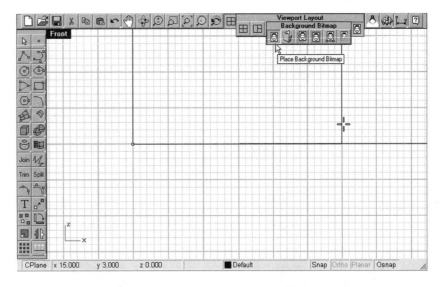

The bitmap image is placed in the background of the Front viewport.

9 Turn off the grid display and zoom the viewport in accordance with figure 4-34.

Fig. 4-34. Bitmap image placed in the Front viewport.

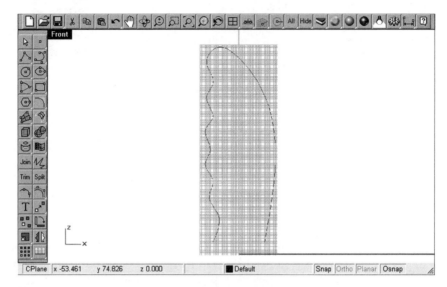

To trace a curve, continue with the following steps. Here, you will use the InterpCrv command. You may try using the Sketch command instead.

10 Select Curve > Free-form > Interpolate Curve, or the Curve/Interpolate Points button from the Curve Tools toolbar.
Command: InterpCrv

11 Select points along the background image and press the Enter key. (See figure 4-35.)

Fig. 4-35. Curve traced.

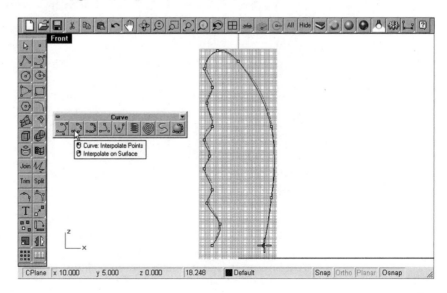

Curve Construction Projects

To place the another bitmap image in the Right viewport, continue with the following steps.

12 Select View > Background Bitmap > Place, or the Place Background Bitmap button from the Background Bitmap toolbar.
Command: PlaceBackgroundBitmap

13 Select the bitmap file (*sketchside.tga*) for the front view of the joystick and select the Open button.

14 Select the points -15,0 and 15,5.

15 Turn off the grid display and zoom the viewport in accordance with figure 4-36.

16 With reference to figure 4-37, construct a free-form curve.

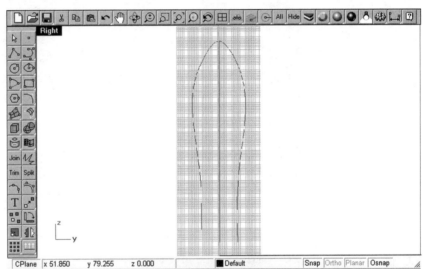

Fig. 4-36. Bitmap placed in the Right viewport.

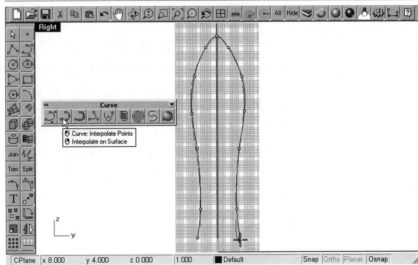

Fig. 4-37. Free-form curve constructed.

Chapter 4: Wireframe Modeling and Curves: Part 3

Now return to a four-viewport display and continue with the following steps.

17 Select the Front viewport.

18 Select View > Background Bitmap > Hide, or the Hide Background Bitmap button from the Background Bitmap toolbar.

19 Select the Right viewport and select View > Background Bitmap > Hide, or the Hide Background Bitmap button from the Background Bitmap toolbar. (See figure 4-38.)

Fig. 4-38. Background bitmap hidden.

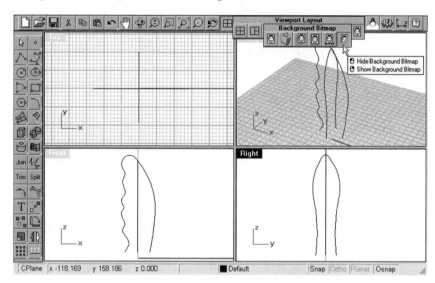

To modify the curves, continue with the following steps.

20 Maximize the Front viewport.

21 Move the curves so that they intersect at the location indicated in figure 4-39.

22 Maximize the Perspective viewport.

23 Select Edit > Split, or the Split button from the Main toolbar.
Command: Split

24 Select a curve. (See figure 4-40.)

25 Select the other curve and press the Enter key.

The first selected curve is split. Note that you would not be able to split the curves if they did not intersect.

26 Repeat the Split command to split the other curve.

Now you should have four curves.

Curve Construction Projects 175

Fig. 4-39. Curve moved.

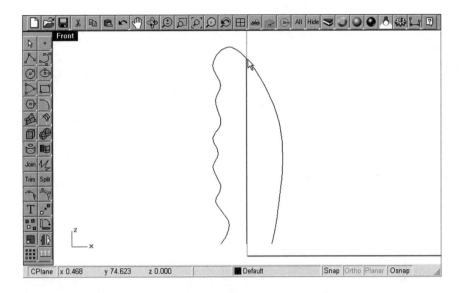

Fig. 4-40. Curves being split into four curves.

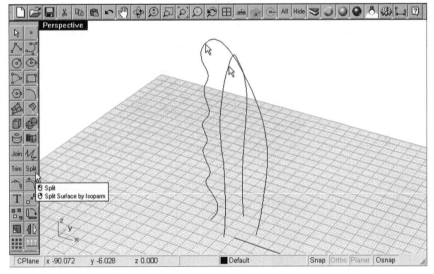

To construct a series of cross-section curves, continue with the following steps.

27 Select Curve > CSec Profiles, or the CSec Profiles button from the Curve Tools toolbar.

Command: CSec

28 Select the curves in the Perspective viewport sequentially (either clockwise or counterclockwise) and press the Enter key. (See figure 4-41.)

Fig. 4-41. Curves selected.

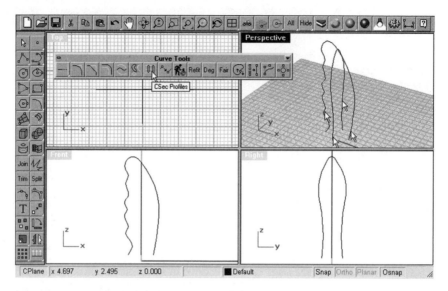

29 Select two points in the Front viewport to indicate the orientation of the cutting plane. (See figure 4-42.)

Fig. 4-42. Cross section specified.

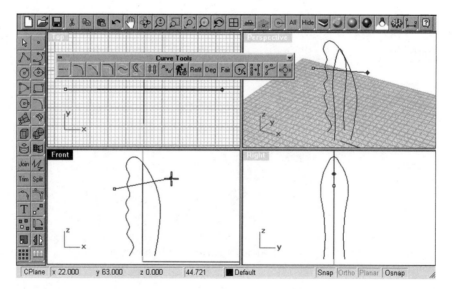

30 Specify five more section planes and press the Enter key.

Six cross-section curves are constructed. (See figure 4-43.) Save your file as *Joystick.3dm*.

Curve Construction Projects

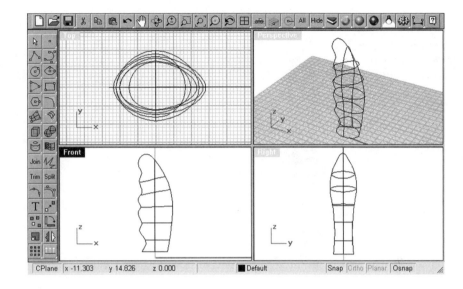

Fig. 4-43. Cross section constructed.

Mobile Phone

In the following, you will construct the curves for making a mobile phone. Figure 4-44 shows the surface model, and figure 4-45 shows the curves.

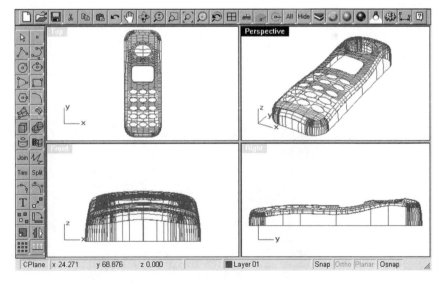

Fig. 4-44. Surface model of a mobile phone.

1 Start a new file. Use the metric (millimeter) template.

2 Maximize the Top viewport.

3 Select Curve > Rectangle > Corner to Corner, or the Rectangle/Corner to Corner button from the Rectangle toolbar.

Command: Rectangle

178   CHAPTER 4: Wireframe Modeling and Curves: Part 3

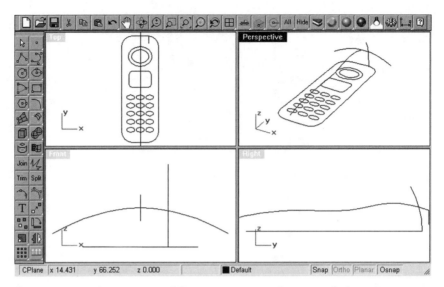

Fig. 4-45. Curves for creating the mobile phone surface model.

4   Type *R* at the command line area to use the Rounded option.

5   Type *-23,-65* at the command line area to specify the first corner.

6   Type *r46,130* at the command line area to specify a relative coordinate.

7   Type *12* at the command line area to specify the corner radius.

8   Select Zoom Extents from the Main toolbar. (See figure 4-46.)

A rectangle is constructed and the display is zoomed. To construct an arc in the Right viewport, continue with the following steps. (See figure 4-47.)

9   Maximize the Right viewport.

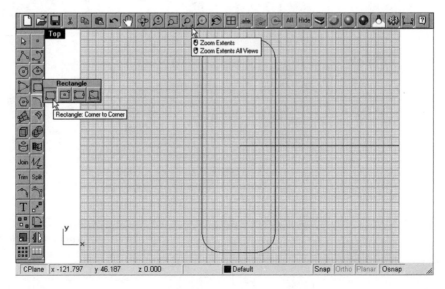

Fig. 4-46. Rectangle constructed.

Curve Construction Projects

Fig. 4-47. Arc constructed.

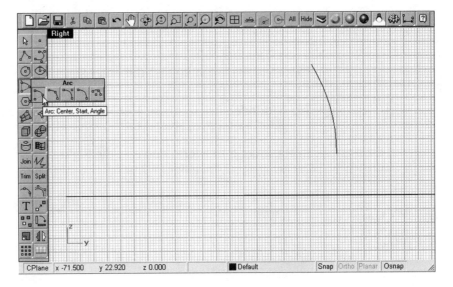

10 Select Curve > Arc > Center, Start, Angle, or the Arc/Center, Start, Angle button from the Arc toolbar.
Command: Arc

11 Select any point on the screen to specify the center.

12 Type *r60,0* at the command line area to specify the start point.

13 Type *32* at the command line area to specify the angle.

To move the arc, continue with the following steps. (See figure 4-48.)

Fig. 4-48. Arc being moved.

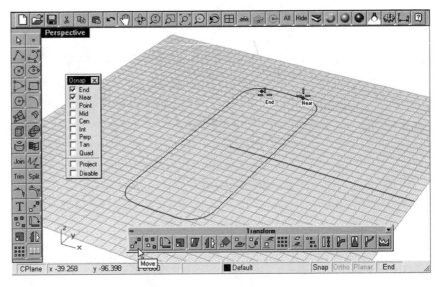

14 Maximize the Perspective viewport.

Chapter 4: Wireframe Modeling and Curves: Part 3

15 Select Transform > Move, or the Move button from the Transform toolbar.

Command: Move

16 Select the arc and press the Enter key.

17 Check the End and Near boxes of the Osnap dialog box.

18 With reference to figure 4-48, select the end point of the arc and then a point near the curve.

To construct a free-form curve in the Right viewport, continue with the following steps. (See figure 4-49.)

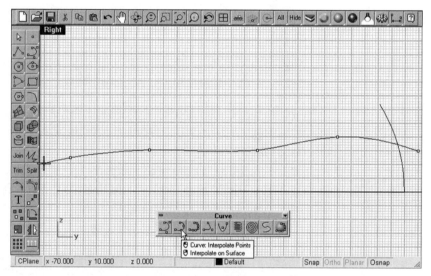

Fig. 4-49. Curve constructed in the Right viewport.

19 Maximize the Right viewport.

20 Construct a free-form curve to interpolate along the following points:

❐ 70,15/40,20/10,15/-30,15/-60,12/-70,10

To construct a free-form curve in the Front viewport, continue with the following steps. (See figure 4-50.)

21 Maximize the Front viewport.

22 Construct a free-form curve to interpolate along the following points:

❐ 35,5/0,15/-35,5

To move the curve constructed in the Front viewport, continue with the following steps.

23 Maximize the Perspective viewport.

Curve Construction Projects 181

Fig. 4-50. Curve constructed in the Front viewport.

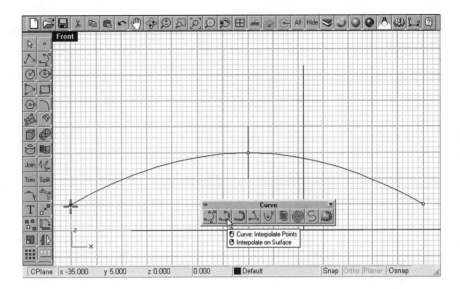

24 Select Transform > Move, or the Move button from the Transform toolbar.

Command: Move

25 Select the curve indicated in figure 4-51 and press the Enter key.
26 Check the End and Mid boxes of the Osnap dialog box.
27 With reference to figure 4-51, select the mid point of the arc and then the end point of another curve.

To turn off the grid and grid axes display and construct three ellipses, continue with the following steps. (See figure 4-52.)

Fig. 4-51. Curve being moved.

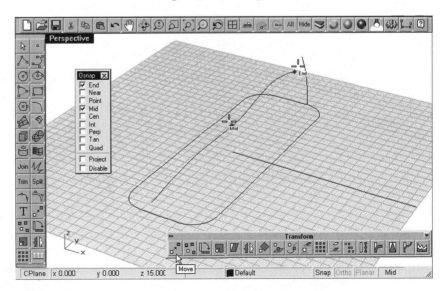

Fig. 4-52. Three ellipses constructed.

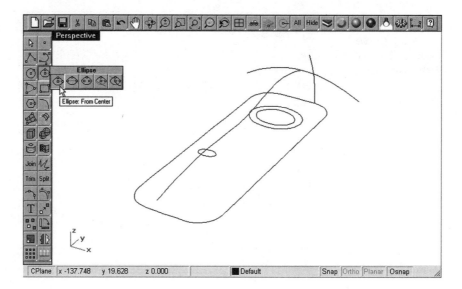

28 Type *Grid* and then *Gridaxe* at the command line area.

29 Select Curve > Ellipse > from Center, or the Ellipse/From Center button on the Ellipse toolbar.

Command: Ellipse

30 Type *0,40* at the command line area to specify the center.

31 Type *r15 < 0* at the command line area to specify the end of the first axis.

32 Type *r12 < 90* at the command line area to specify the end of the second axis.

33 Repeat the Ellipse command to construct two more ellipses, as follows:

- Center: *0,40*/End of first axis: *r11 < 0*/End of second axis: *r8 < 90*

- Center: *-12,-10*/End of first axis: *r5 < 0*/End of second axis: *r3 < 90*

To construct a rectangular array of an ellipse, continue with the following steps. (See figure 4-53.)

34 Select Transform > Array > Rectangular, or the Rectangular Array button from the Array toolbar.

Command: Array

35 Select the ellipse indicated in figure 4-53 and press the Enter key.

36 Type *3* at the command line area to specify three copies in the X direction.

37 Type *5* at the command line area to specify five copies in the Y direction.

Curve Construction Projects

Fig. 4-53. Rectangular array constructed.

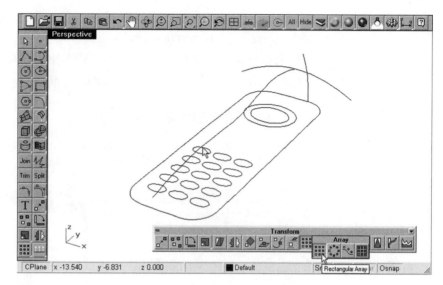

38 Type *1* at the command line area to specify one copy in the Z direction.
39 Type *12* at the command line area to specify the X spacing.
40 Type *-10* at the command line area to specify the Y spacing.

Now you will construct a rectangle and fillet the four corners. (See figure 4-54.)

Fig. 4-54. Rectangle constructed.

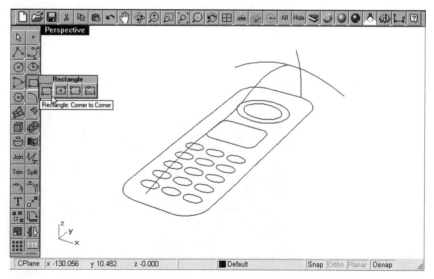

41 Select Curve > Rectangle > Corner to Corner, or the Rectangle/Corner to Corner button from the Rectangle toolbar.
Command: Rectangle

42 Type *R* at the command line area to use the Rounded option.

184 CHAPTER 4: Wireframe Modeling and Curves: Part 3

43 Type *-14,2* at the command line area to specify the first corner.
44 Type *r28,20* at the command line area to specify a relative coordinate.
45 Type *4* at the command line area to specify the corner radius.

The curves are complete. Save your file as *Mphone.3dm*.

Scale Model Car Body

In the following, you will construct the curves for making the surface model of a model car. (See figures 4-55 and 4-56.)

Fig. 4-55. Car surface model.

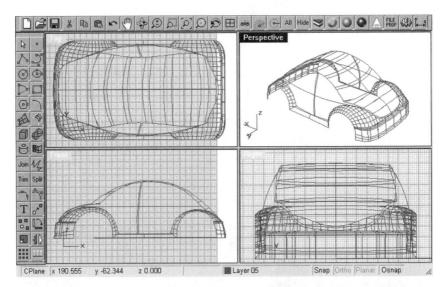

Fig. 4-56. Curves for the model car.

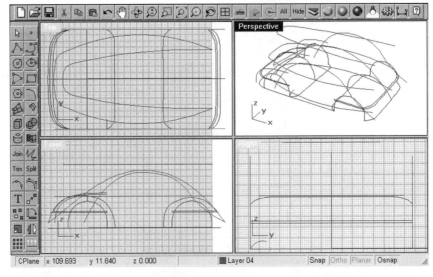

1 Start a new file. Use the metric (millimeter) template.

Curve Construction Projects

2   Set the current layer to *Layer 01*.
3   Maximize the Front viewport.
4   Construct four free-form curves (figure 4-57) that interpolate along the points outlined in table 4-2.

Fig. 4-57. Four free-form curves constructed.

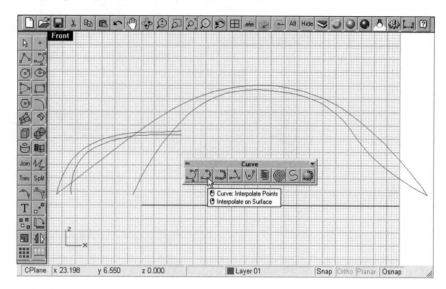

Table 4-2: Points for Constructing Four Free-form Curves

Curve 1	Curve 2	Curve 3	Curve 4
-32,4	-26.5,4	-32,4	-3,4
-28.5,13.5	-25,13.5	-18.5,14.5	4,17
-22,21	-19.5,20.5	-4.8,24.8	13,29
-13.5,25	-11.5,24	9.2,34.4	23.5,38
-4,27	-2.5,26	24.5,41.6	37,43
5.8,27.5	6.5,26.5	41,45	52,43
15.5,27.8	15.5,26.5	58,44	66,40
—	—	74,38.5	78,32
—	—	88,29	87,21
—	—	100.17.5	111,4
—	—	111,4	—

To move and copy two curves, continue with the following steps.

5   Maximize the Perspective viewport.

CHAPTER 4: Wireframe Modeling and Curves: Part 3

6 Select Transform > Move, or the Move button from the Transform toolbar.

Command: Move

7 Select the curves indicated in figure 4-58 and press the Enter key.

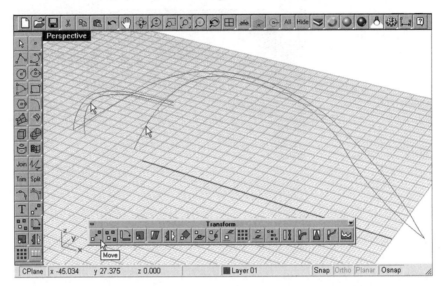

Fig. 4-58. Curves being moved.

8 Select any point in the viewport to specify a point to move from.

9 Type *r39 < 270* at the command line area to specify a relative distance of 39 units in 270 degrees.

10 Select Transform > Copy, or the Copy button from the Transform toolbar.

Command: Copy

11 Select the curves indicated in figure 4-59.

12 Select any point in the viewport to specify a point to copy from.

13 Type *r79 < 90* at the command line area to specify a relative distance of 78 units in 90 degrees, and press the Enter key.

Curves for making the top of the car are complete. To construct the curves for making the side body of the car, continue with the following steps.

14 Maximize the Top viewport.

15 Set the current layer to *Layer 02*.

16 Construct six free-form curves (figure 4-60) that interpolate along the points outlined in table 4-3.

17 Maximize the Perspective viewport.

18 With reference to figure 4-61, construct four fillet curves with a radius of 10 units.

19 Join the curves indicated in figure 4-61 into a single curve.

Curve Construction Projects

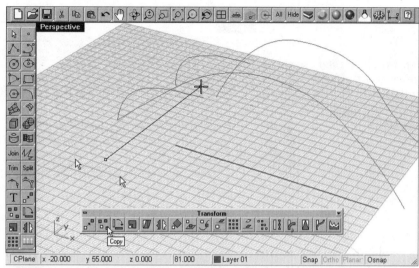

Fig. 4-59. Curves being copied.

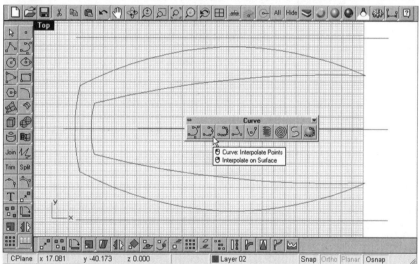

Fig. 4-60. Six free-form curves constructed.

Table 4-3: Points for Constructing Six Free-form Curves

Curve 1	Curve 2	Curve 3	Curve 4	Curve 5	Curve 6
101,23	101,-23	101,23	101,-23	-25,18.5	-18.5,11
37.5,35	37.5,-35	41,21	41,-21	-27,0	-20,0
-25,18.5	-25,-18.5	-18.5,11	-18.5,-11	-25,-18.5	-18.5,-11

20 Select Transform > Move, or the Move button of the Transform toolbar.

Command: Move

Fig. 4-61. Four fillet curves constructed, and five curves joined and being moved.

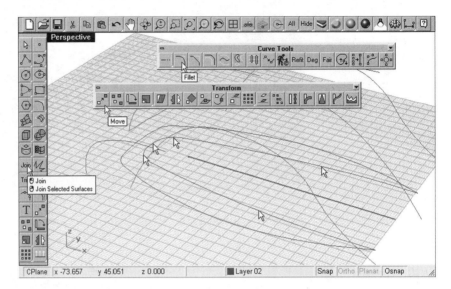

21 Select the joined curve and press the Enter key.
22 Type *V* at the command line area to use the Vertical option.
23 Select any point in the viewport to specify a point to move from.
24 Type *50* at the command line area to move the selected curve a distance of 50 units in a direction vertical to the current construction plane.
25 Join the curves indicated in figure 4-62.

Fig. 4-62. Curves joined and being copied.

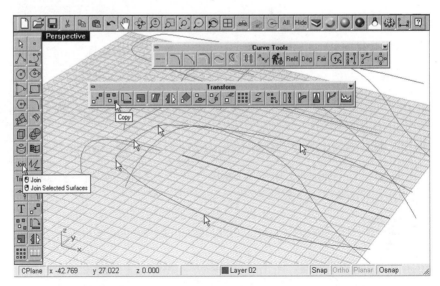

26 Select Transform > Copy, or the Copy button from the Transform toolbar.
Command: Copy

Curve Construction Projects

27 Select the joined curve and press the Enter key.
28 Type *V* at the command line area to use the Vertical option.
29 Select any point in the viewport to specify a point to copy from.
30 Type *25* at the command line area to copy the selected curve a distance of 25 units in a direction vertical to the current construction plane.
31 Press the Enter key to terminate the command.

Curves for making the side body of the car are complete. To construct the curves for making the front part of the car, continue with the following steps.

32 Set the current layer to *Layer 02*.
33 Maximize the Front viewport.
34 Construct a free-form curve (figure 4-63) to interpolate along the following points:

   ❐ 22.5,0/15,19.5/2,25/-5,26/-11.9,25/-18.5,22.7/-24,18.5/-28,13

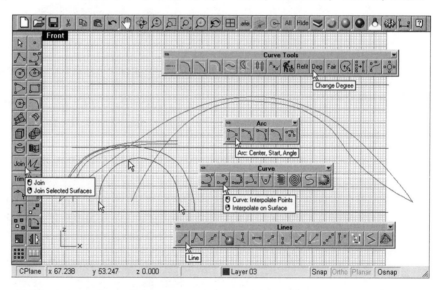

Fig. 4-63. Curves constructed in the Front viewport.

35 Construct an arc, as follows:

   ❐ Center: *0,5*/Start: *r16 < 0*/Angle: *180*

36 Construct two line segments, as follows:

   ❐ Line1 start: *16,0*/Line1 end: *16,5*

   ❐ Line2 start: *-16,0*/Line2 end: *-16,5*

37 Join the lines and arc indicated in figure 4-63 into a single curve.
38 Change the polynomial degree of the joined curve to *3*.

39 Maximize the Perspective viewport.

40 Select Transform > Move, or the Move button from the Transform toolbar.

Command: Move

41 Select the curve indicated in figure 4-64 and press the Enter key.

Fig. 4-64. Curve being moved.

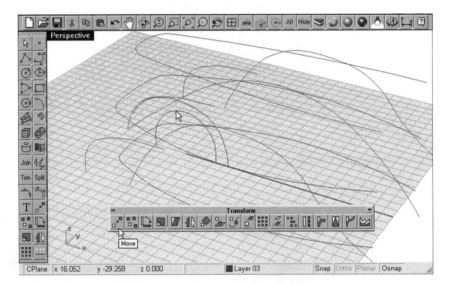

42 Select any point in the viewport to specify the point to move from.

43 Type *r39<270* at the command line area to specify the distance and direction of the move.

44 Select Transform > Copy, or the Copy button from the Transform toolbar.

Command: Copy

45 Select the curve indicated in figure 4-65 and press the Enter key.

46 Select any point in the viewport to specify the point to copy from.

47 Type *r78<90* at the command line area to specify the distance and direction of the copy.

48 Press the Enter key to terminate the command.

49 Construct two free-form curves (figure 4-66) to interpolate along the points outlined in table 4-4.

50 Turn off grid and grid axes display in the Perspective viewport.

51 Construct a line segment joining the end points indicated in figure 4-67.

52 Select Transform > Copy, or the Copy button from the Transform toolbar.

Command: Copy

Curve Construction Projects

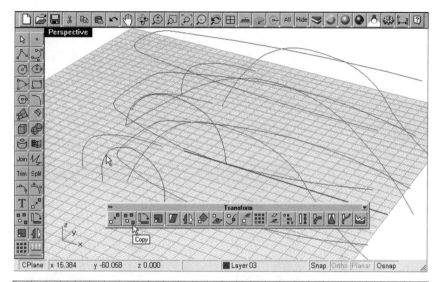

Fig. 4-65. Curve being copied.

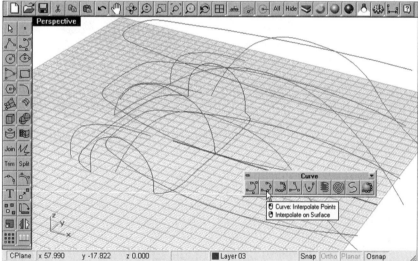

Fig. 4-66. Two curves constructed.

Table 4-4: Points for Constructing Two Free-form Curves

Curve 1	Curve 2
16,39	-16,39
21,35	-26.5,34
22.5,27	-30.5,23.5
22.5,15	-32,0
22.5,0	-30.5,-23.5

Curve 1	Curve 2
22.5,-15	-26.5,-34
22.5,-27	-16,-39
21,-35	—
16,-39	—

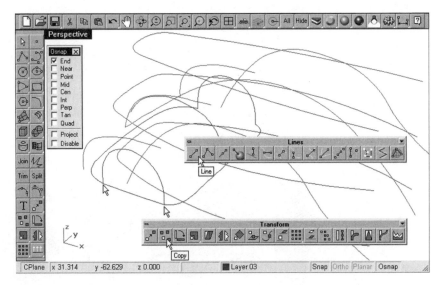

Fig. 4-67. Line constructed and being copied.

53 Select the line you constructed and press the Enter key.

54 Type *V* at the command line area to use the Vertical option.

55 Select any point in the viewport to specify the point to copy from.

56 Type *12* at the command line area to specify a vertical distance for the first copy.

57 Type *13* at the command line area to specify a vertical distance for the second copy.

58 Press the Enter key to terminate the command.

A line segment and two copies of the line segment are constructed.

59 Construct a free-form curve (figure 4-68) to interpolate along the following points:

❒ -36,-39/-41,-37/-47.5,-30/-49.5,-20/-50,-10/-50,0/-50, 10/-49.5,20/-47.5,30/-41,37/-36,39

60 Check the End and Int boxes of the Osnap dialog box.

Curve Construction Projects

Fig. 4-68. Free-form curve constructed.

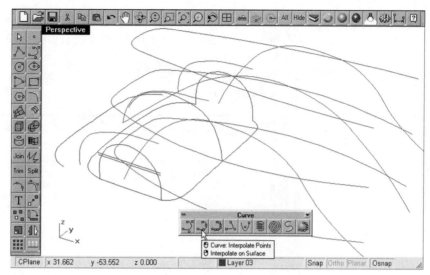

61 Select Transform > Move, or the Move button from the Transform toolbar.

Command: Move

62 Select curve A indicated in figure 4-69 and press the Enter key.

63 Select the end point A indicated in figure 4-69 to specify the point to move from.

64 Select the intersection point B indicated in figure 4-69 to specify the point to move to.

65 Select Transform > Copy, or the Copy button from the Transform toolbar.

Command: Copy

Fig. 4-69. Curve being moved.

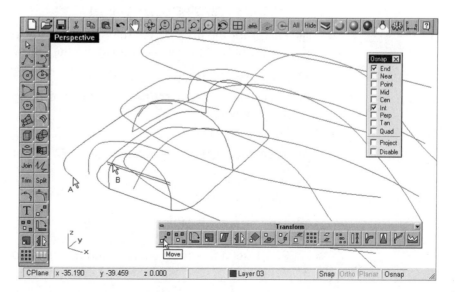

66 Select curve A indicated in figure 4-70 and press the Enter key.

67 Select the end point B indicated in figure 4-70 to specify the point to copy from.

68 Select the intersection point C indicated in figure 4-70 to specify the point to copy to.

Fig. 4-70. Curve being copied.

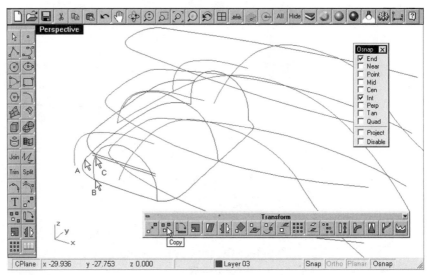

69 Press the Enter key to terminate the command.

70 Maximize the Right viewport.

71 Construct a free-form curve (figure 4-71) to interpolate along the following points:

- -39,21/-35,23.5/-27,24.5/-18,24.5/-9,24.5/0,24.5/9,24.5/ 18,24.5/27,24.5/35,23.5/39,21

The curves for the front part of the car are complete. To construct curves for the rear part of the car, continue with the following steps.

72 Maximize the Perspective viewport.

73 Select Transform > Mirror, or the Mirror button from the Transform toolbar.

Command: Mirror

74 Select the curves indicated in figure 4-72 and press the Enter key.

75 Type *42,0* at the command line area to specify the start of the mirror plane.

76 Type *42,1* at the command line area to specify the end of the mirror plane.

77 Construct a free-form curve to interpolate along the following points:

Curve Construction Projects

Fig. 4-71. Curve constructed.

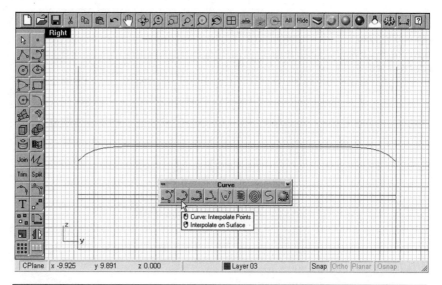

Fig. 4-72. Curves being mirrored.

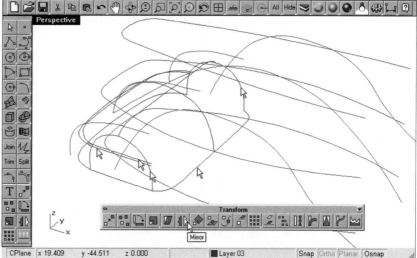

- 128,39/131,35/133.5,27/134,18/134.2,9/134.3,0/134.2, -9/134,-18/133.5,-27/131,-35/128,-39

78 Construct another free-form curve (figure 4-73) to interpolate along the following points:

- 100,39/104,36/106.6,31/107.5,21/107.5,10/107.5,0/107.5, -10/107.5,-21/106.6,-31/104,-36/100,-39

79 Select Transform > Move, or the Move button from the Transform toolbar.

Command: Move

196   CHAPTER 4: Wireframe Modeling and Curves: Part 3

Fig. 4-73. Two free-form curves constructed.

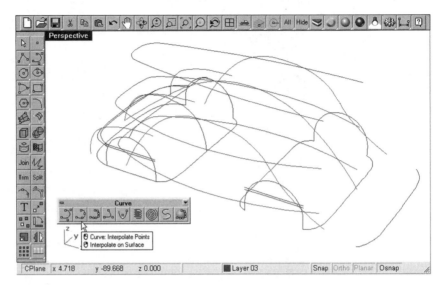

80 Select curve A indicated in figure 4-74 and press the Enter key.
81 Select the end point A indicated in figure 4-74 to specify the point to move from.
82 Select the intersection point B indicated in figure 4-74 to specify the point to move to.
83 Select Transform > Copy, or the Copy button from the Transform toolbar.
    Command: Copy
84 Select curve A indicated in figure 4-75 and press the Enter key.
85 Select the end point B indicated in figure 4-75 to specify the point to copy from.

Fig. 4-74. Curve being moved.

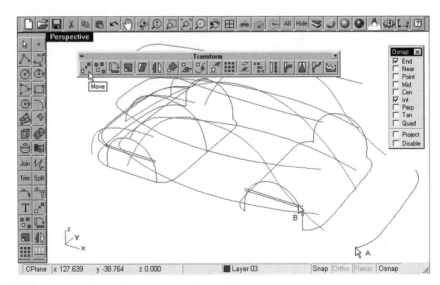

Curve Construction Projects

Fig. 4-75. Curve being copied.

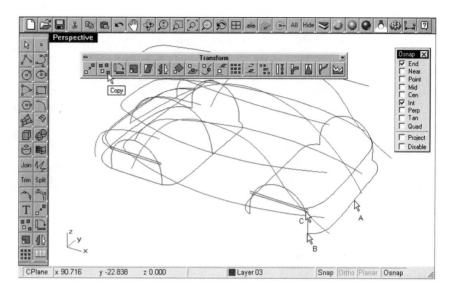

86 Select the intersection point C indicated in figure 4-75 to specify the point to copy to.
87 Press the Enter key to terminate the command.
88 Maximize the Front viewport.
89 Construct a free-form curve (figure 4-76) to interpolate along the following points:

❐ 104.3,13/99,18/92.8,22/84,23.8/78,23/69,19.5/63.5,14/61.5,0

Fig. 4-76. Rear part of car completed.

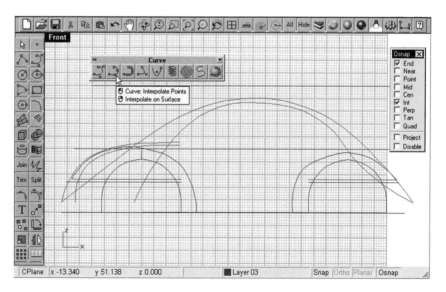

The curves for the rear part of the car are complete. To create the curves for making the side skirt of the car, continue with the following steps.

90 Maximize the Perspective viewport.
91 Set the current layer to *Layer04*.
92 Construct a free-form curve to interpolate the following points:

- 16,-39/16,-36,3/16,-31.5,3.5

93 Construct a line to join the end points indicated in figure 4-77.

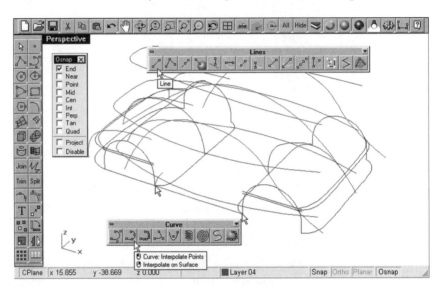

Fig. 4-77. Free-form curve and line constructed.

The curves for the surfaces of the model car are complete. Save your file as *Beetle.3dm*.

■■ Summary

This is the final chapter of a series of chapters on wireframe modeling and curves. In this chapter you learned various analysis procedures and tools: evaluating points, measuring length, measuring distance, measuring angle, evaluating curvature radius, constructing a bounding box, displaying a curvature graph, constructing a curvature circle, checking geometric continuity, finding out deviation between curves, displaying a database list, finding out a curve's direction, and checking bad objects in a file. In addition to learning the analysis tools, you also used Rhino as a tool to construct 3D curves for making surfaces.

■■ Review Questions

1 Explain how to find out the continuity of a joint in a curve.
2 What is the difference between displaying a database list and reporting bad objects in a file?

Chapter 5

Surface Modeling: Part 1

■■ Objectives

The goals of this chapter are to introduce the key concepts of surface modeling, to explain the basic ways of representing a surface in the computer, to outline an approach to surface modeling, to delineate the advantages and limitations of surface modeling, to explain the use of Rhino NURBS surface modeling tools in the construction of primitive surfaces, to discuss basic free-form surfaces, and to delineate the use of the MicroScribe digitizer. After studying this chapter, you should be able to:

❒ Explain the key concepts involved in surface modeling

❒ State the two types of surface representation methods

❒ Explain the advantages and limitations of surface modeling

❒ Use Rhino as a tool to construct primitive and free-form NURBS surfaces

❒ Use a digitizer in NURBS surface construction

■■ Overview

In our daily lives, we encounter many objects with free-form body shapes. Some examples are the handle of a razor, the blade of an electric fan, the casing of a computer pointing device (mouse), the casing of a mobile phone, the handle of a joystick (figure 5-1), and the body panels of an automobile.

Free-form surfaces are used in objects to meet two basic design needs: aesthetic and functional. Aesthetically, an object has to be eye-pleasing and eye-catching to attract customers. Hence, various types of free-form shapes are used in many consumer products.

Fig. 5-1. Free-form body shape.

Functionally, a surface needs to comply with a certain form and shape to serve specific purposes. For example, the blade of an electric fan should conform to aerodynamic requirements, the profile and silhouette of a shoe should match the shape of a human foot, and the joystick handle shown in figure 5-1 has to match the human hand.

In chapters 2 through 4 you learned the basic ways of constructing point objects, as well as how to construct, edit, transform, and analyze curves. In this chapter you will learn various surface construction methods.

■■ Surface Modeling Concepts

Surface modeling is a means of representing the geometric shape of a 3D object in the computer by using a set of surfaces put together to resemble the boundary faces of the object. Each surface is a mathematical expression that represents a 3D shape with no thickness. There are two basic ways of representing a surface in the computer: using polygon meshes to approximate a surface and using complex mathematics to obtain an exact representation of the surface.

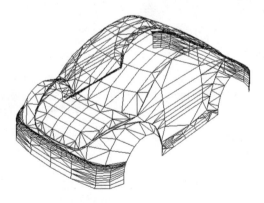

Fig. 5-2. Polygon mesh.

Polygon Mesh

A polygon mesh is a method of approximately representing a surface. It reduces a smooth, free-form surface to a set of planar polygonal faces, curved edges, and silhouettes to sets of straight line segments. Accuracy of representation is inversely proportional to the size of the polygon faces and line segments. A mesh with smaller polygon size represents a surface better, but the memory required to store the mesh is larger. Figure 5-2 shows the polygon mesh of a scale model car body.

A severe drawback of the polygon mesh method is that despite using a very small polygon it can never represent a surface accurately because the surface is always faceted. As a result, this method can be used only for visualization of the real object, and cannot be used in most downstream computerized manufacturing systems.

NURBS Surfaces

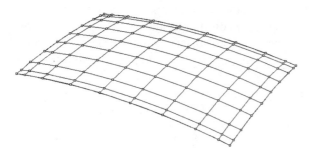

Fig. 5-3. NURBS surface and control vertices.

To accurately represent free-form smooth surfaces in 3D design applications and computerized manufacturing systems, a higher-order spline surface is used. This is the NURBS (non-uniform rational B-spline) surface, which uses NURBS mathematics to define a set of control vertices and a set of parameters (knots).

The distribution of control vertices together with the values of the parameters controls the shape of the surface. The use of NURBS mathematics allows the implementation of multi-patch surfaces with cubic surface mathematics and maintains full continuity control even with trimmed surfaces. Figure 5-3 shows a NURBS surface and control vertices.

Isoparm Curves

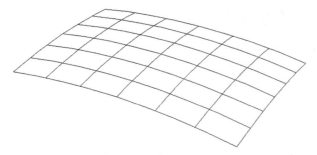

Fig. 5-4. U and V isoparm curves.

Because a NURBS surface is an accurate representation of a smooth surface, there is no facet but only boundary edges and silhouettes. To help visualize a smooth NURBS surface in the computer display, isoparametric curves (isoparm) distributed in two orthogonal directions are placed on the surface. (See figure 5-4.) (To avoid confusion with the X and Y axes of the coordinate system, these directions are called U and V.)

Surface Modeling Approaches

There are many ways to construct NURBS surfaces in the computer. If you want to construct a surface of basic primitive geometric shape, you simply select a primitive surface from the menu and specify the parameters. To meet aesthetic and functional needs, free-form surfaces are required.

The fundamental method of creating a free-form surface is to construct a framework of points and/or curves and let the computer generate the surfaces. If you already have a physical object or mock-up of the physical object, you use the digitizer to obtain appropriate data from the object in constructing the framework of points and/or curves.

Framework of Points and/or Curves

To construct a surface from a framework of points and/or curves, you construct the framework, select a command, and let the computer construct the surface. The outcome of the surface is determined by the location and pattern of the points and/or curves and the command applied. To prepare yourself for making free-form models, you will learn surface construction using various types of points and curves. Figure 5-5 shows a set of curves, a surface constructed from the curves, and a rendered image of the surface.

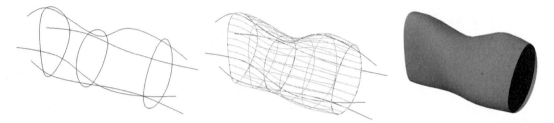

Fig. 5-5. Curve framework and surface constructed from the framework.

Thinking About Surface Modeling

Because a surface model represents an object by using a set of surfaces, surface modeling concerns making a set of surfaces in accordance with the appearance of the object. The sections that follow discuss various aspects of surface modeling with which you need to be familiar.

Deconstruction

To make a surface model, you start thinking about the boundary faces of the object. With careful analysis of the shape of the object, you deconstruct the object's surface into a number of discrete regions of surfaces, with each region following a particular geometric pattern. Then you think about the shape of each individual surface and identify ways of constructing those surfaces.

After this initial thinking process, you construct the surfaces accordingly. Thus, the process of surface modeling consists of two major steps: reduction of the complex 3D object into simple objects (individual surfaces) and making the simple objects to form the complex object. Figure 5-6 shows the surface model of a car body and the individual surfaces exploded.

Surface Modeling Concepts

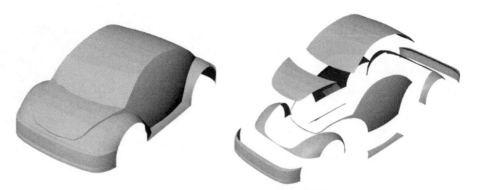

Fig. 5-6. Surface model and individual surfaces exploded.

After you deconstruct the surface of a complex 3D object into a set of surfaces, you start thinking about how to construct each of the surfaces. Among the many ways to construct an individual surface, the fundamental method is to construct 3D curves and/or points as a framework defining the profile and silhouette of the surface and apply NURBS surface modeling tools on the framework.

Thinking About Curves and Points

To construct a surface from a framework of curves and/or points, you should think about the curves and points and reduce the surface to its defining curves and/or points. Because the curves and points of a surface are not readily perceived, reduction of a surface into its defining curves and points requires in-depth analysis of what curves and/or points are needed, where the curves and/or points should be located, and what surface modeling commands are to be applied on them to obtain the required surface.

To be able to identify the locations and types of curves and/or points and the command to apply on them, you need to master various surface construction methods. You also need a good understanding of the shapes of surfaces generated from these methods.

In any design project, you will find that the most tedious job in surface modeling is creating curves and points, and that making surfaces from curves is simple. You need only use the appropriate surface construction commands. Hence, the prime focus of surface modeling is thinking about the curves and points, designing the curves and points, and making the curves and points. Try to determine the curves and/or points for making the surface shown in figure 5-7.

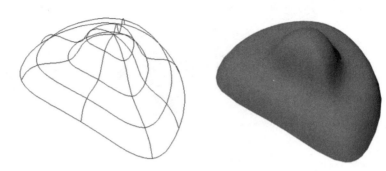

Fig. 5-7. Rail revolve surface.

Curve Data

In Chapter 2 you learned basic techniques of curve construction, editing, and transformation. To guide you through various curve and surface construction techniques, some of the data for the curves are given to you in the tutorials. However, you must realize that this type of data is not readily available when you design a surface. You need to think about the curves and determine the data for the curves on your own.

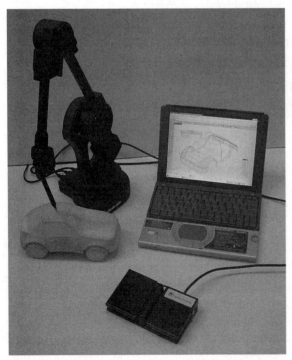

Fig. 5-8. Digitizing a physical model.

One way to determine the curves is to use graph paper and sketches. (See Chapter 3.) After you have discerned the relationships among surface shapes, the defining curves, and the applicable surface modeling commands, you construct freehand sketches on graph paper that depict the curves.

From these sketches, you obtain the coordinates of interpolation points. Using these interpolation points, you construct preliminary curves in the computer. From the preliminary curves, you improvise and construct curves to better represent the surface according to your design intent.

Digitizing a Physical Model

Another way to define curves is to make a physical model of the object and use a digitizer to obtain the coordinates of the control points of the surface. Using the coordinates of the markings, you construct curves. Figure 5-8 shows a 3D digitizer connected to a computer.

Surface Modeling Concepts

Prerequisite Knowledge and Thinking

Bearing in mind that the curves and points for the surface are implicit and imaginary, you may not find them on the object. Thus, a good understanding of the characteristics of various surface modeling commands and the types of curves and points required for these commands is essential. Insight into the types of curves and points needed for a particular type of surface is advantageous.

Modification

There are times when you are not satisfied with the surface constructed from a set of curves. It is probable that the surface you construct does not match the surface you conceive in your mind. To improvise, you have two options: (1) think about the curves and points again, reconstruct the curves and points, and make a new surface from the curves and points, or (2) modify the surface by editing and transforming.

One way to modify a surface is to manipulate control points. It is important to note that when you translate a control point on a surface, not only is the control point moved but quite a large area of the surface affected. Figure 5-9 shows a surface modified by translating two control points.

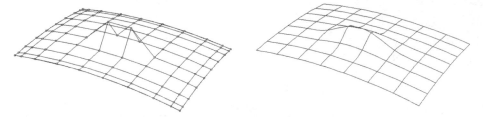

Fig. 5-9. Surface modified by moving its control points.

Advantages of Surface Models

Because a NURBS surface accurately represents smooth, free-form surfaces in the computer, it is the most appropriate tool for aesthetic and engineering design. A surface model contains all surface data of a 3D object, and you can retrieve the coordinates of any point on the surface of the object. Hence, you can use a surface model in most downstream computerized manufacturing operations, as well as in visualization of the object.

For example, you generate stereolithographic data for making rapid prototypes and cross sections for CNC machining. In visualization, you generate hidden-line projection views, shaded images, and photo-realistic rendered images. Figure 5-10 shows a rapid prototype constructed by using the 3D printing machine.

Fig. 5-10. Rapid prototype constructed from a surface model.

Limitations of Surface Models

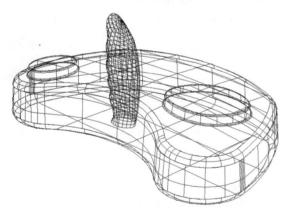

A surface has no thickness. Hence, a surface model is simply a thin shell with no volume, and therefore represents no information other than surface data. As a result, you cannot evaluate the mass properties of a surface model. This produces difficulties such as detecting collision between two surface models in situations in which one surface model lies completely inside another. Figure 5-11 shows the surface model of a joystick lying partly inside the surface model of a joypad.

Fig. 5-11. One surface model lying partly inside another surface model.

■■ Rhino Surface Modeling Tools

Basically, Rhino uses NURBS mathematics to accurately represent surfaces in the computer. In addition, Rhino enables you to construct and manipulate polygon meshes. (Polygon meshes are discussed further in Chapter 9.) By joining a set of surfaces, you obtain a polysurface. If the set of polysurfaces encloses a volume without any gap or opening, a solid is implied. (You will learn solid modeling in Chapter 8.) Figure 5-12 shows a set of surfaces joined to form a polysurface, and figure 5-13 shows an implied solid.

Rhino Surface Modeling Tools

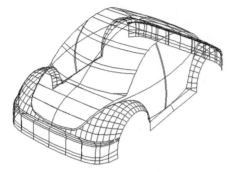

Fig. 5-12. Polysurface.

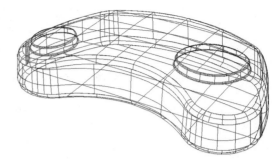

Fig. 5-13. Polysurface enclosing a volume.

Command Menu

Using Rhino as a surface modeling tool, you construct, edit, transform, and analyze NURBS surfaces as well as manipulate polygon meshes. The NURBS surface construction tool is available in the Surface pull-down menu. To edit and transform a NURBS surface, you use the Edit and Transform pull-down menu. Figure 5-14 shows Surface pull-down menu items for NURBS surface modeling, the Polygon Mesh cascading menu of the Tools pull-down menu, and the Surface, Surface Tools, and Mesh toolbars.

Fig. 5-14. NURBS surface modeling pull-down menu items.

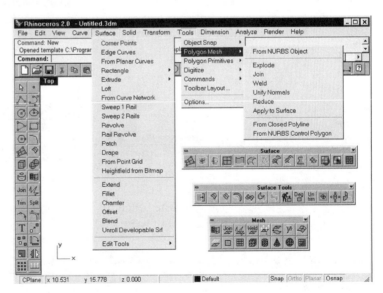

NURBS Surfaces

Because of their importance in design and manufacture, this chapter focuses mainly on NURBS surfaces. Basically, you can classify NURBS surfaces into three main categories in accordance with how they are constructed. The first category of surfaces is the primitives. To construct these types of surfaces, you select a surface type from the menu and specify the parameters of the surface.

The second category of surfaces is the basic free-form surfaces constructed from a framework of curves and/or points. The third category derives from existing surfaces. Their shape bears a relationship to surfaces you have already constructed. After you have constructed these surfaces, you can modify them in various ways and analyze them to determine various surface data.

Constructing Primitive Surfaces

Primitive surfaces are surfaces you specify without having to make a framework of points or curves. To produce a primitive surface, you select a command and specify the parameters. Corner-points surfaces and rectangular surfaces are primitive surfaces. A corner-points surface is a planar surface constructed by specifying four corner points. A rectangular surface, as the name implies, is a planar rectangular surface.

In essence, both are quadrilateral planar surfaces. Figure 5-15 shows the primitive surface commands in the Surface toolbar. Apart from these planar surfaces, you may also construct other types of primitive surfaces, such as box, cylinder, and sphere. Because these primitive objects are closed polysurfaces with no gap or opening, they are treated as solids (which you will learn about in Chapter 8).

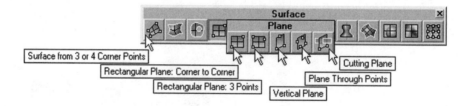

Fig. 5-15. Primitive surface construction commands.

Table 5-1 outlines the functions of primitive surface commands and their location in the pull-down menu.

Table 5-1: Primitive Surface Commands and Their Functions

Command Name	Pull-down Menu	Function
SrfPt	Surface > Corner Points	For constructing a planar surface by specifying four corner points

Rhino Surface Modeling Tools

Command Name	Pull-down Menu	Function
Plane	Surface > Rectangle > Corner to Corner	For constructing a rectangle by specifying its diagonal corners
Plane3Pt	Surface > Rectangle > 3 Points	For constructing a rectangle by specifying three points
PlaneV	Surface > Rectangle > Vertical	For constructing a rectangle perpendicular to the construction plane
CutPlane	Surface > Rectangle > Cutting Plane	For constructing a rectangle wide enough to pass selected objects through it
PlaneThroughPt	Surface > Rectangle > Through Points	For constructing a rectangle through a set of selected points

Quadrilateral Planar Surface

A quadrilateral planar surface is represented by four corner points lying on a plane. To make the surface, you specify four corner points.

1 Start a new file. Use the metric (millimeter) template.

2 Select Surface > Corner Points, or the Surface from 3 or 4 Corner Points button from the Surface toolbar.
   Command: SrfPt

3 Select the four corner points indicated in figure 5-16.

A planar surface is constructed on the Top viewport construction plane. Save your file as *Rectangle.3dm*.

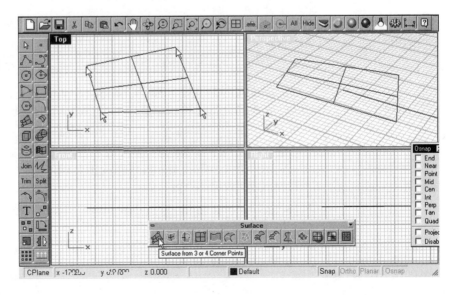

Fig. 5-16. Quadrilateral surface constructed from four corner points.

Rectangular Surfaces

There are five ways to construct a rectangular surface. These methods are discussed in the sections that follow.

Diagonal Corners. Rectangular surfaces with edges parallel to the X and Y axes can be constructed by specifying two diagonal corners, as follows.

1 Select Surface > Rectangle > Corner to Corner, or the Rectangular Plane/Corner to Corner button from the Plane toolbar.
   Command: Plane

2 Select the point in the Front viewport indicated in figure 5-17.

3 Select the point in the Top viewport indicated in figure 5-17.

A rectangular surface parallel to the Top viewport and at an elevation defined by a point specified in the Front viewport is constructed.

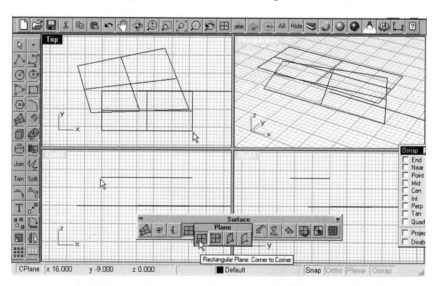

Fig. 5-17. Rectangular surface constructed.

Edge and Width. If the edges of the rectangular surface are not parallel to the main axes, you specify two points to depict an edge and specify the third point to depict the width, as follows.

1 Select Surface > Rectangle > 3 Points, or the Rectangular Plane/3 Points button from the Plane toolbar.
   Command: Plane3Pt

2 Select a point in the Front viewport (figure 5-18) to indicate the start of the edge.

3 Select a point in the Top viewport (figure 5-18) to indicate the end of the edge.

Rhino Surface Modeling Tools

4 Select a point in the Top viewport to indicate the width of the rectangle.

A rectangular surface is constructed from three selected points.

Fig. 5-18. Rectangle constructed from three points.

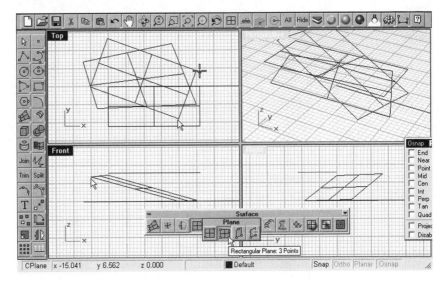

Perpendicular to the Construction Plane. To construct a rectangular surface perpendicular to the construction plane, perform the following steps.

1 Select Surface > Rectangle > Vertical, or the Vertical Plane button from the Plane toolbar.
   Command: PlaneV

2 Select the two points indicated in the Front viewport shown in figure 5-19 to specify the start and end points of the edge of the rectangle.

3 Select the point indicated in the Right viewport shown in figure 5-19 to specify the width of the rectangle.

A rectangle perpendicular to the Front viewport is constructed.

Cutting Plane. To construct a rectangular surface that cuts across a set of selected objects, perform the following steps.

1 Select Surface > Rectangle > Cutting Plane, or the Cutting Plane button from the Plane toolbar.
   Command: CutPlane

2 Select the objects indicated in figure 5-20 and press the Enter key.

3 Select two points to specify the location of a cutting plane and press the Enter key.

Fig. 5-19. Rectangle perpendicular to the construction plane.

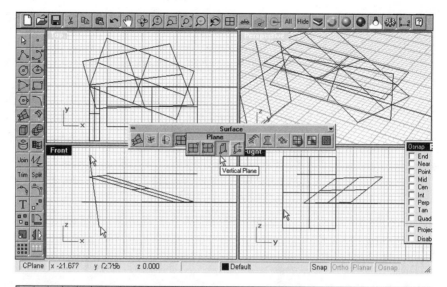

Fig. 5-20. Cutting plane constructed.

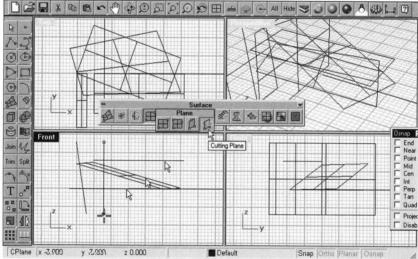

A rectangular surface that is wide enough to cut through using the selected objects is constructed.

Through Points. To construct a rectangular surface through a number of point objects, perform the following steps.

1 Construct a number of point objects.

2 Select Surface > Rectangle > Through Points, or the Plane Through Points button from the Plane toolbar.
Command: PlaneThroughPt

3 Select all of the point objects.

A surface is constructed. Save your file as *Rectangle.3dm*.

Constructing Basic Free-form Surfaces

As explained earlier, the fundamental way to construct a free-form surface is to construct a set of curves and/or point objects and let the computer make the surface. The surface profile is determined by the location and pattern of the curves and/or points and the method used to construct the surface. Figure 5-21 shows the free-form surface construction commands of the Surface toolbar.

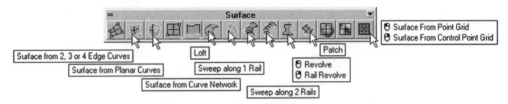

Fig. 5-21. Free-form surface commands.

Table 5-2 outlines the functions of basic free-form surface commands and their location in the pull-down menu.

Table 5-2: Basic Free-form Surface Commands and Their Functions

Command Name	Pull-down Menu	Function
EdgeSrf	Surface > Edge Curves	For constructing a surface from four edge curves
PlanarSrf	Surface > From Planar Curve	For constructing a trimmed planar surface from a planar closed-loop curve
Extrude	Surface > Extrude > Straight	For constructing a surface by extruding a curve perpendicular to the construction plane
ExtrudeAlongCrv	Surface > Extrude > Along Curve	For constructing a surface by extruding a shape curve along a path curve
ExtrudeToPt	Surface > Extrude > To Point	For constructing a surface by extruding a curve toward a point
Ribbon	Surface > Extrude > Ribbon	For constructing a surface by offsetting a curve
Loft	Surface > Loft	For constructing a surface that passes through a set of selected curves
NetworkSrf	Surface > From Curve Network	For constructing a surface from a network of curves
Sweep1	Surface > Sweep 1 Rail	For constructing a surface by sweeping a set of section curves along a rail curve
Sweep2	Surface > Sweep 2 Rails	For constructing a surface by sweeping a set of section curves along two rail curves

Command Name	Pull-down Menu	Function
Revolve	Surface > Revolve	For constructing surfaces by revolving curves about an axis
RailRevolve	Surface > Rail Revolve	For constructing a surface by revolving a curve along a path curve and about an axis
Patch	Surface > Patch	For constructing a patch surface from selected curves and/or points
SrfPtGrid	Surface > From Point Grid	For constructing a surface on a matrix of point objects

From-edge-curves Surface

A from-edge-curves surface is a 3D free-form surface that interpolates among four edges. To construct this type of surface, perform the following steps.

1 Start a new file. Use the metric (millimeter) template. Double click on the Front viewport to maximize the viewport.

2 Construct a free-form curve in accordance with figure 5-22.

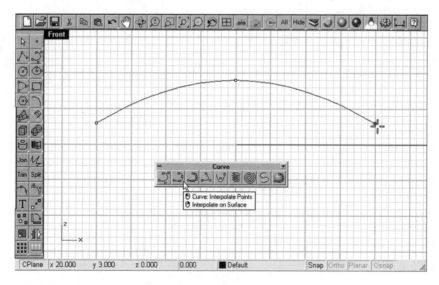

Fig. 5-22. Free-form curve constructed on the Front viewport construction plane.

In the following, you will set up a construction plane that is parallel to the Front viewport and 30 mm away. You will construct a curve on the new construction plane. Continue with the following steps.

3 Select View > Set CPlane > Elevation, or the Set Cplane/Elevation button from the Set CPlane toolbar.

Command: CPlaneElevation

4   Type *30* at the command line area to specify a distance of 30 mm.
5   Construct a free-form curve in accordance with figure 5-23.

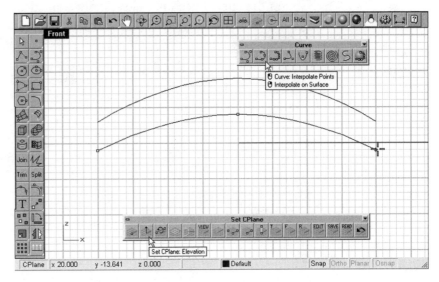

Fig. 5-23. Second curve constructed on a plane parallel to the Front viewport construction plane.

6   Double click on the Front viewport title to return to a four-viewport display.
7   Select the Zoom Extents All Views button from the Standard toolbar.

To construct the third and fourth curves on the Right viewport, continue with the following steps.

8   Construct two free-form curves on the Right viewport in accordance with figure 5-24.

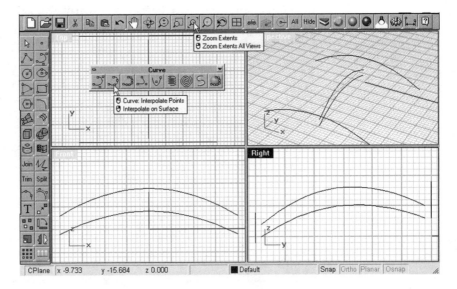

Fig. 5-24. Two free-form curves constructed on the Right viewport.

9 Maximize the Perspective viewport.

10 Move the curves indicated in figure 5-25. Note that the end points of the curves need not necessarily coincide.

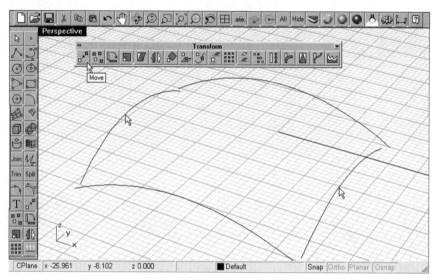

Fig. 5-25. Curves translated.

Four curves in two orthogonal directions are complete. To construct a surface, continue with the following steps.

11 Select Surface > Edge Curves, or the Surface from 2, 3 or 4 Edge Curves button from the Surface toolbar.
Command: EdgeSrf

12 Select the four curves.

A surface from four edge curves is constructed. (See figure 5-26.)

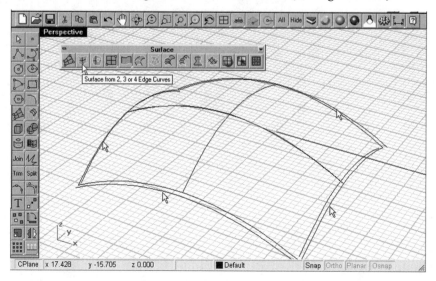

Fig. 5-26. Surface interpolating four edges constructed.

Rhino Surface Modeling Tools

To better control the shape of the surface, you will modify the curves so that adjacent end points meet. To hide the surface, edit the curves, and construct another surface, continue with the following steps.

13 Select Edit > Visibility > Hide, or the Hide Objects button from the Visibility toolbar.

Command: Hide

14 Select the surface and press the Enter key.

15 Select Curve > Edit Tools > Match, or the Match button from the Curve Tools toolbar.

Command: Match

16 Select the curves indicated in figure 5-27.

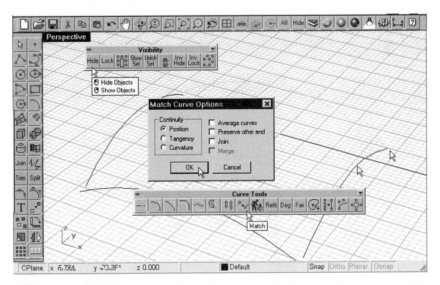

Fig. 5-27. Surface hidden and curves being matched.

17 In the Match Curve Options dialog box, select Position and the OK button.

18 Repeat the Match command three more times to modify the other ends of the curves. (See figure 5-28.)

19 Select Surface > Edge Curves, or the Surface from 2, 3 or 4 Edge Curves button from the Surface toolbar.

Command: EdgeSrf

20 Select four edges.

A surface from the modified curves is constructed. To compare the two surfaces, unhide the hidden surface, as follows. (See figure 5-29.)

21 Select Edit > Visibility > Show, or the Show Objects button from the Visibility toolbar.

Command: Show

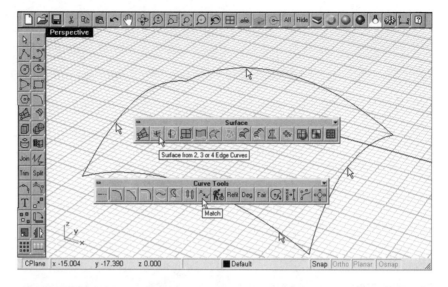

Fig. 5-28. Curves matched and surface being constructed.

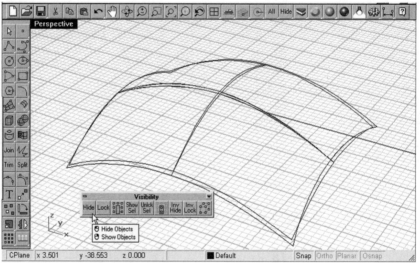

Fig. 5-29. Surfaces constructed.

Save your file as *EdgeSrf.3dm*.

Trimmed Planar Surface

A surface constructed from planar curves is, in essence, a rectangular planar curve trimmed by the selected planar curves. You construct planar curves (letting the computer construct a planar surface large enough to encompass the curves) and use the curves to trim the surface. To construct a trimmed planar surface, perform the following steps.

1 Start a new file. Use the metric (millimeter) template. Maximize the Top viewport.

2   Construct three closed-loop free-form curves in accordance with figure 5-30.

Fig. 5-30. Curves constructed.

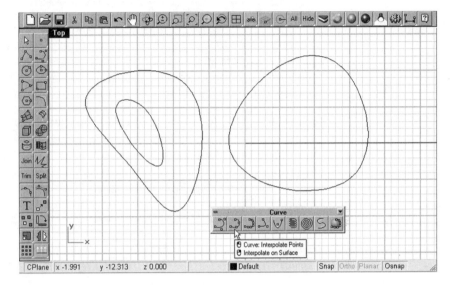

3   Select Surface > From Planar Curves, or the Surface from Planar Curves button from the Surface toolbar.
Command: PlanarSrf

4   Select the curves and press the Enter key.

Two surfaces are constructed. As shown in figure 5-31, two curves with one of them completely inside the other produces a surface with an opening. Save your file as *PlanarSrf.3dm*.

Fig. 5-31. Surfaces constructed.

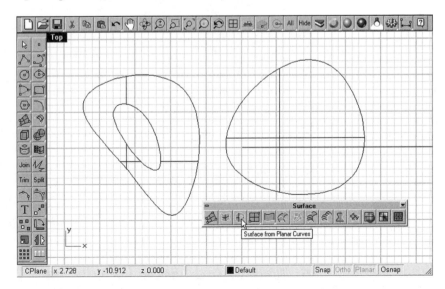

Extruded Surfaces

There are four ways to extrude a curve to a surface. These methods are discussed in the sections that follow.

Straight Extrude. The simplest way to extrude a curve to a surface is to extrude it along a straight line perpendicular to the current construction plane. In the following, you will construct a 3D curve from two planar curves and extrude the curve in a straight line.

1 Start a new file. Use the metric (millimeter) template.
2 Construct two free-form curves in the Front viewport in accordance with figure 5-32.

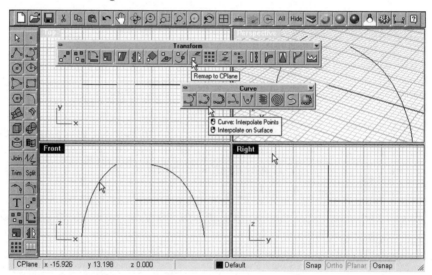

Fig. 5-32. Planar free-form curves constructed.

Two curves depicting two orthogonal views of a 3D curve are constructed. Continue with the following steps.

3 Select Transform > Orient > Remap to Cplane, or the Remap to CPlane button from the Transform toolbar.
Command: RemapCPlane

4 Select the curve indicated in figure 5-32 and select the Right viewport.

One of the curves is remapped to the Right viewport. Continue with the following steps.

5 Select Curve > From 2 Views, or the Curve From 2 Views button from the Curve Tools toolbar.
Command: Crv2View

6 Select the curves indicated in figure 5-33.

Rhino Surface Modeling Tools

Fig. 5-33. 3D curve constructed from planar curves in two views.

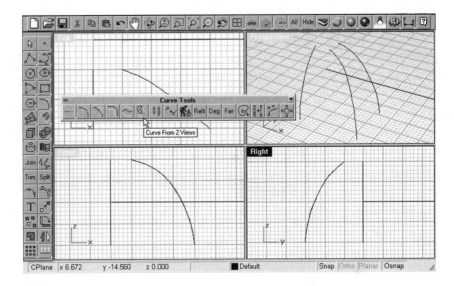

A 3D curve is constructed. Continue with the following steps.

7   Select Surface > Extrude > Straight, or the Extrude Straight button from the Extrude toolbar.

    Command: Extrude

8   Select the 3D curve in the Top viewport.

9   Select a point in the Front viewport to indicate the extrude distance.

A surface is extruded in a direction perpendicular to the Top viewport. (See figure 5-34.) Save your file as *Extrude.3dm*.

Fig. 5-34. 3D curve being extruded.

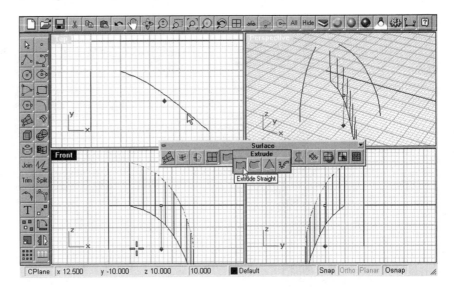

Chapter 5: Surface Modeling: Part 1

Extrude Along a Path. The second way to make an extruded surface is to extrude a curve along a path, as follows.

1. Select File > Save As and specify a new file name of *ExtrudeAlongCrv.3dm*. Maximize the Perspective viewport.
2. Select Surface > Extrude > Along Curve, or the Extrude Along Curve button from the Extrude toolbar.
   Command: ExtrudeAlongCrv
3. Select the curve indicated in figure 5-35.

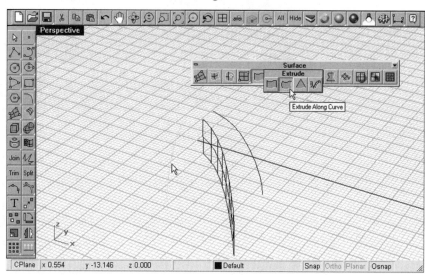

Fig. 5-35. Shape curve being selected.

4. Select a point near an end of the other curve indicated in figure 5-36 to indicate a path and direction.

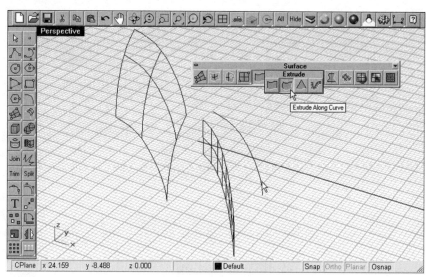

Fig. 5-36. Path curve being selected.

Rhino Surface Modeling Tools

A surface is extruded in a direction guided by a curve. (See figure 5-36.) Save your file.

Extrude to a Point. The third way to make an extruded surface is to extrude a curve toward a point, as follows.

1 Select File > Save As and specify a new file name of *ExtrudeToPt.3m*.

2 Select Surface > Extrude > To Point, or the Extrude to Point button from the Extrude toolbar.
Command: ExtrudeToPt

3 Select the curve and indicate a point. (See figure 5-37.)

A surface is extruded from a curve to a point. Save your file.

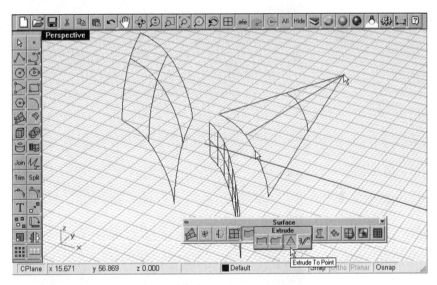

Fig. 5-37. Curve extruded to a point.

Extrude Ribbon. The fourth way to construct an extruded surface is to extrude a curve sideways, as follows.

1 Select File > Save As and specify a new file name of *Ribbon.3dm*.

2 Select Surface > Extrude > Ribbon, or the Ribbon button of the Extrude toolbar.
Command: Ribbon

3 Select the edge indicated in figure 5-38.

Because the edge you selected has two curves (the edge you use to extrude and an edge of the extruded surface), you need to select one from the menu.

4 Select Curve.

Fig. 5-38. Curve being extruded to form a ribbon.

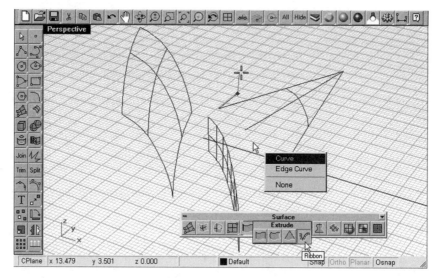

5 Type *D* at the command line area and type *10* to specify the width of the ribbon.

6 Select the point indicated in figure 5-38 to specify the side to be offset.

Extruding a curve sideways is like making a ribbon, especially if you extrude a wavy 3D curve. (See figure 5-39.) Save your file.

Fig. 5-39. Ribbon constructed.

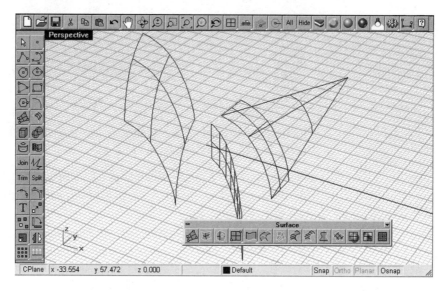

Lofted Surface

A lofted surface is constructed from two or more section profile curves. The surface interpolates from the first section curve to the second, and

Rhino Surface Modeling Tools 225

to the next curve. To construct a loft surface, perform the following steps.

1 Start a new file. Use the metric (millimeter) template.

2 Construct three free-form curves in accordance with figure 5-40.

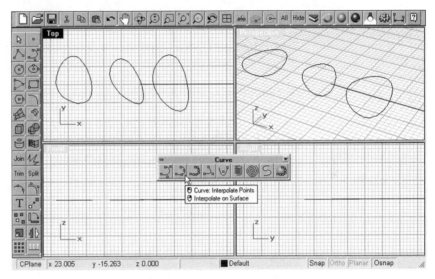

Fig. 5-40. Curves constructed.

Curves for making a lofted surface need not be a closed-loop curve. You may use open-loop curves as well.

3 Select the curves in the Front viewport one by one, and drag them to new positions in accordance with figure 5-41.

4 Select Surface > Loft, or the Loft button from the Surface toolbar.
Command: Loft

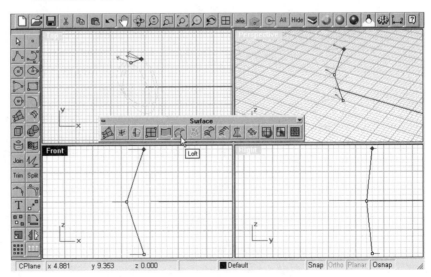

Fig. 5-41. Curves translated and being selected to construct a lofted surface.

Chapter 5: Surface Modeling: Part 1

5 Select the curves in the Front viewport (top to bottom) and press the Enter key.

Note the direction arrows indicated in the seam points of the curves.

6 If the directions of the arrows are not congruent, type *F* and select the non-congruent seam point. If the arrows are all in the same direction, press the Enter key.

7 In the Loft Options dialog box, accept the default setting and press the OK button.

A loft surface is constructed. (See figure 5-42.) Save your file as *Loft.3dm*.

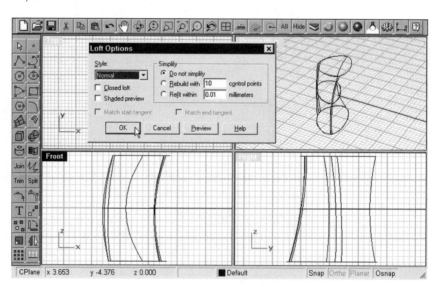

Fig. 5-42. Lofted surface constructed.

From-Curve-Network Surface

A from-curve-network surface is similar to the from-edge-curves surface, with additional curves in two orthogonal directions. To construct this type of surface, perform the following steps.

1 Start a new file. Use the metric (millimeter) template.

2 Construct three free-form curves in the Front viewport and three free-form curves in the Right viewport in accordance with figure 5-43.

3 Translate the curves so that they form a network of curves in two orthogonal directions. (See figure 5-44.)

4 Select Surface > From Curve Network, or the Surface from Curve Network button from the Surface toolbar.

Command: NetworkSrf

Fig. 5-43. Free-form curves constructed.

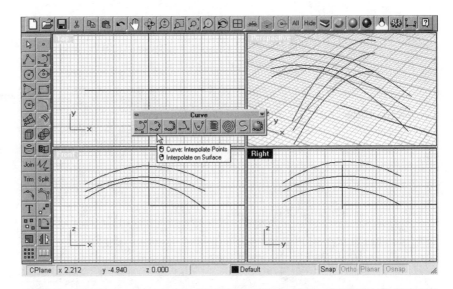

Fig. 5-44. Curves translated and surface being constructed.

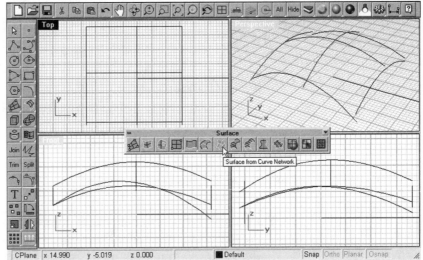

5 Select all curves and press the Enter key.

In essence, valid curves will be automatically sorted into two sets of orthogonal curves, regardless of the sequence of selection. However, if one of the curves is invalid or ambiguously defined, the system will prompt you to select the curve sets manually. Then you should select one set of curves, press the Enter key, continue to select the second set of orthogonal curves, and press the Enter key again.

After selecting the curves, the Surface From Curve Network dialog box displays. Here, you specify edge tolerance and continuity of matching edges, as follows.

6 In the dialog box, accept the default and click on the OK button.

A surface is constructed from a curve network. (See figure 5-45.) Save your file as *NetworkSrf.3dm*.

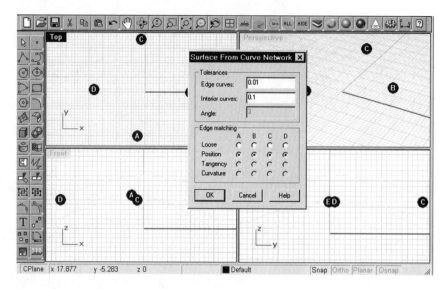

Fig. 5-45. Surface From Curve Network dialog box.

Sweeping Along a Rail

In making a lofted surface, the surface profile interpolates smoothly from the first curve to the second curve, and to the next curve. To impose control on the flow of a surface profile from a section curve to another section curve, you may add a guiding rail. The surface is called a sweep surface. To construct a surface by sweeping cross-section curves along a rail, perform the following steps.

1 Start a new file. Use the metric (millimeter) template.

2 Construct a free-form curve, a circle around the free-form curve, and an ellipse around the free-form curve in accordance with figure 5-46.

3 Select Surface > Sweep 1 Rail, or the Sweep along 1 Rail button from the Surface toolbar.

Command: Sweep1

4 Select the free-form curve as the rail curve, select the circle and ellipse as the cross-section curves, and press the Enter key.

5 In the Sweep 1 Rail Options dialog box, accept the default and click on the OK button.

A surface is constructed. (See figure 5-47.) Save your file as *Sweep1.3dm*.

Fig. 5-46. Free-form curve, and circle and ellipse around the free-form curve, constructed.

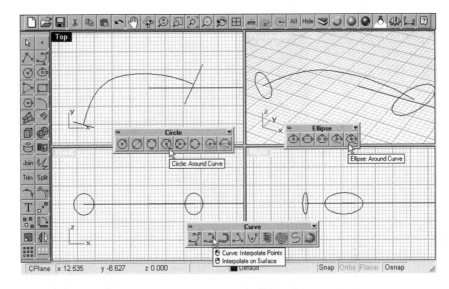

Fig. 5-47. Sweep surface constructed.

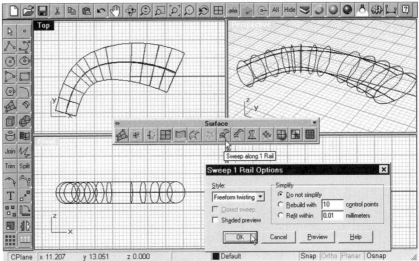

Sweeping Along Two Rails

To add further control to the surface profile while sweeping, you may use two guiding rails. To construct a surface by sweeping cross-section curves along two rails, perform the following steps.

1 Start a new file. Use the metric (millimeter) template.

2 With reference to figure 5-48, construct a free-form curve on the Top viewport, a free-form curve on the Front viewport, and a circle and an ellipse vertical to the Top viewport.

3 Check the End and Quad boxes in the Osnap dialog box.

Fig. 5-48. Curves, circle, and ellipse constructed.

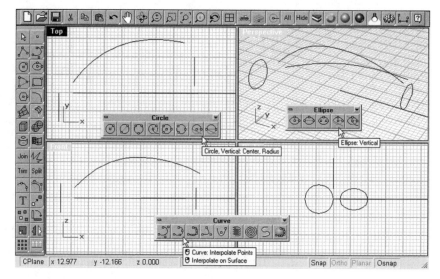

4 Select Transform > Orient > 2 Points, or the Orient/2 Points button from the Transform toolbar.

Command: Orient

5 Select the circle and press the Enter key.

6 Select the quadrant locations as the reference points.

7 Select the end points of the free-form curves as the target points. (See figure 5-49.)

Fig. 5-49. Circle being oriented.

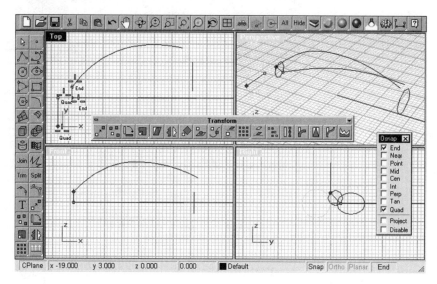

8 Repeat the Orient command to orient the ellipse. (See figure 5-50.)

To construct a surface by sweeping two cross-section curves along two rail curves, continue with the following steps.

Fig. 5-50. Ellipse being oriented.

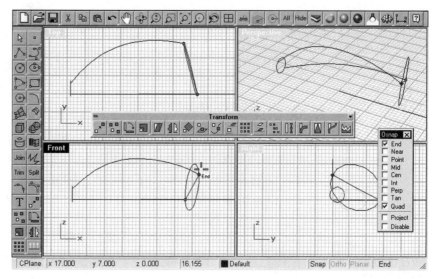

9 Select Surface > Sweep 2 Rails, or the Sweep along 2 Rails button from the Surface toolbar.

Command: Sweep2

10 Select the free-form curves as rails.

11 Select the circle and ellipse as cross-section curves and press the Enter key.

12 If the direction of the seam point is congruent, press the Enter key. Otherwise, type *F* and select the connection point to be flipped. (See figure 5-51.)

Fig. 5-51. Sweep surface being constructed.

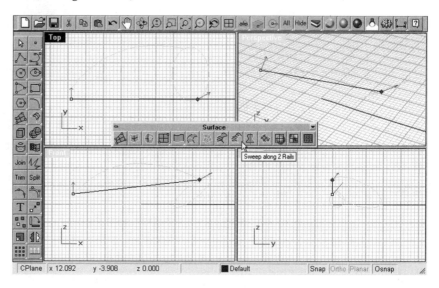

13 Press the Enter key to accept the direction.

14 In the Sweep 2 Rails dialog box, accept the default and click the OK button.

A surface is constructed by sweeping the circle and ellipse along two rails. Save your file as *Sweep2.3dm*.

Revolved Surface

A revolved surface is constructed by revolving a curve about an axis. To construct a revolved surface, perform the following steps.

1 Start a new file. Use the metric (millimeter) template.

2 Construct a free-form curve in accordance with figure 5-52.

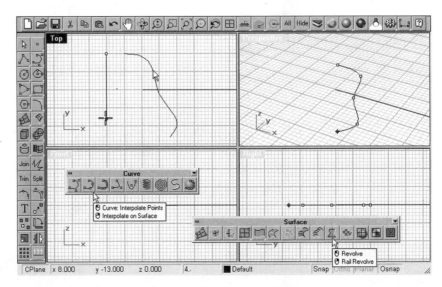

Fig. 5-52. Revolve surface being constructed.

3 Select Surface > Revolve, or the Revolve button from the Surface toolbar.

Command: Revolve

4 Select the curve and press the Enter key.

5 Select two points indicated in figure 5-52 as the start and end points of the revolved axis.

6 In the Revolve Options dialog box, accept the default and click on the OK button.

A revolved surface is constructed. Save your file as *Revolve.3dm*.

Revolved Surface Guided by a Rail

To control the surface profile while it is being revolved about an axis, you may add a guide rail. To construct a revolved surface guided by a rail, perform the following steps.

1 Start a new file. Use the metric (millimeter) template.

2 Construct a curve in the Top viewport and another curve in the Front viewport in accordance with figure 5-53.

3 Select Surface > Rail Revolve, or the Rail Revolve button from the Surface toolbar.

Command: RailRevolve

4 Select the curve in the Front viewport as the profile curve and select the curve in the Top viewport as the path curve.

5 Select two points indicated in figure 5-53 to specify the origin and direction of the revolve axis.

A surface revolved about an axis and guided by a rail is constructed. Save your file as *RailRevolve*.

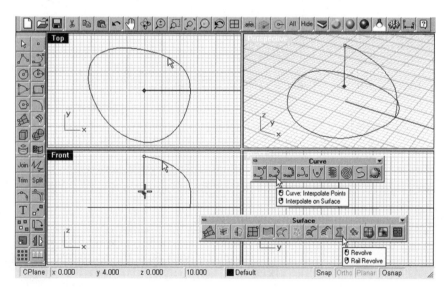

Fig. 5-53. Revolve rail surface being constructed.

Patch Surface

To construct a patch surface, the input data can be curves and/or points. You will construct five patch surfaces: a patch surface from a closed-loop curve, a patch surface from point objects, a patch surface from a closed-loop curve and point objects, a patch surface from an open-loop curve and point objects, and a patch surface from a number of curves.

234   CHAPTER 5: Surface Modeling: Part 1

To construct two curves and a set of point objects, perform the following steps.

1 Start a new file. Use the metric (millimeter) template.

2 With reference to figure 5-54, construct two free-form curves (an open-loop curve and a closed-loop curve) in the Top viewport, three point objects in the Front viewport, and three point objects in the Right viewport.

3 Set the current layer to *Layer 01*.

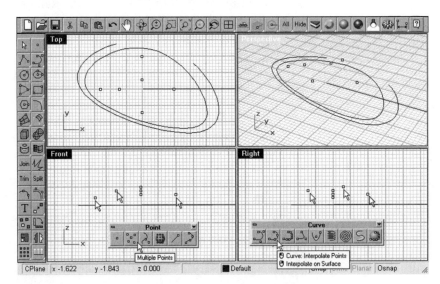

Fig. 5-54. Curve and point objects constructed.

Patch Surface from Closed-loop Curve. To construct a patch surface from a curve, perform the following steps.

1 Select Surface > Patch, or the Patch button from the Surface toolbar.

   Command: Patch

2 Select the closed-loop curve and press the Enter key.

3 In the Patch Options dialog box, accept the default and click on the OK button.

A patch surface is constructed from the closed-loop curve. (See figure 5-55.)

> **NOTE:** *Do not confuse this command with the PlanarSrf command in which the curves for making a surface have to be planar.*

Fig. 5-55. Patch surface constructed from the closed-loop curve.

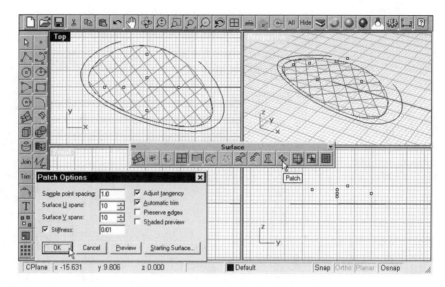

Patch Surface from Point Objects. To construct a patch surface from point objects, perform the following steps.

1 Turn off *Layer 01* and set the current layer to *Layer 02*.
2 Use the Patch command, select the points, and press the Enter key.
3 Click on the OK button in the Patch Options dialog box.

A surface is constructed. (See figure 5-56.)

Fig. 5-56. Patch surface constructed from the point objects.

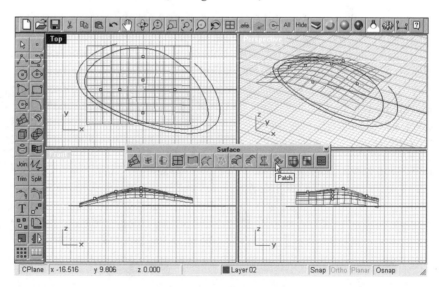

Patch Surface from Closed-loop Curve and Point Objects. To construct a patch surface from the closed-loop curve and point objects, perform the following steps.

1. Turn off *Layer 02* and set the current layer to *Layer 03*.
2. Use the Patch command, select the closed-loop curve and the points, press the Enter key, and click on the OK button of the Patch Options dialog box. (See figure 5-57.)

Fig. 5-57. Patch surface constructed from the point objects and the closed-loop curve.

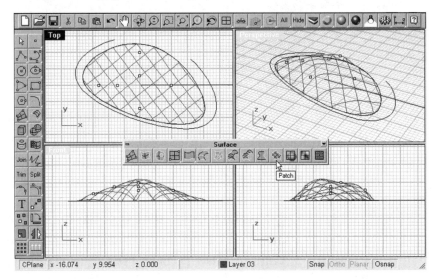

Patch Surface from Open-loop Curve and Point Objects. To construct a patch surface from the open-loop curve and the point objects, perform the following steps.

1. Turn off *Layer 03* and set the current layer to *Layer 04*.
2. Use the Patch command, select the open-loop curve and points, press the Enter key, and click on the OK button of the Patch Options dialog box. (See figure 5-58.)

Fig. 5-58. Patch surface constructed from the point objects and the open-loop curve.

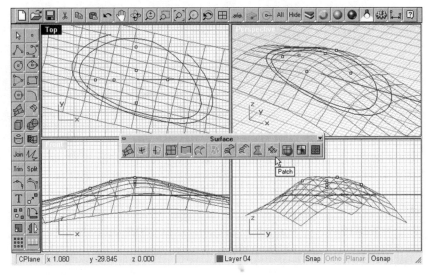

Patch Surface from Curves. To construct two curves and a patch surface from the curves, perform the following steps.

1. Turn off *Layer 04* and set the current layer to *Layer 05*.
2. Construct a free-form curve in the Front viewport and another free-form curve in the Right viewport. (See figure 5-59.)
3. Use the Patch command, select the curves indicated in figure 5-59 and the closed-loop curve, press the Enter key, and click on the OK button of the Patch Options dialog box.

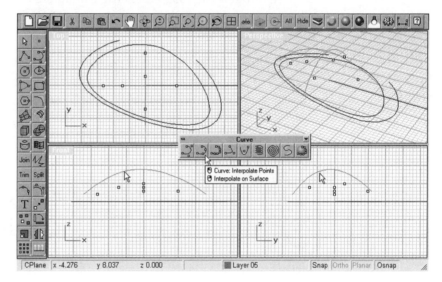

Fig. 5-59. Curves constructed.

A patch surface from the curves is constructed. (See figure 5-60.)

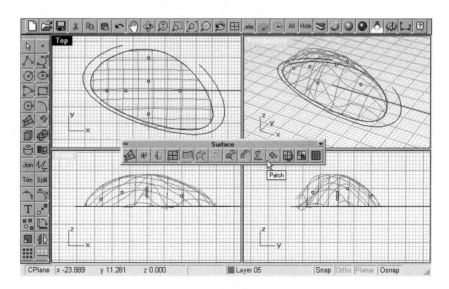

Fig. 5-60. Patch surface constructed from curves.

238   CHAPTER 5: Surface Modeling: Part 1

Five patch surfaces are complete. Try to relate the results with the input points and/or curves. Save your file as *Patch.3dm*.

Surface from a Pattern of Point Grids

A surface constructed from a pattern of point grids requires a rectangular pattern of point grids. To construct this type of surface, perform the following steps.

1 Start a new file. Use the metric (millimeter) template.

2 Construct three point objects in the Top viewport in accordance with figure 5-61.

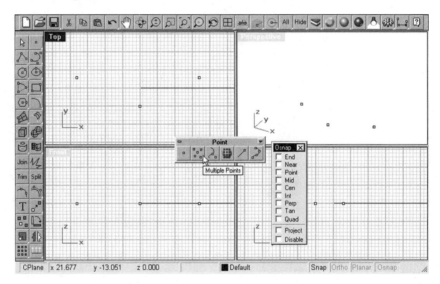

Fig. 5-61. Point objects constructed.

Three point objects are constructed. To construct three point objects on a construction plane parallel to the Top viewport and at an elevation of 15 mm, perform the following steps.

3 Select View > Set Cplane > Elevation, or the Set Cplane/Elevation button from the Set CPlane toolbar.

Command: CPlaneElevation

4 Type *15* and press the Enter key.

5 Clear all boxes in the Osnap dialog box.

6 Construct the three point objects indicated in figure 5-62.

To construct three more point objects and a surface from the point objects, continue with the following steps.

7 Use the Cplane/Elevation option again to set the elevation to 15. (The elevation height is cumulative.)

Rhino Surface Modeling Tools

Fig. 5-62. Point objects constructed on a construction plane at an elevation of 25 mm.

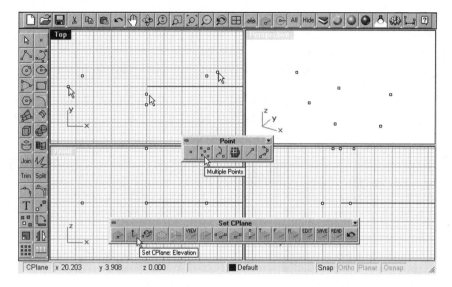

8 Construct the three point objects indicated in figure 5-63.

Fig. 5-63. More point objects constructed on a new construction plane.

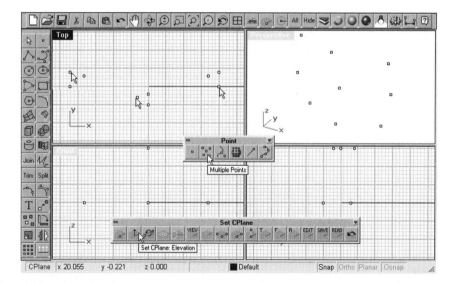

9 Select View > Cplane > To View, or the Set CPlane/To View button from the Set CPlane toolbar, to reset the construction plane to the Top viewport construction plane.
Command: CPlaneToView

10 Maximize the Perspective viewport.

11 Select Surface > From Point Grid, or the Surface From Point Grid button from the Surface toolbar.
Command: SrfPtGrid

240 CHAPTER 5: Surface Modeling: Part 1

12 Type *3* to specify three rows of a point grid.
13 Type *3* to specify three columns of a point grid.
14 Check the Point box in the Osnap dialog box.
15 Select the point objects, left to right, top to bottom.

A surface from a rectangular pattern of point objects in three rows and three columns is constructed. (See figure 5-64.) Save your file as *SrfPtGrid.3dm*.

Fig. 5-64. Surface constructed from a rectangular pattern of point objects.

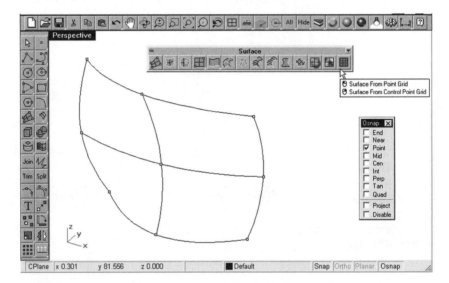

Digitizing

Prior to constructing a basic free-form surface, you construct points and/or curves. In making the point objects and curves, you have to specify locations. To specify locations, you can extract coordinate information from sketches that depict your design idea, use the grid mesh background or bitmap background and construct free-form sketches directly by selecting locations in the viewport, or use a digitizer if you already have a physical object.

By using a digitizer, you obtain coordinates of selected points on the surface of a physical object. If the physical object is a hand-made mock-up of the real object, chances are the surfaces are not smooth enough. For example, the silhouette and profiles of a physical model you digitize may deviate from the model's ideal shape and may even be wavy.

If the waviness and irregularities on the mock-up are not intended in the final computer model, you need to edit the curves you obtained from the digitizer and/or the basic free-form surfaces you construct from the digitized data. You already learned curve editing in Chapter 3. You will learn

Rhino Surface Modeling Tools

surface editing after you learn how to digitize and how to construct derived surfaces in Chapter 6.

There are many types of digitizers. The following section discusses digitizing using the MicroScribe digitizer. If you do not have a MicroScribe digitizer connected to your computer, you may skip the following tutorial. However, the following has general value in terms of understanding how digitizers operate and how they are employed.

Using the MicroScribe Digitizer

Figure 5-65 shows the MicroScribe digitizer. The MicroScribe digitizer has eight key components: base, shoulder, counterweight, upper arm, elbow, lower arm, wrist, and stylus.

The base unit is the foundation of the digitizer. You place it on any flat surface. At the back of the base unit are connections to the foot pedal, electric power supply, and data transmission cord to the computer. On the pedal unit are two buttons that function similarly to the buttons of a mouse (freeing your hands to hold the stylus and the object to be digitized).

On top of the base unit is a swiveling, upright shoulder unit that can freely rotate horizontally. At the upper end of the shoulder unit is a pivot joint that connects to the upper arm of the digitizer. The upper arm then connects to the elbow unit that, in turn, connects to the lower arm. At the other end of the lower arm is a wrist unit that links to the stylus. To balance the weight of the elbow and other objects connected to it, a counterweight is placed at the lower end of the upper arm.

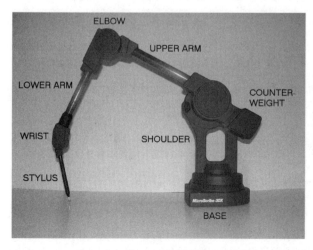

Fig. 5-65. MicroScribe digitizer and foot pedal.

To operate the digitizer, you select a command, hold the stylus, and place the stylus tip on the surface of the object you want to digitize. You then depress the right-hand foot pedal to pick a point. Information regarding the position of the stylus tip relative to the base unit is transmitted to the computer.

MicroScribe Digitizer Commands

A set of built-in commands is available if your computer is connected to the 3D MicroScribe digitizer. These commands are found in the MicroScribe toolbar, shown in figure 5-66.

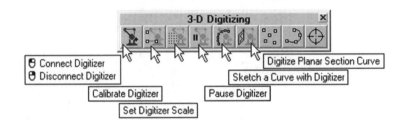

Fig. 5-66. MicroScribe toolbar.

Table 5-3 outlines the functions of MicroScribe digitizer commands and their location in the pull-down menu.

Table 5-3: Digitizer Commands and Their Functions

Command Name	Pull-down Menu	Function
Dig	Tools > Digitizer > Connect	For activating the connected MicroScribe digitizer
DigDisconnect	Tools > Digitizer > Disconnect	For deactivating the digitizer
DigCalibrate	Tools > Digitizer > Calibrate	For calibrating the origin and the X, Y, and Z axes of the digitizer
DigScale	Tools > Digitizer > Set Scale	For setting the scale of digitized output
DigPause	Tools > Digitizer > Pause	For pausing the digitizer
DigSketch	Tools > Digitizer > Sketch Curve	For sketching
DigSection	Tools > Digitizer > Planar Section Curve	For constructing a series of cross-section curves

Activating and Calibrating the MicroScribe Digitizer

Prior using the MicroScribe digitizer, you need to connect it to an electric power supply and to your computer. You also need a flat surface on

Rhino Surface Modeling Tools 243

which you place the base unit of the digitizer and the physical object you want to digitize. To activate the digitizer, perform the following steps.

1 Select Tools > Digitize > Connect, or the Connect Digitizer button from the MicroScribe toolbar.

Command: Dig

2 In the Select Digitizer dialog box, select Immersion MicroScribe and click on the OK button. (See figure 5-67.)

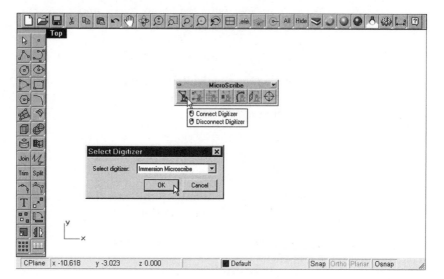

Fig. 5-67. Select Digitizer dialog box.

3 In the MicroScribe Port and Baud dialog box, accept the default settings and click on the OK button. (See figure 5-68.)

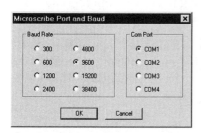

Fig. 5-68. MicroScribe Port and Baud dialog box.

The digitizer is activated. After activating the digitizer, you need to calibrate it. First, you will specify three points to depict a reference plane. The first point is the origin, the second point (together with the first point) defines the X axis, and the third point (together with the first and second points) defines a plane and the Y axis. Continue with the following steps.

4 Hold the stylus, move the stylus tip to a point on the flat surface to use the point as the reference origin, and depress to right-hand foot pedal. (See figure 5-69.)

5 Bring the stylus tip to a second point on the flat surface to define the X axis direction and depress the right-hand foot pedal.

6 Bring the stylus tip to a third point to define a plane and depress the right-hand foot pedal.

Fig. 5-69. Specifying the origin, X axis, and Y axis.

7 Press the Enter key to use the world origin.

The MicroScribe digitizer is connected to your computer and is calibrated. If you wish to recalibrate the digitizer, you use the DigCalibrate command, as follows.

8 Select Tools > Digitize > Calibrate, or the Calibrate Digitizer button from the MicroScribe toolbar.

Command: DigCalibrate

9 Select three points to indicate the origin, X axis, and the reference plane, and press the Enter key.

Digitizing Sketched Curves

In essence, you use the digitizer as a 3D mouse. In contrast to an ordinary mouse, with which you can select only 2D points on a construction plane, the digitizer enables you to select points in 3D, regardless of the current construction plane. Using the digitizer with appropriate commands, you construct point objects and curves.

Before you start to digitize, you have to decide the types of surfaces to be constructed, because they need different patterns of points and curves. (Refer to the previous section on basic free-form surfaces.) To construct a curve, perform the following steps.

1 Select Tools > Digitize > Sketch Curve, or the Sketch a Curve with Digitizer button of the MicroScribe toolbar.

Command: DigSketch

2 In the Digitize Options dialog box, check only the Curves box and click on the OK button. (See figure 5-70.) This way, open-loop curves will be constructed.

3 Place the stylus tip on the surface of the model at the location you want to start sketching, and depress and hold down the right-hand foot pedal.

Fig. 5-70. Digitize Options dialog box.

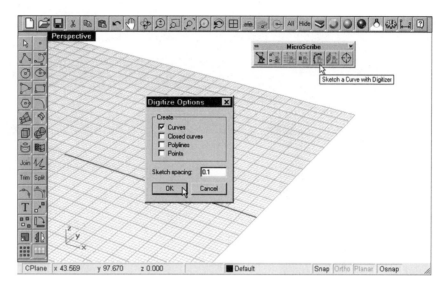

4 Move the stylus tip across the face of the physical model and then release the foot pedal.

A free-form curve is constructed along the path of the stylus movement.

5 Repeat the command by depressing the left-hand foot pedal, and repeat steps 3 and 4 to digitize.

When you have sufficient curves, you may use a free-form surface command to construct a surface. Figure 5-71 shows a set of free-form curves constructed for making a patch surface.

Fig. 5-71. Free-form sketch curves.

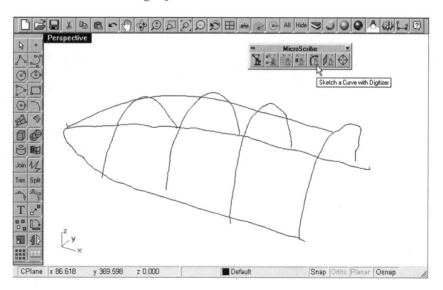

Because the curves may not be smooth enough for a making a smooth surface, you need to edit the curve, as follows.

6 Select Curve > Edit Tools > Fair, or the Fair button from the Curve Tools toolbar. (See figure 5-72.)

Command: Fair

Fig. 5-72. Curves faired.

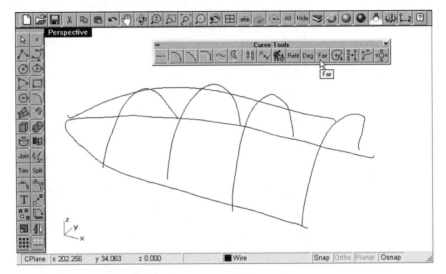

7 Select all curves and press the Enter key.

8 Select Curve > Edit Tools > Match, or the Match button from the Curve Tools toolbar.

Command: Match

9 Select the curves indicated in figure 5-73 to match the end points.

Fig. 5-73. Curves being matched.

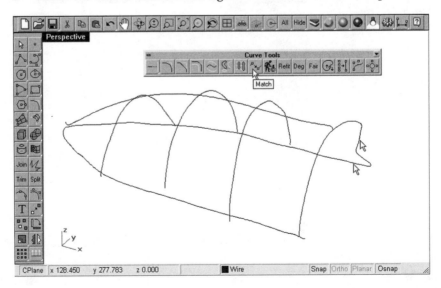

10 Repeat the Match command to match the other end points.

To construct a patch surface, continue with the following steps.

11 Select Surface > Patch, or the Patch button from the Surface toolbar.
Command: Patch

12 Select the curves and press the Enter key.

A patch surface is constructed. (See figure 5-74.) Save your file as *DigitizePatch.3dm*.

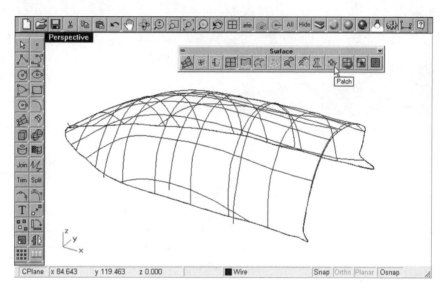

Fig. 5-74. Patch surface constructed.

Digitizing Cross-section Profiles

In the following, you will construct a set of cross-section curves using the digitizer. Figure 5-75 shows the physical object placed on a flat horizontal board with two vertical boards attached that along with the base board serve as digitizing reference planes.

1 Start a new file. Use the metric (millimeter) template.
2 Maximize the Perspective viewport.
3 Select Tools > Digitize > Planar Section Curve, or the Digitize Planar Section Curve button from the MicroScribe toolbar.
   Command: DigSection
4 In the Digitize Options dialog box, select Curve and click on the OK button. (See figure 5-76.)

To define a vertical plane and the X axis, continue with the following steps.

Fig. 5-75. Physical object to be digitized.

Fig. 5-76. Digitize Options dialog box.

> **5** Select a point and press the right-hand foot pedal to define the origin of the vertical plane. (See figure 5-77.)
>
> **6** Select a point and press the right-hand foot pedal to define the second point of the vertical plane.
>
> **7** Select a point and press the right-hand foot pedal to define the third point of the vertical plane.
>
> A vertical plane is defined. (See figure 5-78.)

Fig. 5-77. Defining a vertical plane by selecting three points on a vertical board.

> **8** Select a point and depress the right-hand foot pedal to define the start point of the X axis. (See figure 5-79.)
>
> **9** Select a point and depress the right-hand foot pedal to define the end point of the X axis.
>
> The X axis is defined. (See figure 5-80.)

Rhino Surface Modeling Tools

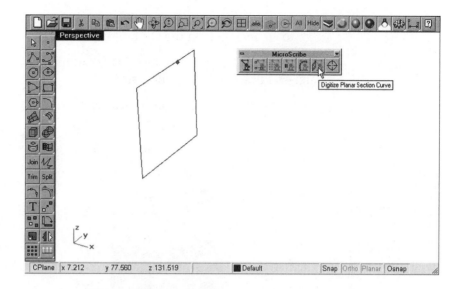

Fig. 5-78. Vertical plane defined.

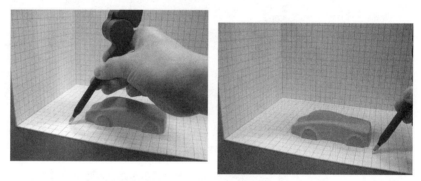

Fig. 5-79. Defining the X axis.

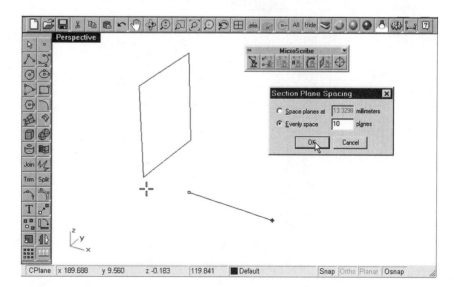

Fig. 5-80. X axis defined.

10 In the Section Plane Spacing dialog box, set the number of planes to *10* and click on the OK button.

Ten section planes are defined. They are parallel to the vertical plane and evenly spaced along the X axis.

To construct a set of section curves, continue with the following steps.

11 Bring the stylus onto the surface of the physical object, and depress and hold down the right-hand foot pedal. (See figure 5-81.)

Fig. 5-81. Describing a curve along the surface.

12 Move the stylus along the surface to describe a curve and release the right-hand foot pedal.

Point objects are displayed at the intersection between this curve and the vertical planes. (See figure 5-82.)

Fig. 5-82. Intersection points displayed.

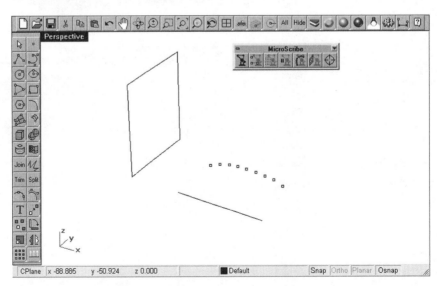

13 Repeat steps 11 and 12 to construct more points. (See figure 5-83.)

Fig. 5-83. More intersection points displayed.

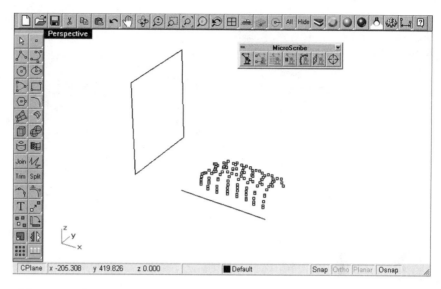

14 Press the Enter key.

A set of cross-section curves is constructed. (See figure 5-84.)

15 Construct a lofted surface. (See figure 5-85.)

The surface is constructed. Save your file as *DigitLoft.3dm*.

Fig. 5-84. Section curves constructed.

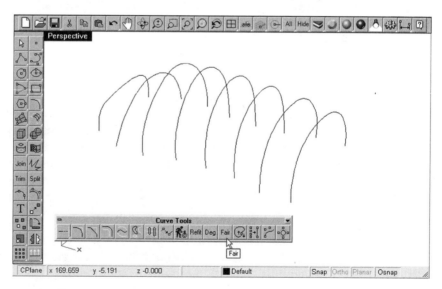

Apart from the DigSketch and DigSection commands, you can use the MicroScribe digitizer to replace your mouse to select 3D points for constructing the point objects and curves you learned about in chapters 2 through 4.

Fig. 5-85. Surface constructed

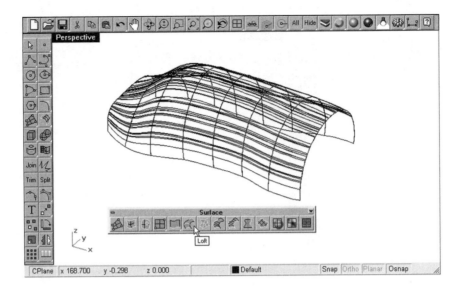

■■ Summary

A surface model is a set of 3D surfaces joined in 3D space to represent a 3D object. Each surface of a surface model is a mathematical expression depicting the profile and silhouette of the surface. There are two basic methods of representing a surface in the computer: using planar polygon meshes to approximate a surface and using complex spline surfaces to exactly represent a surface.

Using Rhino, you construct and manipulate both polygon meshes and accurate surfaces. A polygon mesh approximates a smooth surface by employing a set of small planar polygonal faces. Accuracy of representation is inversely proportional to the size of the polygon. The smaller the polygon size, the more accurate the surface. One major disadvantage of the polygon mesh is that file size increases tremendously with a decrease in polygon size. Another disadvantage is that a surface can never be accurately represented with a mesh, no matter how small the polygon size.

Hence, the polygon mesh is only useful in visualization and is not appropriate for downstream computerized manufacturing operations. A better way to represent a continuously smooth surface is to use spline mathematics. In most contemporary computer-aided design (CAD) applications, NURBS surface mathematics is used. In this chapter you worked mainly on NURBS surfaces. NURBS surfaces can be classified into three categories according to the manner in which they are constructed: primitive surfaces, basic free-form surfaces, and derived surfaces.

In this chapter you learned primitive surfaces and basic free-form surfaces. Primitive surfaces are fundamental geometric shapes. To construct a primitive surface, you select a shape and specify the parameters. You also learned about quadrilateral and rectangular planar surfaces in this chapter. You will learn other primitives in Chapter 8.

Basic free-form surfaces are those you use most often in constructing 3D surface models. This category of surface includes the following: extruded, revolved, rail revolved, from-edge-curves, lofted, one-rail sweep, two-rail sweep, curve network, patch, and point grid.

To construct free-form surfaces, you define a set of curves and/or points. Creating free-form surfaces is simple after the curves and points are constructed. However, making the curves and points can be a tedious job. In addition to learning primitive and free-form surface construction, you also learned how to use a digitizer to digitize a physical model to obtain appropriate curve data for creating surfaces.

■■ Review Questions

1. Explain the concept of surface modeling.
2. Differentiate the NURBS surface and the polygon mesh.
3. Outline the advantages and limitations of surface modeling.
4. List NURBS primitive surfaces and basic free-form surfaces.
5. Use simple sketches to illustrate the types of points and/or curves required to construct various types of basic free-form surfaces.

Chapter 6

Surface Modeling: Part 2

■■ Objectives

The goals of this chapter are to delineate the methods of constructing derived surfaces, and to explain how to edit and transform surfaces. After studying this chapter, you should be able to:

❒ Construct various types of derived surfaces
❒ Edit and transform surfaces

■■ Overview

This chapter is a continuation of Chapter 5. Here you will learn how to construct various types of derived surfaces from existing surfaces and bitmaps. You will also learn how to edit and transform surfaces. In the next chapter you will learn how to analyze surfaces and work on a number of surface modeling projects.

■■ More Surface Modeling Tools

The sections that follow discuss various types of derived surfaces, surface editing techniques, and numerous operations related to transforming surfaces. These sections contain practice exercises associated with the methods discussed.

Constructing Derived Surfaces

Derived surfaces, as the name implies, are surfaces derived from existing objects. To construct a derived surface, you do not need to construct any point objects or curves. Instead, you use existing objects to derive new surfaces.

There are six basic types of derived surfaces (extended, filleted, chamfered, offset, blended, and draped) and two special types of derived surfaces (heightfield and unrolled). Figure 6-1 shows the derived surface commands from the Surface and Surface Tools toolbars.

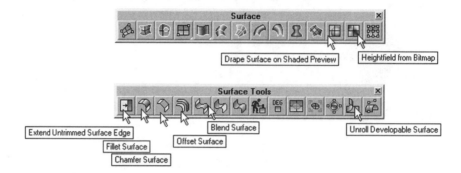

Fig. 6-1. Surface and Surface Tools toolbars.

Table 6-1 outlines the functions of derived surface commands and their location in the pull-down menu.

Table 6-1: Derived Surface Commands and Their Functions

Command Name	Pull-down Menu	Function
Drape	Surface > Drape	For deriving a surface by draping a rectangular sheet over selected objects
Heightfield	Surface > Heightfield from Bitmap	For deriving a surface from a bitmap in accordance with a color value
ExtendSrf	Surface > Extend	For extending the untrimmed edge of a surface
FilletSrf	Surface > Fillet	For constructing a filleted surface at the intersection of two surfaces
ChamferSrf	Surface > Chamfer	For constructing a chamfer surface at the intersection of two surfaces
OffsetSrf	Surface > Offset	For deriving surfaces offset from selected surfaces
BlendSrf	Surface > Blend	For constructing a blended surface connecting two sets of surface edges
UnrollSrf	Surface > Unroll Developable Srf	For constructing a flat pattern of a developable surface

In the following, you will construct a set of surfaces. Based on these surfaces, you will derive extended, filleted, chamfered, offset, and blended surfaces.

More Surface Modeling Tools

1 Start a new file. Use the metric (millimeter) template. Maximize the Front viewport.

2 Construct two free-form curves in accordance with figure 6-2.

3 Maximize the Perspective viewport and construct two extruded surfaces. (See figure 6-3.)

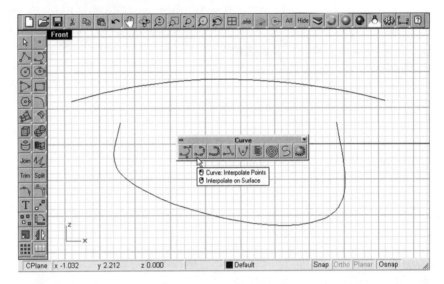

Fig. 6-2. Free-form curves.

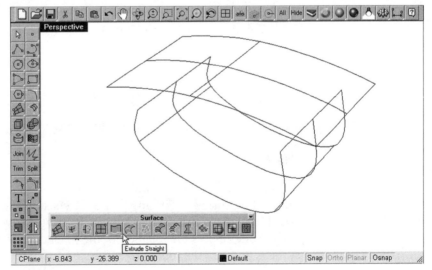

Fig. 6-3. Extruded surfaces.

Extend

In the following, you will extend a surface. Note that you can only extend the untrimmed edge of a surface.

1. Select Surface > Extend, or the Extended Untrimmed Surface Edge button from the Surface Tools toolbar.

   Command: ExtendSrf

2. There are two types of extension: smooth and linear. Type *T* at the command line and press Enter. If Type is Smooth, change type to Linear and select the edge indicated in figure 6-4.

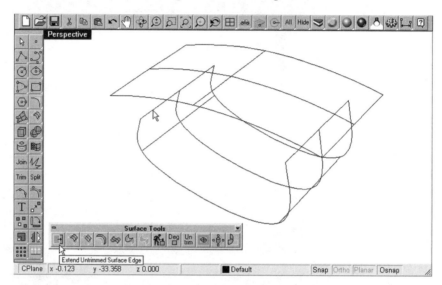

Fig. 6-4. Surface edge selected.

3. Select two points to indicate the extension length.

An edge of the surface is extended. (See figure 6-5.)

4. Repeat the ExtendSrf command and extend the edge indicated in figure 6-5.

The edges of the surface are extended. Save your file as *ExtendSrf.3dm*.

Fillet

A filleted surface has an arc-shaped cross section. It derives from two intersecting surfaces. For any pair of intersecting surfaces, there are four possible filleted surfaces. The location at which you select the original surface determines the location of the filleted surface. To derive a filleted surface, perform the following steps.

1. Select File > Save As and specify a file name of *FilletSrf.3dm*.
2. Select Surface > Fillet, or the Fillet Surface button from the Surface Tools toolbar.

   Command: FilletSrf

More Surface Modeling Tools

Fig. 6-5. Untrimmed edge extended.

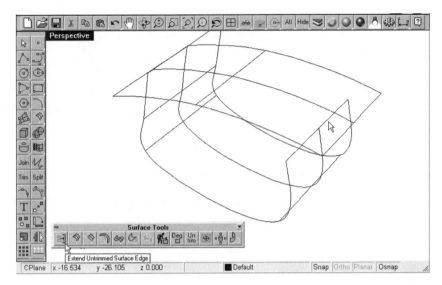

3   Type *R* at the command line area and select two points to specify a radius. Note that you can type in a radius value for more precision.

4   Select the surfaces indicated in figure 6-6.

The filleted surface is constructed. Save your file.

Fig. 6-6. Filleted surface being constructed.

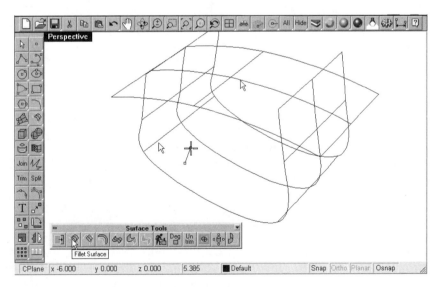

Chamfer

A chamfered surface also derives from two intersecting surfaces. It forms a beveled edge between two surfaces. Similar to filleting, there are four possible chamfered surfaces on the intersecting surfaces. To create a chamfered surface, perform the following steps.

1. Select File > Save As and specify a file name of *ChamferSrf.3dm*.
2. Select Surface > Chamfer, or the Chamfer Surface button from the Surface Tools toolbar.

   Command: ChamferSrf

3. Type *D* at the command line and select four points to specify the chamfer distances. Note that you can type in a radius value for more precision.
4. Select the surfaces indicated in figure 6-7.

A chamfered surface is constructed. Save your file.

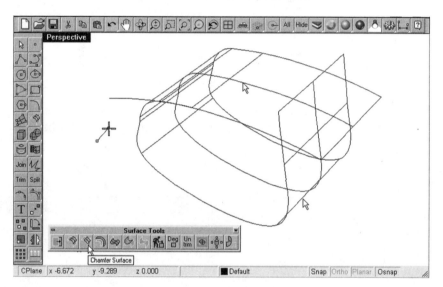

Fig. 6-7. Chamfered surface being constructed.

Offset

An offset surface is a surface that offsets from an existing surface. Every point on the offset surface is equal in distance from the original surface. To create an offset surface, perform the following steps.

1. Select File > Save As and specify a file name of *OffsetSrf.3dm*.
2. Select Surface > Offset, or the Offset Surface button from the Surface Tools toolbar.

   Command: OffsetSrf

3. Select the surface indicated in figure 6-8 and press the Enter key.
4. Type *D* and then *4* to specify an offset distance.

An offset surface is constructed.

5. Repeat the OffsetSrf command to construct one more offset surface.

More Surface Modeling Tools

Fig. 6-8. Offset surface constructed.

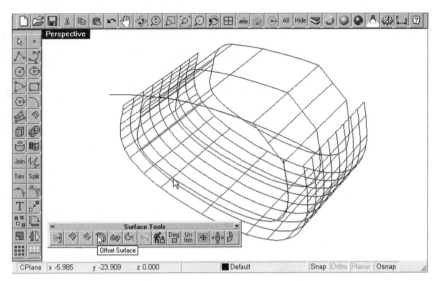

6 Type *F* to flip the direction of offset. (See figure 6-9.) (This command has a number of options: FlipAll, Solid, Loose, and Tolerance. In particular, the Solid option constructs a solid. You will learn more about solid modeling using Rhinoceros in Chapter 8.)

Another offset surface is constructed. Save your file.

Fig. 6-9. Second offset surface being constructed.

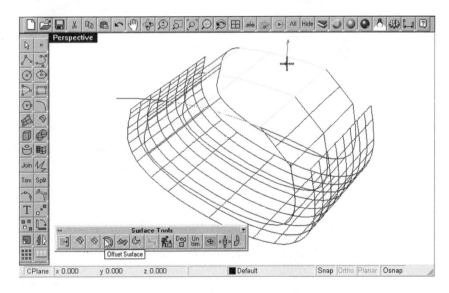

Blend

A blended surface joins two non-intersecting surfaces. It is a smooth transition between the two original surfaces. To construct a blended surface, perform the following steps.

1 Select File > Save As and specify a file name of *BlendSrf.3dm*.
2 Select Surface > Blend, or the Blend Surface button from the Surface Tools toolbar.

Command: BlendSrf

3 Select the surface edges indicated in figure 6-10 and press the Enter key. (Pick near the same end to avoid getting a twisted surface.)
4 In the Blend Bulge dialog box, accept the default and click on the OK button. (The sliders determine the distance of influence the surface has on the edge curve.)

A blended surface is constructed. (See figure 6-11.) Save your file.

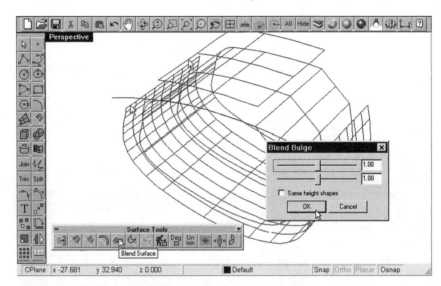

Fig. 6-10. Blended surface being constructed.

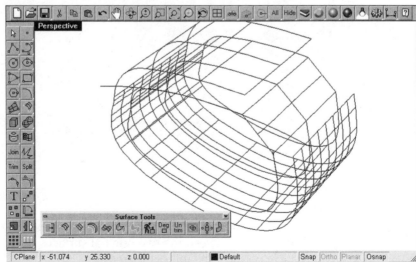

Fig. 6-11. Blended surface constructed.

Continuity of Contiguous Surfaces

Chamfered, filleted, and blended surfaces have one thing in common: they bridge two surfaces. Chamfered and filleted surfaces bridge two intersecting surfaces, and the blended surface bridges two non-intersecting surfaces. At the "joint" between the surfaces, three types of surface continuity are manifested, which are discussed in the sections that follow.

Positional Continuity (G0 Continuity).

In a G0 (positional) continuity joint, the control points at the edges of contiguous surfaces coincide. The chamfered surface edges indicated in figure 6-12 have G0 continuity.

Tangent Continuity (G1 Continuity).

In a G1 (tangent) continuity joint, the tangent directions of the control points at the edges of the contiguous surfaces are the same. The filleted surface edges indicated in figure 6-13 have G1 continuity.

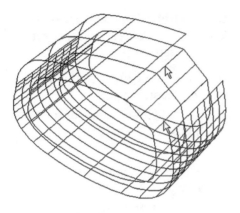

Fig. 6-12. G0 (positional) continuity.

Curvature Continuity (G2 Continuity).

In a G2 (curvature) continuity joint, the curvature and tangent direction of the control points at the edges of contiguous surfaces are the same. The blended surface edges indicated in figure 6-14 have G2 continuity.

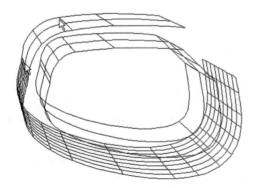

Fig. 6-13. G1 (tangent) continuity.

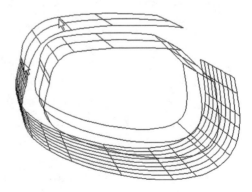

Fig. 6-14. G2 (curvature) continuity.

Drape

Making a draped surface is analogous to warping a rectangular piece of elastic sheet on a set of 3D objects in a way similar to vacuum forming. Vacuum forming is a type of plastics forming process. You heat up plastic sheeting, warp the sheet onto a 3D object (the mold), and apply a vacuum to deform the sheet. Figure 6-15 shows a vacuum forming machine.

Fig. 6-15. Vacuum forming machine.

To construct draped surfaces from a cone and a sphere, perform the following steps.

1. Start a new file. Use the metric (millimeter) template.
2. Select Solid > Cone, or the Cone button from the Solid toolbar.
   Command: Cone
3. Select two points in the Top viewport to specify the center and the radius.
4. Check the Ortho box in the status bar.
5. Select a point in the Front viewport.

A cone is constructed. (See figure 6-16.)

6. Select Solid > Sphere > Center, Radius, or the Sphere/Center, Radius button from the Solid toolbar.
   Command: Sphere
7. Select two points in the Front viewport to indicate the center and radius. (See figure 6-17.)
8. Select Surface > Drape, or the Drape Surface on Shaded Preview button from the Surface toolbar.
   Command: Drape

To provide a better visual aid, the objects are automatically shaded in all viewports.

More Surface Modeling Tools

Fig. 6-16. Cone constructed.

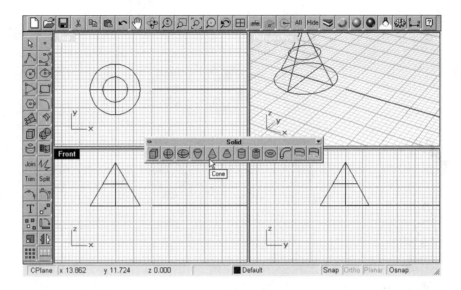

Fig. 6-17. Sphere constructed.

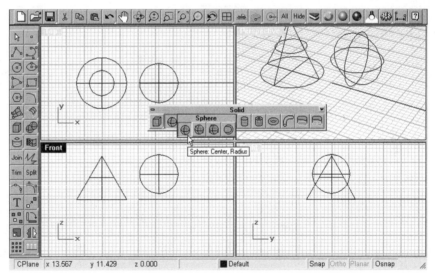

9 Select two diagonal points in the Top viewport indicated in figure 6-18.

A draped surface is constructed. (See figure 6-19.)

Now hide the draped surface.

10 Select Edit > Visibility > Hide, or the Hide Objects button from the Visibility toolbar.
Command: Hide

11 Select the draped surface and press the Enter key.

Fig. 6-18. Draped surface being constructed.

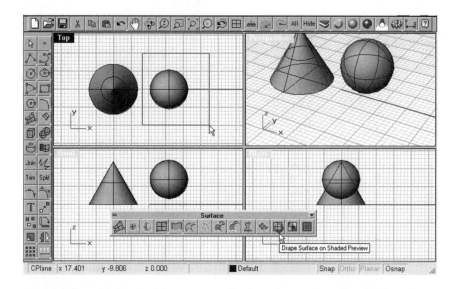

Fig. 6-19. Draped surface constructed.

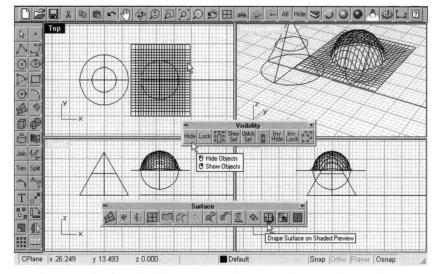

The draped surface is hidden. Now construct another draped surface.

12 Select Surface > Drape, or the Drape Surface on Shaded Preview button from the Surface toolbar.

Command: Drape

13 Select two diagonal points in the Top viewport to include the cone and the sphere.

Another draped surface is constructed. (See figure 6-20.)

Unhide the hidden surface and compare the draped surfaces to discover the difference in the location of the flat portion of the surfaces.

More Surface Modeling Tools

Fig. 6-20. Second draped surface being constructed.

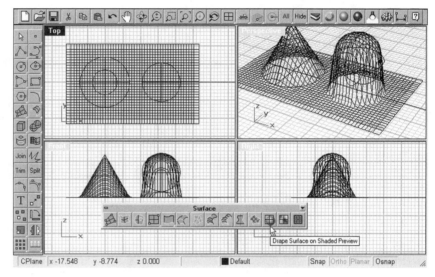

14 Select Edit > Visibility > Show, or the Show Objects button from the Visibility toolbar.

The surfaces are complete. Save your file as *Drape.3dm*.

Deriving Surfaces from Bitmaps

Using the color value of a bitmap, you can derive a surface. Basically, the bitmap can be color or black and white. However, it is more appropriate to use a black-and-white image because only the brightness value is considered when a surface is derived. If you use a black-and-white image, you can better perceive the outcome before making the surface.

Naturally, you need a bitmap to make a surface of this type. Figure 6-21 shows digital black-and-white images being taken via digital camera.

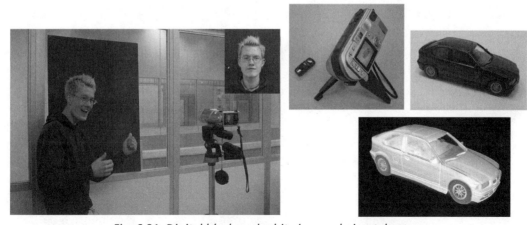

Fig. 6-21. Digital black-and-white images being taken.

1. Start a new file. Use the metric (millimeter) template. Maximize the Top viewport.
2. Select Surface > Heightfield from Bitmap, or the Heightfield from Bitmap button from the Surface toolbar.
   Command: Heightfield.
3. In the Select Bitmap dialog box, select the image file *Car.tga*.
4. Select two points in the Top viewport to indicate the size of the surface to be derived from the bitmap. (See figure 6-22.)

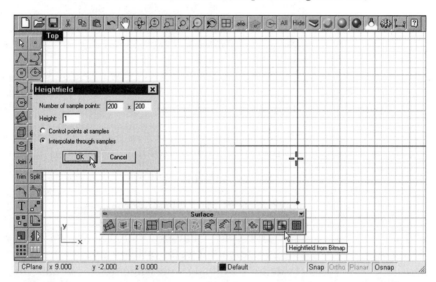

Fig. 6-22. Heightfield surface being constructed.

5. In the Heightfield dialog box, set the number of sample points to 200 times 200, set Height to 1, select *Interpolate through samples*, and click on the OK button.

A surface is derived. (See figure 6-23.) The surface is complete. Save your file as *Heightfield.3dm*.

Unrolling Developable Surfaces

Before you make a sheet metal object, you need a development (flat) pattern of the object as a 2D sheet. You then roll or fold the 2D sheet to the 3D object. An object constructed this way can be unrolled into a flat sheet. A cylindrical surface, for example, can be unrolled to a rectangle. On the other hand, you cannot unroll a sphere. To unroll a developable surface, perform the following steps.

1. Start a new file. Use the metric (millimeter) template.
2. Construct a free-form curve in the Front viewport and extrude the curve in a straight line to form a surface. (See figure 6-24.)

More Surface Modeling Tools

Fig. 6-23. Heightfield surface.

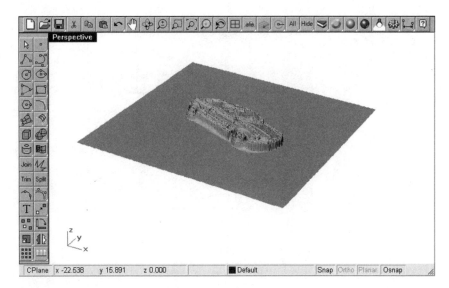

Fig. 6-24. Curve constructed and extruded.

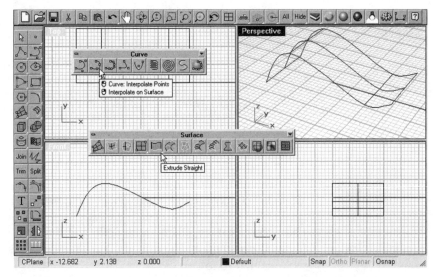

3 Select Surface > Unroll Developable Srf, or the Unroll Developable Surface button from the Surface Tools toolbar.
   Command: UnrollSrf

4 Select the surface and press the Enter key.

The surface is unrolled. (See figure 6-25.) Save your file as *UnrollSrf.3dm.*

Fig. 6-25. Surface unrolled.

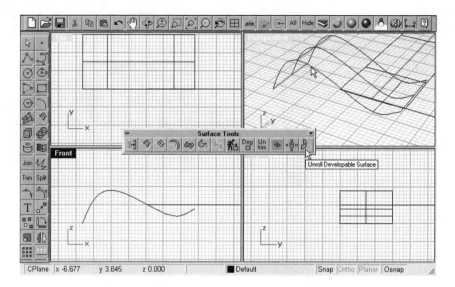

Surface Editing

Similar to editing curves, there are three categories of surface editing, as follows. These types of surface editing are discussed in the sections that follow.

- ❑ Edit by joining, exploding joined surfaces, trimming, and splitting
- ❑ Edit by manipulating points of a surface
- ❑ More advanced editing

Joining, Exploding, Trimming, and Splitting

The Join, Explode, Trim, and Split commands you learned about in Chapter 2 also apply to surfaces. These commands are discussed in the sections that follow.

Joining Surfaces. Joining two or more contiguous surfaces, you get a polysurface. To join surfaces, perform the following steps. (Note that the surfaces will not join unless they meet edge to edge within tolerance. In Chapter 7 you will learn how to merge edges before joining.)

1 Open the file *Blend.3dm*.

2 Select Edit > Join, or the Join button from the Main toolbar.
    Command: Join

3 Select the surfaces indicated in figure 6-26 and press the Enter key.

The surfaces are joined to become a polysurface.

More Surface Modeling Tools

Fig. 6-26. Surfaces being joined.

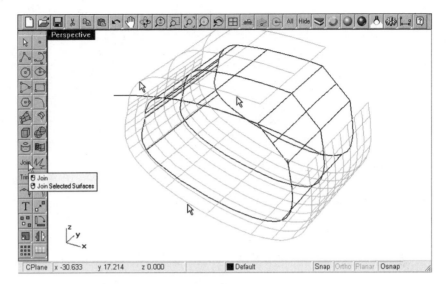

Exploding Polysurfaces. Exploding a polysurface renders a set of individual surfaces. To explode a polysurface, perform the following steps.

1 Select Edit > Explode, or the Explode button from the Main toolbar.
   Command: Explode

2 Select the polysurface indicated in figure 6-27 and press the Enter key.

The polysurface is exploded. (In Chapter 8, you will learn how to extract individual surfaces from a polysurface without exploding the polysurface.)

Fig. 6-27. Polysurface being exploded.

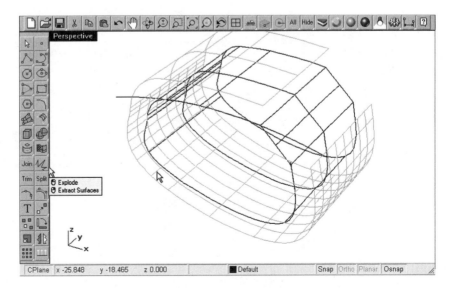

Trimming and Splitting Surfaces. To produce a smooth surface, it is necessary to use smooth defining wires and smooth boundary lines. However, most of the surfaces you will use to compose a design do not necessarily have smooth boundaries, although they have smooth profiles. To construct a smooth surface with an irregular boundary, you construct a large smooth surface and then trim the smooth surface with a cutting object. The resulting surface is a smooth, trimmed surface.

The splitting operation is very similar to trimming in that you need a cutting object. The difference is that the cutting object cuts the target object in two. In the following, you will trim and split surfaces.

1 Start a new file. Use the metric (millimeter) template.

2 With reference to figure 6-28, construct three free-form curves in the Front viewport, extrude two curves at different lengths, and construct a circle in the Top viewport.

Because you will use the surfaces in trimming and splitting, you will save two files and use one file for trimming and another file for splitting. Continue with the following steps.

3 Save your file as *SplitSrf.3dm*.

4 Select File > Save As and specify a file name of *TrimSrf.3dm*.

5 Select Edit > Trim, or the Trim button from the Main toolbar.
Command: Trim

6 Select the surfaces as the cutting objects and press the Enter key.

7 Select the surfaces indicated in figure 6-28.

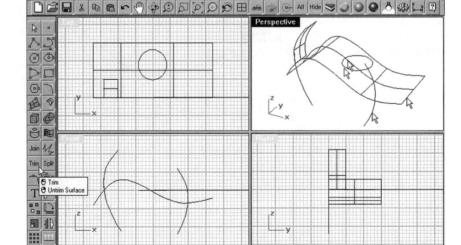

Fig. 6-28. Surfaces being selected as cutting objects.

More Surface Modeling Tools

The narrower surface is trimmed by the wider surface, but not vice versa. (See figure 6-29.) You can also use curves for cutting operations. You can trim surfaces and polysurfaces using curves. You can also split surfaces using curves, but not polysurfaces.

Fig. 6-29. Surface trimmed.

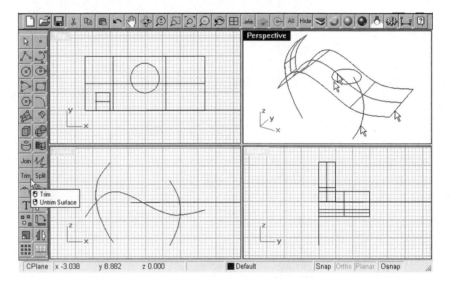

8 Use the Trim command again.

9 Select the circle and the curve as cutting objects and press the Enter key.

10 Select the surface indicated in figure 6-29.

To use the surface to trim a curve, continue with the following steps.

11 Repeat the Trim command.

12 Select the surface indicated in figure 6-30 and press the Enter key.

13 Select the curve indicated in figure 6-30 and press the Enter key.

The curve is trimmed. (See figure 6-31.) Save your file.

14 Open the file *SplitSrf.3dm*.

15 Select Edit > Split, or the Split button from the Main toolbar.
Command: Split

16 Select the surface indicated in figure 6-32 as the object to be split.

17 Select the other surface and click on the OK button.

Fig. 6-30. Surfaces trimmed by the curves.

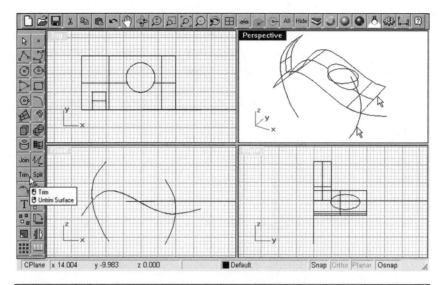

Fig. 6-31. Curve trimmed by the surface.

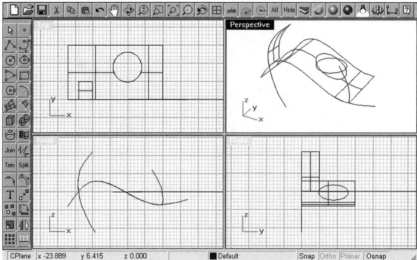

There is no effect because the narrower surface cannot split the wider surface.

18 Repeat the Split command.

19 Select the surface indicated in figure 6-32 again as the object to be split.

20 Select the circle and the curve, and press the Enter key.

The surface is split into three surfaces. (See figure 6-33.)

More Surface Modeling Tools

Fig. 6-32. Surface being split by a narrower surface, a circle, and a curve.

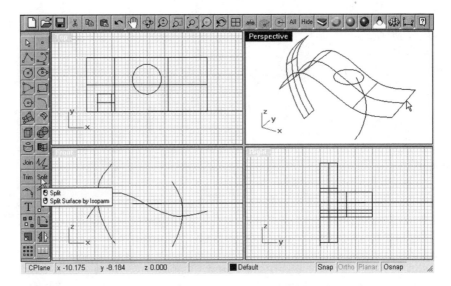

Fig. 6-33. Surface split by the circle and the curve.

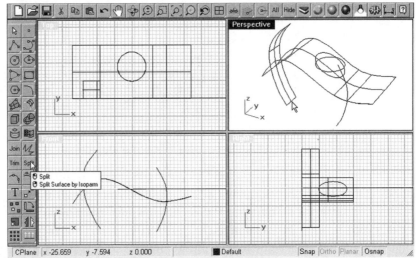

21 Repeat the Split command.
22 Select the surface indicated in figure 6-33 as the object to be split.
23 Select the wider surface and press the Enter key.

The selected surface is split. (See figure 6-34.)

24 Repeat the Split command.
25 Select the curve indicated in figure 6-34 as the object to be split.
26 Select the surface indicated in figure 6-34 and press the Enter key.

The curve is split. Save your file.

Fig. 6-34. Surface split.

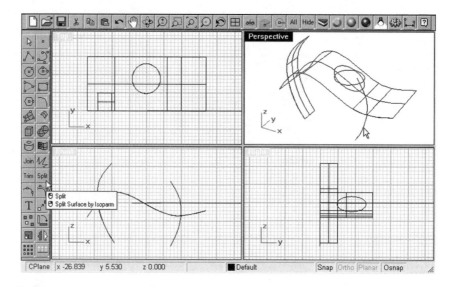

Point Editing

Similar to curve editing, you edit a surface by manipulating its control point location and weight, using a handlebar, and adding or removing knots. However, there is no edit point available on a surface and there is no kink point because any kink point on a curve will result in multiple surfaces.

Control Point Manipulation. To manipulate the control points of a surface, perform the following steps.

1 Start a new file. Use the metric (millimeter) template.

2 Construct a free-form curve in the Front viewport and a free-form curve in the Right viewport. (See figure 6-35.)

3 Maximize the Perspective viewport and move the curves in accordance with figure 6-36.

4 Select Surface > Sweep1, or the Sweep Along 1 Rail button from the Surface toolbar.

Command: Sweep1

5 Select the curve indicated in figure 6-36.

6 Select the other curve and press the Enter key.

A swept surface is constructed.

7 Select Edit > Point Editing > Control Points On, or the Control Points On button from the Point Editing toolbar.

Command: PtOn

8 Select the surface and click on the OK button.

More Surface Modeling Tools

Fig. 6-35. Curve constructed.

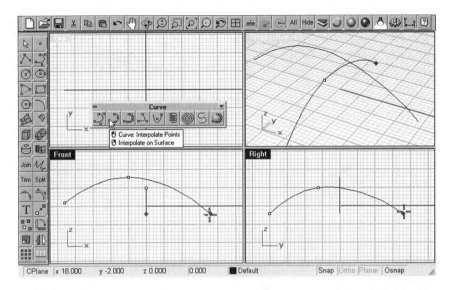

Fig. 6-36. Curve translated.

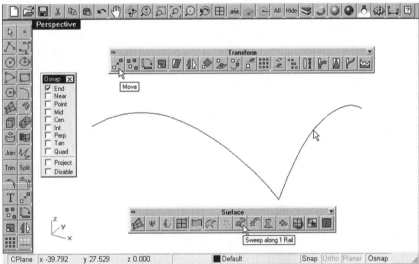

The control points are turned on. (See figure 6-37.)

Now you will modify the surface's shape by moving a control point. This is done in one of three ways.

❒ By selecting the control point, holding down the mouse button, and dragging the control point to a new location.

❒ By selecting the control point and using the nudge keys. (By default, the nudge keys are the Alt and arrow keys.)

❒ Using the *MoveUVN* command.

Fig. 6-37. Control points turned on.

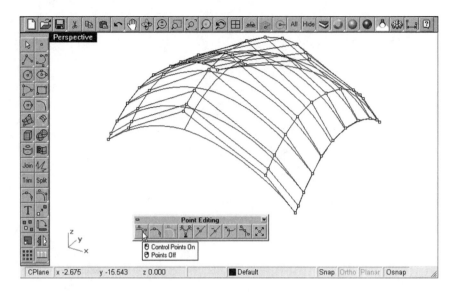

To use the *MoveUVN* command, perform the following.

9 Type *MoveUVN* at the command line area. Select the control point indicated in figure 6-38 and press the Enter key. In the MoveUVN dialog box, use the slider bars to move the control point in the U and V directions, as well as in a direction normal to the surface.
Command: Move UVN

The surface is modified. (See figure 6-38.)

Fig. 6-38. Control point translated.

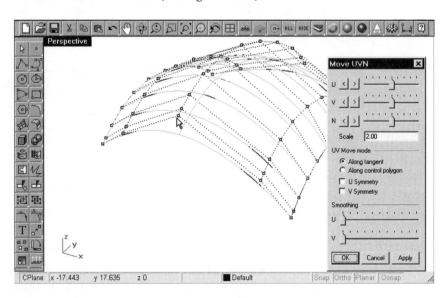

More Surface Modeling Tools

Now you will change the control point's weight. Like the control points of a spline, the higher the weight, the closer the surface will be pulled to the control point.

10 Select Edit > Point Editing > Edit Weight, or the Edit Control Point Weight button from the Point Editing toolbar.

Command: Weight

11 Select a control point and press the Enter key.

12 In the Set Control Point Weight dialog box, set the weight to 6 and click on the OK button. (See figure 6-39.)

The control point weight is modified.

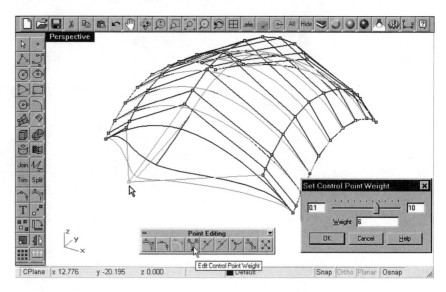

Fig. 6-39. Control point weight adjusted.

Handlebar Editor. To edit the surface using the handlebar, perform the following steps.

1 Select Edit > Point Editing > Handlebar Editor, or the Handlebar Editor button from the Point Editing toolbar.

Command: HBar

2 Select the surface. (See figure 6-40.)

3 Select a node of the handlebar to modify the surface. (See figure 6-41.)

Fig. 6-40. Handlebar selected.

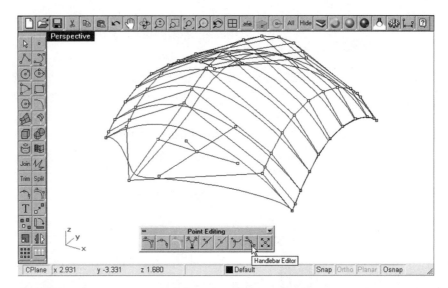

Fig. 6-41. Handlebar being manipulated.

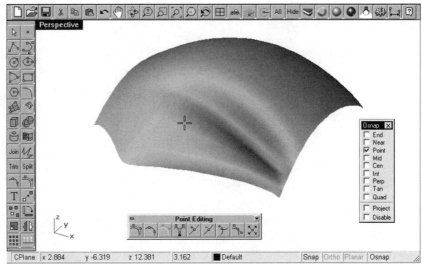

Adding and Removing Knots. In the following, you will add knots to the surface. Adding knots increases the number of control points. The surface profile will not change until the next time you manipulate its control points. The number of control points of a surface has a significant effect on how the surface will change in shape if a control point is moved.

1 Select Edit > Point Editing > Insert Knot, or the Insert Knot button from the Point Editing toolbar.

Command: InsertKnot

2 Select the surface and a point along the surface and press the Enter key.

More Surface Modeling Tools

A knot is inserted. Compare figures 6-40 and 6-42 to note the difference in control point number.

Fig. 6-42. Knot added.

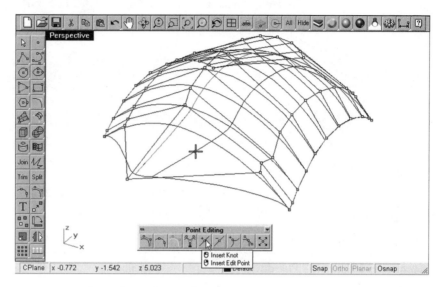

To remove a knot from the surface, continue with the following steps.

3 Select Edit > Point Editing > Remove Knot, or the Remove Knot button from the Point Editing toolbar.

Command: RemoveKnot

4 Select the surface and a knot on the surface and press the Enter key. (See figure 6-43.)

The knot is removed. Save your file as *PointEditSrf.3dm*.

Fig. 6-43. Knot being removed.

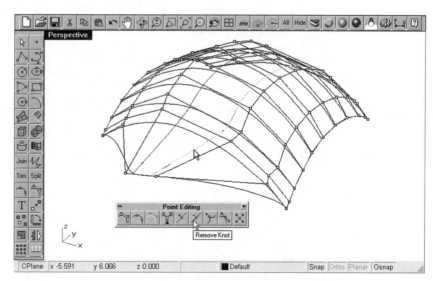

Advanced Editing

The more advanced commands for editing a surface are indicated in the Surface Tools and Main toolbars, shown in figure 6-44.

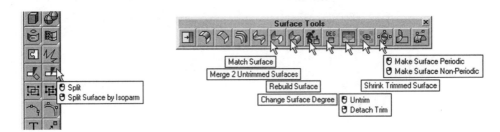

Fig. 6-44. Surface editing tools.

Table 6-2 outlines the functions of surface editing commands and their location in the pull-down menu.

Table 6-2: Surface Editing Commands and Their Functions

Command Name	Pull-down Menu	Function
MatchSrf	Surface > Edit Tools > Match	For matching an untrimmed edge of a surface with an untrimmed surface of another surface by modifying the position, tangency, and/or curvature of the selected surfaces
MergeSrf	Surface > Edit Tools > Merge	For merging and joining the untrimmed edges of two surfaces
RebuildSrf	Surface > Edit Tools > Rebuild	For reconstructing a selected surface with a new degree setting and number of control points in the U and V directions
ChangeDegreeSrf	Surface > Edit Tools > Change Degree	For modifying a surface by changing its degree setting
Untrim	Surface > Edit Tools > Untrim	For removing a selected trimmed edge of a surface to restore it to its untrimmed state
DetachTrim	Surface > Edit Tools > Detach Trim	For removing a selected trimmed edge of a surface and converting the trimmed edge to a curve
SplitSrf	Surface > Edit Tools > Split by Isoparm	For splitting a surface into multiple surfaces along the isoparm
ShrinkTrimmedSrf	Surface > Edit Tools > Shrink Trimmed Surface	For reducing the size of the underlying untrimmed surface of a trimmed surface
MakeSrfPeriodic	Surface > Edit Tools > Make Periodic	For constructing a periodic surface from a surface

More Surface Modeling Tools

Matching Two Surfaces. To construct two surfaces and match their edges, perform the following steps.

1 Start a new file. Use the metric (millimeter) template.

2 With reference to figure 6-45, construct two free-form curves in the Front viewport and extrude them to two surfaces.

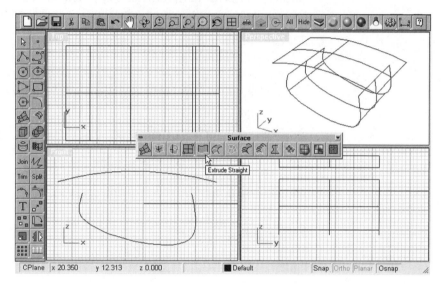

Fig. 6-45. Extruded surfaces constructed.

3 Select Surface > Edit Tools > Match, or the Match Surface button from the Surface Tools toolbar.

Command: MatchSrf

4 Select the edges indicated in figure 6-46.

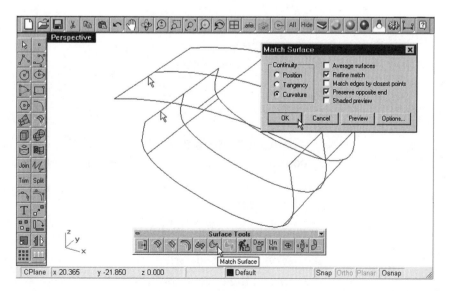

Fig. 6-46. Surface edges being matched.

5 In the Match Surface dialog box, accept the default and click on the OK button.

The surface edges are matched. Save your file as *MatchSrf.3dm*.

Merging Surfaces. To merge two contiguous surfaces into a single surface, perform the following steps. The edges of the contiguous surfaces have to be untrimmed edges for the merge to work.

1 Select File > Save As and specify a new file name of *MergeSrf.3dm*.

2 Select Surface > Edit Tools > Merge, or the Merge 2 Untrimmed Surface button from the Surface Tools toolbar.
Command: MergeSrf

3 Select the surfaces indicated in figure 6-47.

The surfaces are merged into a single surface. Save your file. Note that a merged surface is a single surface, not a polysurface. Therefore, you cannot explode it into the surfaces you merged.

Fig. 6-47. Surfaces being merged.

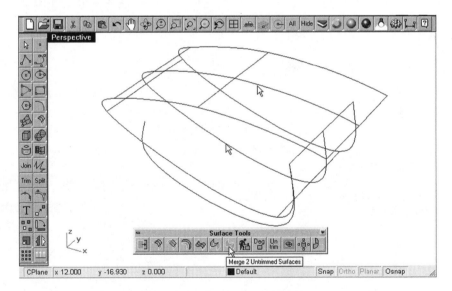

Making a Surface Periodic. A periodic surface is a smooth, closed-loop surface. To make a surface periodic, perform the following steps.

1 Select File > Save As and specify a file name of *MakeSrfPeriodic.3dm*.

2 Select Surface > Edit Tools > Make Periodic, or the Make Surface Periodic button from the Surface Tools toolbar.
Command: MakeSrfPeriodic

3 Select the edge of the surface.

More Surface Modeling Tools

The surface is made periodic. (See figure 6-48.) Save your file.

Fig. 6-48. Surface made periodic.

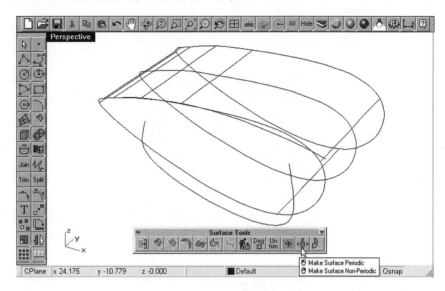

Rebuilding. To rebuild a surface, you change the number of control points and the degree of the polynomial. To rebuild a surface, perform the following steps.

1 Select File > Save As and specify a file name of *RebuildSrf.3dm*.

2 Select Surface > Edit Tools > Rebuild, or the Rebuild Surface button from the Surface Tools toolbar.
   Command: RebuildSrf

3 Select the surface indicated in figure 6-49 and press the Enter key.

4 In the Rebuild Surface dialog box, accept the default point counts and degree of the polynomial and click on the OK button.

The surface is rebuilt. Save your file.

Changing Polynomial Degree. To change the polynomial degree of a surface, perform the following steps.

1 Select File > Save As and specify a file name of *ChangeDegreeSrf.3dm*.

2 Select Surface > Edit Tools > Change Degree, or the Change Surface Degree button from the Surface Tools toolbar.
   Command: ChangeDegreeSrf

3 Select the surface. (See figure 6-50.)

4 Type *4* to change the U degree to 4.

5 Type *4* to change the V degree to 4.

Fig. 6-49. Surface being rebuilt.

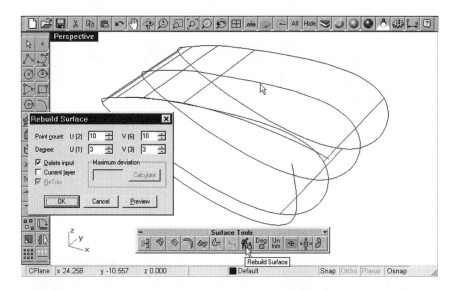

Fig. 6-50. Polynomial degree being changed.

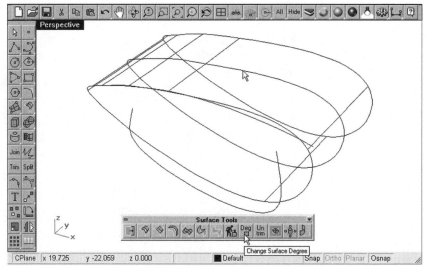

The polynomial degree is changed. Increasing the degree adds more control points to the surface, making it more deformable. Save your file.

Splitting a Surface Along an Isoparm. To split a surface along an isoparm curve, perform the following steps.

1 Open the file *MakeSrfPeriodic.3dm*.

2 Select File > Save As and specify a file name of *SplitIsoparm.3dm*.

3 Select Surface > Edit Tools > Split by Isoparm, or the Split Surface by Isoparm button from the Main toolbar.

Command: SplitSrf

More Surface Modeling Tools

4 Select the surface. (See figure 6-51.)
5 Select a point on the surface and press the Enter key.

The surface is split along the isoparm. Save your file.

Fig. 6-51. Surface being split.

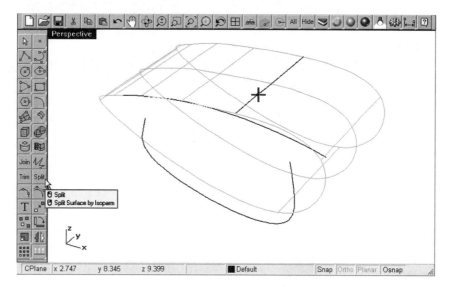

Untrimming a Trimmed Surface. To construct a curve for trimming a surface and learn how to untrim it, perform the following steps.

1 Open the file *MakeSrfPeriodic.3dm*.
2 Select File > Save As and specify a file name of *Untrim.3dm*.
3 Return the display to a four-viewport layout.
4 Construct a free-form curve in the Top viewport. (See figure 6-52.)

Fig. 6-52. Free-form curve constructed.

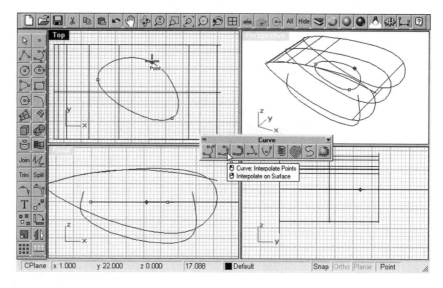

5 Select Edit > Trim, or the Trim button from the Main toolbar.
Command: Trim

6 Select the curve and press the Enter key. (See figure 6-53.)

7 Select the surface and press the Enter key.

The surface is trimmed. Note that there are two surfaces.

Fig. 6-53. Surface being trimmed.

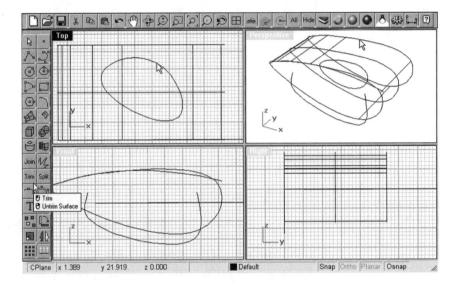

Now you will move one of the trimmed surfaces and untrim it.

8 Maximize the Perspective viewport.

9 Move one of the trimmed surfaces to a new location. (See figure 6-54.)

10 Select Surface > Edit Tools > Untrim, or the Untrim button from the Surface Tools toolbar.
Command: Untrim

11 Select a surface.

The surface is untrimmed. (See figure 6-55.)

Detaching a Trimmed Boundary. To untrim a trimmed surface and retain the trimmed boundary curve, perform the following steps.

1 Select Surface > Edit Tools > Detach Trim, or the Detach Trim button from the Surface Tools toolbar.
Command: DetachTrim

2 Select the surface indicated in figure 6-56.

The surface is untrimmed and the trimmed boundary is retained.

Fig. 6-54. Trimmed surface being moved to a new location.

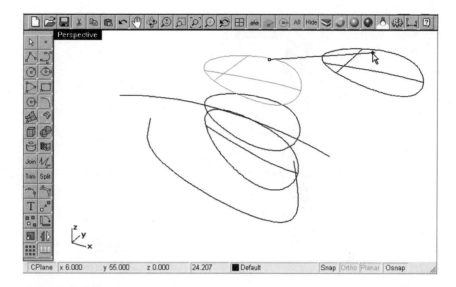

Fig. 6-55. Surface untrimmed.

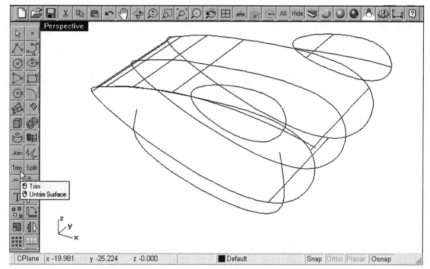

Shrinking a Trimmed Surface. After you untrim or detach a trimmed boundary, the trimmed surface returns to its original shape because the original surface is kept in the database. To reduce the memory required to store a trimmed surface, you can shrink the trimmed surface, as follows.

1 Undo the DetachTrim command and Untrim command. (See figure 6-57.)

2 Select Surface > Edit Tools > Shrink Trimmed Surface, or the Shrink Trimmed Surface button from the Surface Tools toolbar.
Command: ShrinkTrimmedSrf

290 CHAPTER 6: Surface Modeling: Part 2

Fig. 6-56. Trimmed boundary detached.

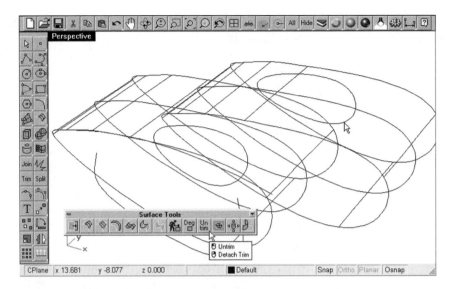

Fig. 6-57. DetachTrim and Untrim commands undone.

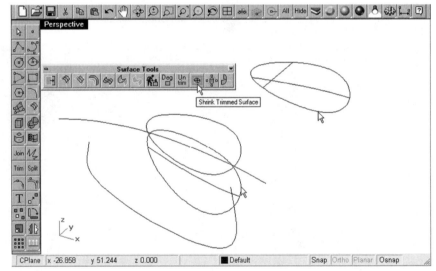

3 Select the surface indicated in figure 6-57.

To appreciate the change, use the Untrim command to untrim the surfaces. (See figure 6-58.) Apart from a reduction in memory space, the topology of the control points of a shrunk, trimmed surface is different from that of the same trimmed surface before it is shrunk. Save your file.

Fig. 6-58. Shrunk surfaces untrimmed.

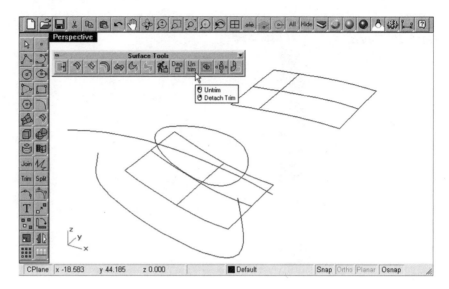

Transforming Surfaces

Transformation of surfaces involves translation and deformation. That is, in the transformation of surfaces you translate surfaces and deform surfaces and control points of surfaces. Some of the transform tools you learned about in Chapter 2 in regard to curves also apply to the transformation of surfaces. The sections that follow discuss various aspects and procedures related to the transformation of surfaces.

Move, Copy, Rotate, Scale, Mirror, Orient, and Array

The operation of these commands as applied to surfaces is similar to that for curves. See the techniques delineated for these commands in Chapter 2.

Setting Points

Like aligning the control points of curves, you can align the control points of a surface. To align the Y coordinates of the control points of a surface, perform the following steps.

1   Open the file *PointEditSrf.3dm*.
2   Select File > Save As and specify a file name of *SetPtSrf.3dm*.
3   Delete the surface, leaving only two curves. (See figure 6-59.)
4   Construct a swept surface. (See figure 6-60.)

Fig. 6-59. Curves.

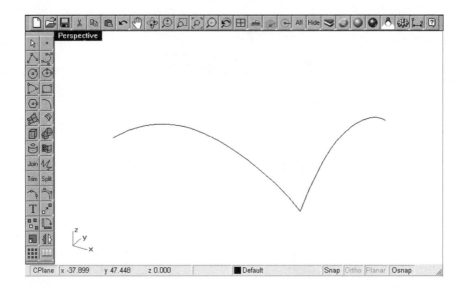

Fig. 6-60. Swept surface constructed.

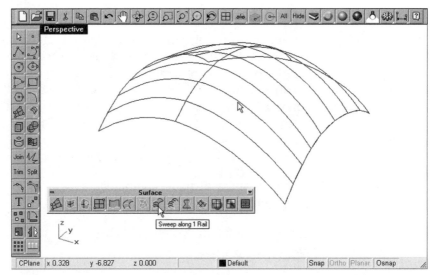

5 Set the display to a four-viewport layout and turn on the control points. (See figure 6-61.)

6 Select Transform > Set Points, or the Set Point button from the Transform toolbar.
Command: SetPt

7 Select the control points indicated in figure 6-62 and press the Enter key.

8 In the Set Points dialog box, check the Set Y box and Align to CPlane box and click on the OK button.

More Surface Modeling Tools

Fig. 6-61. Control points turned on.

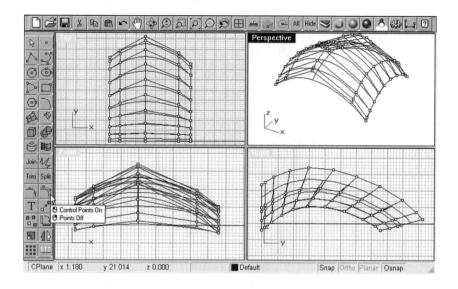

Fig. 6-62. Y coordinate of the selected control points being set.

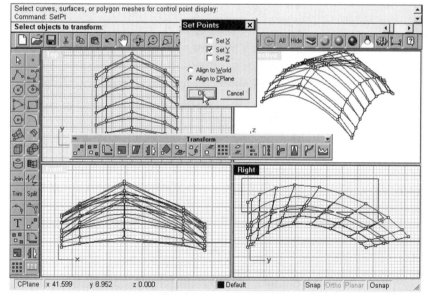

9 Select a point in the Front viewport.

The Y coordinates of the selected control points are aligned. (See figure 6-63.) Save your file.

Projection

By projecting the control points of a surface or the entire surface to the current construction plane, you transform the surface, as follows.

Fig. 6-63. Control points aligned.

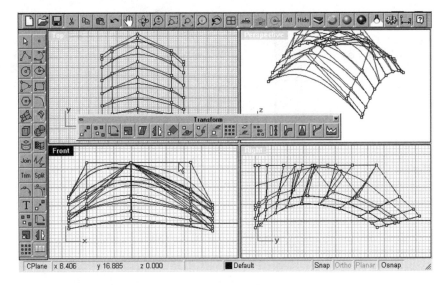

1. Select File > Save As and specify a file name of *ProjectSrf.3dm*.
2. Turn on the control points if you already turned them off.
3. Select Transform > Project to Cplane, or the Project to CPlane button from Transform toolbar.
   Command: ProjectToCPlane
4. Select the control points indicated in figure 6-64 and press the Enter key.
5. In the ProjectToCPlane dialog box, select the Yes button.

The selected control points are projected to the construction plane corresponding to the Top viewport. (See figure 6-65.)

Fig. 6-64. Control points selected.

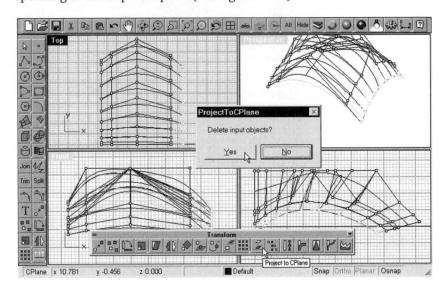

More Surface Modeling Tools

Fig. 6-65. Control points projected.

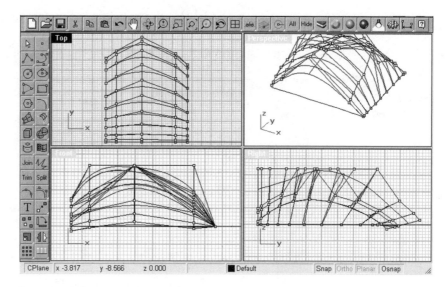

6 Turn off the control points.
7 Repeat the ProjectToCPlane command.
8 Select the surface in the Top viewport and press the Enter key.
9 Select the Yes button in the ProjectToCPlane dialog box.

The entire surface is projected. (See figure 6-66.) Save your file.

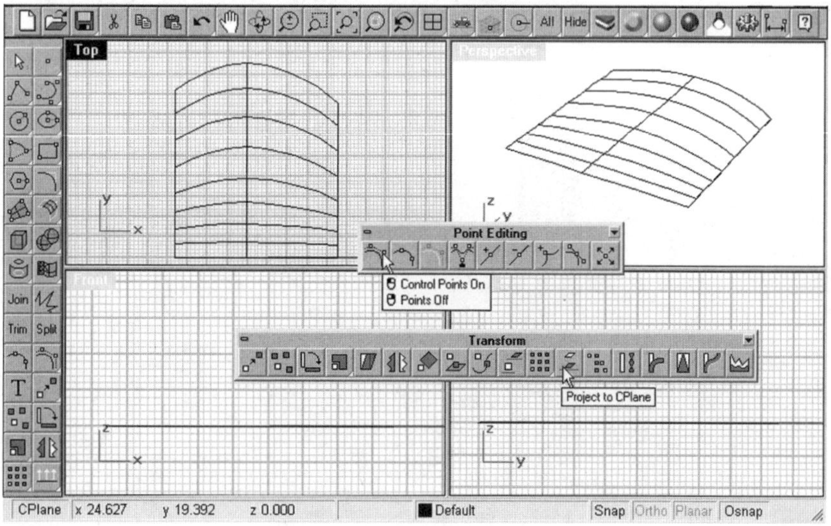

Fig. 6-66. Surface projected

Twisting

In the following, you will construct a rail-revolved surface and deform it in several ways. First, you will twist it, as follows.

296   CHAPTER 6: Surface Modeling: Part 2

1. Start a new file. Use the metric (millimeter) template.
2. Construct a free-form curve in the Front viewport and a free-form curve in the Right viewport. (See figure 6-67.)

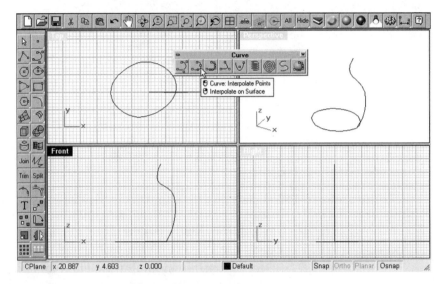

Fig. 6-67. Curves constructed.

3. Construct a rail-revolved surface. (See figure 6-68.)

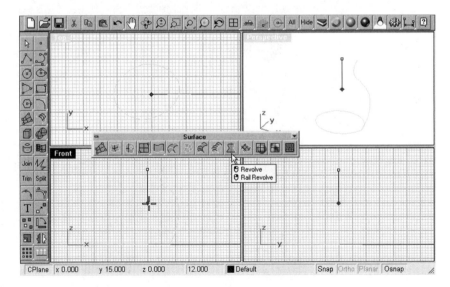

Fig. 6-68. Rail-revolved surface being constructed.

4. Select Transform > Twist, or the Twist button from the Transform toolbar.
   Command: Twist
5. Select the surface and press the Enter key. (See figure 6-69.)

More Surface Modeling Tools

Fig. 6-69. Surface selected.

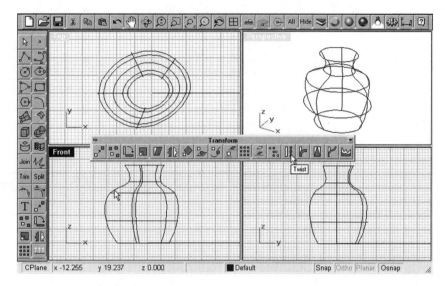

6 With reference to figure 6-70, select two points in the Front viewport to indicate the twist axis and the portion of the surface to be twisted, and select two points in the Top viewport to specify the twist angle.

The surface is twisted. Save your file as *TwistSrf.3dm*.

Fig. 6-70. Twist portion, twist axis, and twist angle being specified.

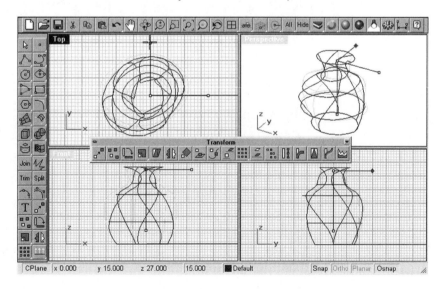

Tapering

To taper the surface, perform the following steps.

1 Select File > Save As and specify a file name of *TaperSrf.3dm*.

2 Select Transform > Taper, or the Taper button of the Transform toolbar.

Command: Taper

3 Select the surface and press the Enter key.

4 With reference to figure 6-71, select two points to indicate the start and end of taper axis, and select two more points to indicate the start and end distance.

The surface is tapered. Save your file.

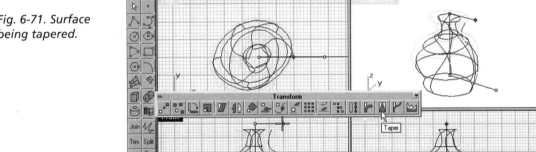

Fig. 6-71. Surface being tapered.

Bending

To bend the surface, perform the following steps.

1 Select File > Save As and specify a file name of *BendSrf.3dm*.

2 Select Transform > Bend, or the Bend button of the Transform toolbar.

Command: Bend

3 Select the surface and press the Enter key.

4 With reference to figure 6-72, select two points in the Front viewport to indicate the start and end of the spine, and select a point to bend through.

The surface is bent. Save your file.

More Surface Modeling Tools

Fig. 6-72. Surface being bent.

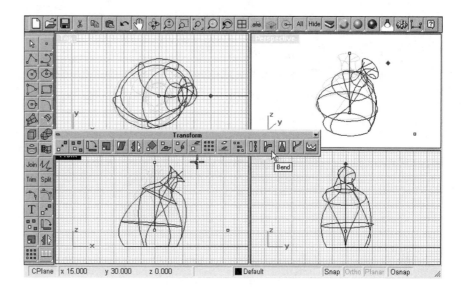

Flowing Along a Curve

To flow the surface along a curve, perform the following steps.

1 Select File > Save As and specify a file name of *FlowSrf.3dm*.

2 Maximize the Front viewport.

3 With reference to figure 6-73, construct a line segment and a free-form curve.

Fig. 6-73. Line and curve constructed.

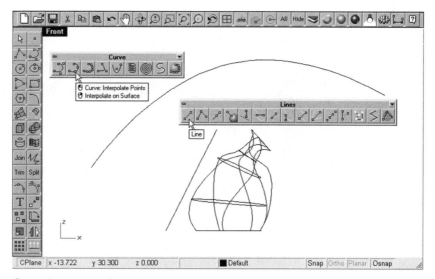

4 Select Transform > Flow along Curve, or the Flow along Curve button from the Transform toolbar.

Command: Flow

5 Select the surface and press the Enter key.
6 Select the lower end of the line as the original backbone. (See figure 6-74.)
7 Select the left end of the curve as the new backbone.

The surface flows along the new backbone curve. Save your file.

Smoothing

Smoothing removes irregularities from the surface. To smooth the surface, perform the following steps.

1 Select File > Save As and specify a file name of *SmoothSrf.3dm*.
2 Select Transform > Smooth, or the Smooth button from the Transform toolbar.
   Command: Smooth
3 Select the surface and press the Enter key.
4 In the Smooth dialog box, accept the default and click on the OK button. (See figure 6-75.)

The surface is smoothed. Save your file.

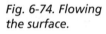
Fig. 6-74. Flowing the surface.

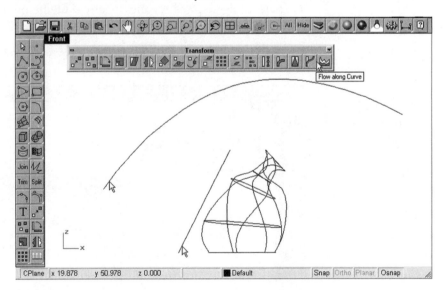

Shearing

To shear a surface, perform the following steps.

1 Open the file *TwistSrf.3dm*.
2 Select File > Save As and specify a file name of *ShearSrf.3dm*.

Fig. 6-75. Surface being smoothed.

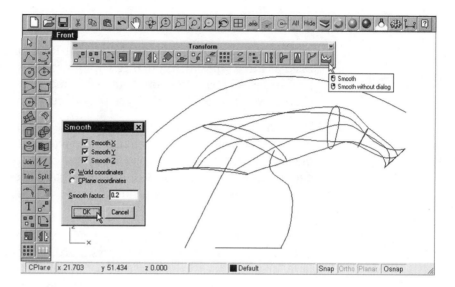

3 Select Transform > Shear, or the Shear button of the Transform toolbar.

Command: Shear

4 Select the surface and press the Enter key.

5 With reference to figure 6-76, select three points in the Front viewport to indicate the origin point, reference point, and angle of shear.

The surface is sheared. Save your file.

Fig. 6-76. Surface being sheared.

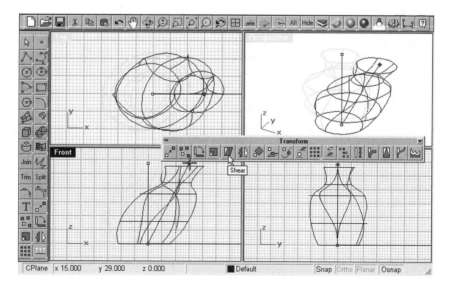

Summary

As mentioned in the previous chapter, NURBS surfaces can be categorized as three types in accordance with how they are constructed: primitive surfaces, basic free-form surfaces, and derived surfaces. In this chapter, you learned the third type of surface: derived surfaces.

There are two categories of derived surfaces. The first derives from existing surfaces. This category includes extended, filleted, chamfered, offset, blended, and draped surfaces. The second category consists of surfaces derived from the heightfield of a bitmap and a developable surface. In addition to learning derived surfaces, you also learned three types of surface continuity.

After you construct a NURBS surface, you can modify it in many ways. You can join surfaces to form a polysurface, explode a polysurface into individual surfaces, trim surfaces to remove unwanted portions, and split surfaces. To change a surface profile, you manipulate control points, insert or remove knots, use the handlebar editor, and use various advanced tools. To transform a surface, you use translate and deform tools.

Review Questions

1. List of the various derived surfaces.
2. List the methods used to edit and transform surfaces.

Chapter 7

Surface Modeling: Part 3

■■ Objectives

The goals of this chapter are to explain methods of analyzing surfaces and to familiarize you with the use of Rhino NURBS surface modeling tools in constructing surface models. After studying this chapter, you should be able to:

❒ Use surface analysis tools

❒ Use Rhino as a tool to construct free-form NURBS surface models

■■ Overview

In chapters 5 and 6 you learned surface modeling concepts; methods of constructing primitive, free-form, and derived surfaces; and methods of editing and transforming surfaces. In this chapter you will learn how to analyze surfaces and enhance your surface modeling techniques by working on a number of surface modeling projects.

■■ Final Surface Modeling Tools

The sections that follow discuss the last of the surface modeling tools you will examine. First, however, let's take a look at how Rhino's analysis tools work with surfaces.

To analyze the surfaces you construct, you use the analysis tools. Basically, most of the tools you use on curves are also applicable to surfaces. The surface analysis tools are located on the Analyze toolbar, shown in figure 7-1.

Fig. 7-1. Analyze toolbar.

Table 7-1 outlines the functions of the commands specific to surface analysis and their location in the pull-down menu.

Table 7-1: Surface Analysis Commands and Their Functions

Command Name	Pull-down Menu	Function
BoundingBox	Analyze > Bounding Box	For constructing a bounding box of a selected object
Area	Analyze > Mass Properties > Area	For evaluating the area of selected surfaces or polysurfaces
AreaCentroid	Analyze > Mass Properties > Area Centroid	For constructing a point at the area centroid of selected surfaces or polysurfaces
AreaMoments	Analyze > Mass Properties > Area Moments	For evaluating area, area centroid, first area moments, second area moments, product moments, area moments of inertia about world coordinate axes, area radii of gyration about world coordinate axes, area moments of inertia about centroid coordinate axes, and area radii of gyration about centroid coordinate axes
CurvatureAnalysis	Analyze > Surface > Curvature Analysis	For displaying a color rendering to indicate the curvature values of selected surfaces
DraftAngleAnalysis	Analyze > Surface > Draft Angle Analysis	For displaying a color rendering to indicate the draft angle values of selected surfaces
Emap	Analyze > Surface > Environment Map	For displaying a color rendering with a bitmap mapped to selected surfaces
Zebra	Analyze > Surface > Zebra	For displaying the zebra pattern on selected surfaces
CurvatureSrf	Analyze > Surface > Curvature Curve	For constructing curvature circles on selected points of a surface
PointFromUV	Analyze > Surface > Point from UV Coordinates	For constructing points on a surface by entering U and V coordinates
EvaluateUVPt	Analyze > Surface > UV Coordinates of a Point	For evaluating UV coordinates of points on a surface

Command Name	Pull-down Menu	Function
PointDeviation	Analyze > Surface > Point Set Deviation	For measuring the deviation of points from surfaces and curves
ShowEdges	Analyze > Edge Tools > Show Edges	For displaying the edges of selected surfaces, including the edge of a closed-loop surface
ShowNakedEdges	Analyze > Edge Tools > Show Naked Edges	For displaying the edges of selected surfaces
PointsAtNakedEdges	Analyze > Edge Tools > Create Points at Naked Edges	For constructing points at the end points of naked edges
SplitEdge	Analyze > Edge Tools > Split Edge	For splitting an edge in two
MergeEdge	Analyze > Edge Tools > Merge Edge	For merging split edges
JoinEdge	Analyze > Edge Tools > Join 2 Naked Edges	For joining edges of contiguous surfaces
RebuildEdges	Analyze > Edge Tools > Rebuild Edges	For restoring the original edges of surfaces
ShowBrokenEdges	Analyze > Diagnostics > Show Broken Edges	For diagnosing problem edges
Dir	Analyze > Direction	For displaying the direction of selected objects

Construction of a Bounding Box

The bounding box of a surface or polysurface consists of a rectangular box enclosing the surface or polysurface. To construct a bounding box for a surface, perform the following steps.

1 Open the file *TwistSrf.3dm*, select File > Save As, and specify a file name of *AnalyzeSrf.3dm*.

2 Select Analyze > Bounding Box, or the Curve Bounding Box button from the Analyze toolbar.
   Command: BoundingBox

3 Select the surface and press the Enter key.

A bounding box is constructed. (See figure 7-2.)

4 Because this bounding box is not required in the following tutorial, select it and press the Delete key to delete it.

Fig. 7-2. Bounding box constructed.

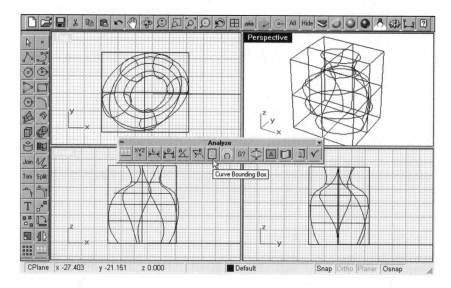

Area, Area Centroid, and Area Moments Inquiry

To assess the area, area centroid, and area moments of a surface, perform the following steps.

1 Select Analyze > Mass Properties > Area, or the Area button from the Mass Properties toolbar.
   Command: Area

2 Select the surface and press the Enter key.

The area of the surface is displayed in the command line area. Continue with the following steps.

3 Select Analyze > Mass Properties > Area Centroid, or the Area Centroid button from the Mass Properties toolbar.
   Command: AreaCentroid

4 Select the surface and press the Enter key.

A point is constructed at the area centroid of the surface. (See figure 7-3.)

5 Select Analyze > Mass Properties > Area Moments, or the Area Moments button from the Mass Properties toolbar.
   Command: AreaMoments

6 Select the surface and click on the OK button.

Area moment information is displayed in the command line area.

Fig. 7-3. Point constructed at area centroid.

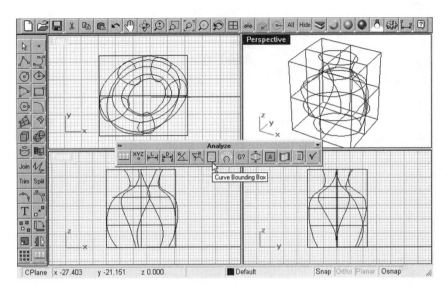

Analyzing Surface Profiles

Because a NURBS surface is a smooth surface with no facets, the profile and silhouette of a NURBS surface are displayed using U and V isoparm curves. To analyze the profile of the surface, you use curvature rendering, draft angle rendering, environmental map rendering, and zebra stripes rendering and construct curvature arcs.

In regard to individual points on a surface, you construct a point on a surface by specifying the U and V coordinates of the point. Given a point on the surface, you evaluate its U and V coordinates. In addition, you evaluate the deviation of any point from a surface.

Curvature Rendering

To display a color rendering to indicate the curvature values of the surface, perform the following steps. In the rendering, different curvature values are represented by different colors.

1 Maximize the Perspective viewport.

2 Select Analyze > Surface > Curvature Analysis, or the Curvature Analysis button from the Surface Analyze toolbar.

Command: CurvatureAnalysis

3 Select the surface and press the Enter key.

A color rendering showing the curvature values in various colors is displayed. (See figure 7-4.)

Fig. 7-4. Curvature values displayed in various colors.

Draft Angle Rendering

Draft angles are an essential consideration in mold and die design. To display a color rendering to illustrate the draft angle value of a surface with respect to the construction plane, perform the following steps.

1 Select Analyze > Surface > Draft Angle Analysis, or the Draft Angle Analysis button from the Surface Analyze toolbar.
   Command: DraftAngleAnalysis

2 Select the surface and press the Enter key.

Draft angle values on the surface are displayed in various colors. (See figure 7-5.)

Environment Map Rendering

To simply inspect the smoothness of a surface, you place a bitmap image on the surface. To place a bitmap on a surface and display a rendering of the surface, perform the following steps.

1 Select Analyze > Surface > Environment Map, or the Environment Map button from the Surface Analyze toolbar.
   Command: EMap

2 Select the surface and press the Enter key.

3 In the Environment Map Options dialog box, select an image.

The selected image is mapped onto the surface. (See figure 7-6.)

Final Surface Modeling Tools

Fig. 7-5. Draft angle values.

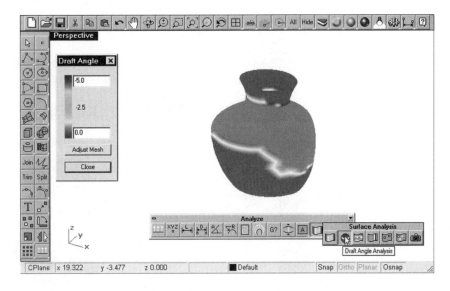

Fig. 7-6. Bitmap mapped on surface.

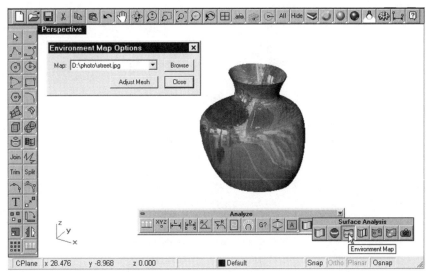

Zebra Stripes Rendering

To effectively view the smoothness of a surface, you use a set of zebra stripes instead of a bitmap. To use zebra stripes to help visualize the smoothness of a surface, perform the following steps.

1 Select Analyze > Surface > Zebra, or the Zebra button from the Surface Analysis toolbar.

Command: Zebra

2 Select the surface and press the Enter key.

3 In the Zebra Options dialog box, set the stripe direction and stripe size.

Zebra stripes are placed on the surface. (See figure 7-7.) This patterning helps you check the continuity between surfaces.

Fig. 7-7. Zebra stripes.

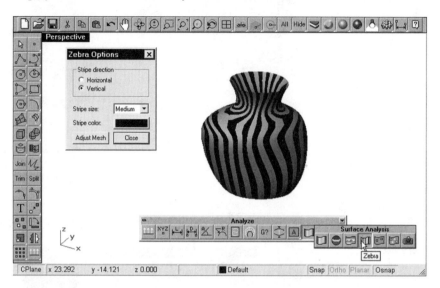

Curvature Arcs

To find out the curvature of a selected point on a surface, you construct two curvature arcs. To construct curvature circles at selected locations of a surface, perform the following steps.

1 Select Analyze > Surface > Curvature Circle.
   Command: CurvatureSrf
2 Select the surface.
3 Select a point on the surface and press the Enter key.

Curvature arcs are constructed. (See figure 7-8.)

Points on a Surface

A surface has two sets of isoparm curves in two directions, U and V. Correspondingly, there are two axis directions, U and V. You can represent any point on the surface by a set of U and V coordinates. To construct a point on a surface by specifying the point's U and V coordinate values, perform the following steps.

1 Select Analyze > Surface > Point from UV Coordinates, or the Point from UV Coordinates button of the Surface Analyze toolbar.
   Command: PointsFromUV

Final Surface Modeling Tools

Fig. 7-8. Curvature arcs constructed.

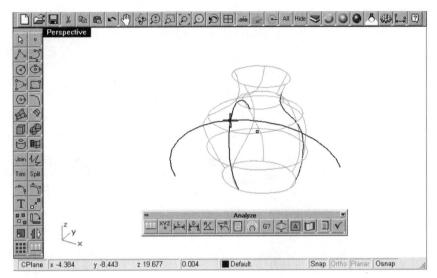

2   Select the surface.
3   Enter a U coordinate value and a V coordinate value, and press the Enter key.

A point is constructed on the surface. (See figure 7-9.)

Fig. 7-9. Point constructed.

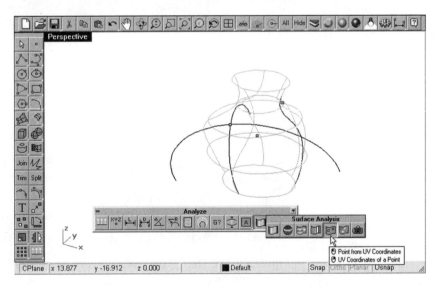

Evaluating the U and V Coordinates of a Point

Contrary to constructing a point on the surface by specifying the U and V coordinates, in the following you will evaluate the U and V coordinates of a point on a surface.

Chapter 7: Surface Modeling: Part 3

1 Select Analyze > Surface > UV Coordinates of a Point, or the UV Coordinates of a Point button from the Surface Analysis toolbar.
   Command: EvaluateUVPt
2 Select the surface.
3 Select the point and press the Enter key.

The U and V coordinates of the selected point on the surface are displayed at the command line area. (See figure 7-10.)

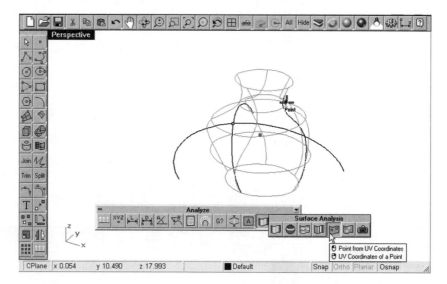

Fig. 7-10. U and V coordinates.

Checking the Deviation of a Point from a Surface

To check the deviation of a point from a surface, perform the following steps.

1 Select Analyze > Surface > Point Set Deviation, or the Point Set Deviation button from the Surface Analyze toolbar.
   Command: PointDeviation
2 Select the point indicated in figure 7-11 and press the Enter key.
3 Select the surface indicated in figure 7-11 and press the Enter key.

A Point/Surface Deviation dialog box is displayed.

Manipulation of Surface Edges

The tools described in the sections that follow enable you to manipulate surface edges. These tools include Show Edges, Show Naked Edges, Create Points at Naked Edges, Split Edge, Merge Edge, Join 2 Naked Edges, Rebuild Edges, and Show Broken Edges.

Final Surface Modeling Tools

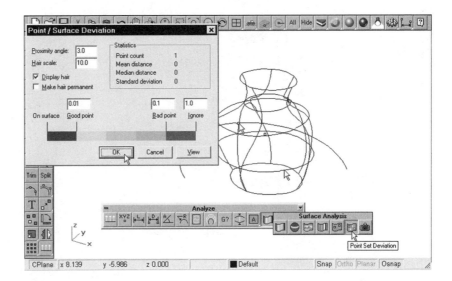

Fig. 7-11. Point/ Surface Deviation dialog box.

Show Edges

In the computer display, both surface edges and isoparm curves are displayed as curves. To see clearly the edges of a surface, you highlight them, as follows.

1 Select Analyze > Edge Tools > Show Edges, or the Show Edges button from the Edge Tools toolbar.

   Command: ShowEdges

2 Select the surface and press the Enter key.

The surface edge is highlighted. In figure 7-12, three edges of the surface (including the seam edge of the closed-loop surface) are highlighted.

Show Naked Edges

In a closed surface such as that shown in figure 7-12, there are four edges. Two of them meet and form a seam. Therefore, you see only three edges. Among them, two are naked edges and one is the seam edge. To highlight the naked edges of a surface, perform the following steps.

1 Select Analyze > Edge Tools > Show Naked Edges, or the Show Naked Edges button from the Edge Tools toolbar.

   Command: ShowNakedEdges

2 Select the surface and click on the OK button.

The naked edges are highlighted. (See figure 7-13.)

Fig. 7-12. Edges highlighted.

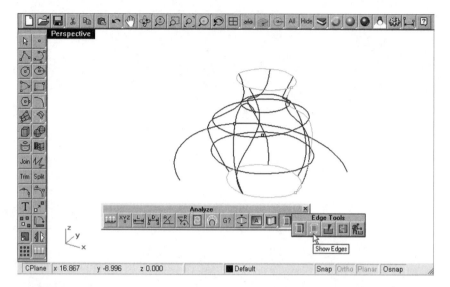

Fig. 7-13. Naked edges highlighted.

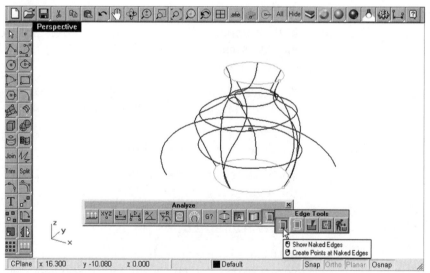

Creating Points at End Points of Naked Edges

The naked edges of the closed surface shown in figure 7-13 form two closed loops. However, each of them has a start point and an end point. To construct a point object at the end points of the naked edges of a surface, perform the following steps.

1 Select Analyze > Edge Tools > Create Points at Naked Edges, or the Create Points at Naked Edges button from the Edge Tools toolbar.
Command: PointsAtNakedEdges

2 Select the surface and press the Enter key.

Final Surface Modeling Tools 315

Points are constructed at the end points of the naked edges. (See figure 7-14.)

Fig. 7-14. Points constructed at the end points of naked edges.

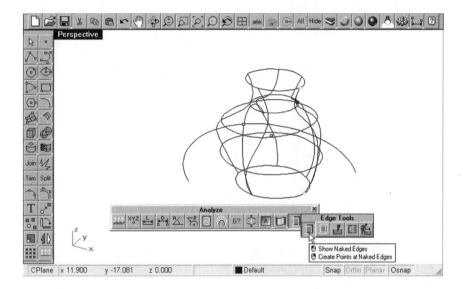

Splitting an Edge

To extrude a surface from an edge, you extrude the entire edge. By splitting an edge in two, you extrude one of the edges, as follows.

1 Select Analyze > Edge Tools > Split Edge, or the Split Edge button from the Edge Tools toolbar.

   Command: SplitEdge

2 With reference to figure 7-15, select an edge and a location along the curve, and press the Enter key.

An edge is split into two edges.

To extrude one of the split edges, continue with the following steps.

1 Select Surface > Extrude > Straight, or the Extrude Straight button from the Extrude toolbar.

   Command: Extrude

2 Select the edge indicated in figure 7-16 and select a location to specify the extrude height.

An edge is extruded. To see how surfaces separated a small distance can be joined (see latter part of this tutorial), you need to move the extruded surface, as follows.

3 Move the extruded surface as indicated in figure 7-17.

Fig. 7-15. Edge being split.

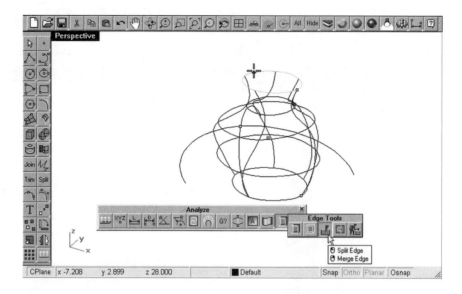

Fig. 7-16. An edge being extruded.

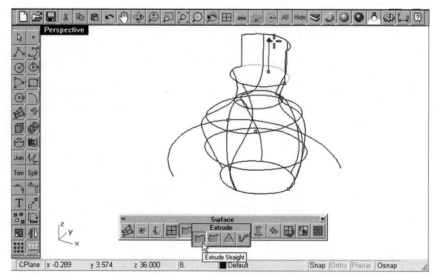

Merging Edges

The opposite of splitting an edge is merging split edges, as follows.

1 Select Analyze > Edge Tools > Merge Edge, or the Merge Edge button from the Edge Tools toolbar.
Command: MergeEdge

2 Select the edge indicated in figure 7-18.

The split edges are merged into a single edge.

Final Surface Modeling Tools

Fig. 7-17. Extruded surface moved.

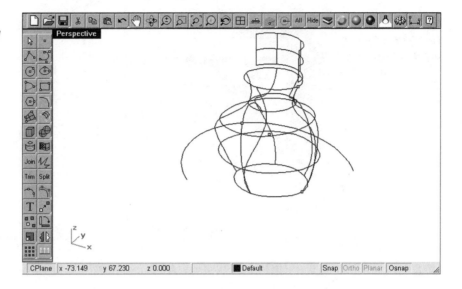

Fig. 7-18. Split edges being merged.

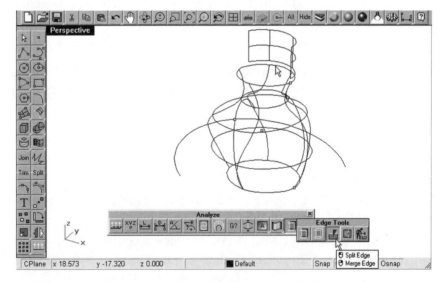

Joining Edges

Naked edges of two surfaces, although separated by a small distance, can be merged to become a single edge. Joining naked edges "heals" any small gaps between contiguous surfaces. To join the naked edges of two surfaces, perform the following steps.

1 Select Analyze > Edge Tools > Join 2 Naked Edges, or the Join 2 Naked Edges button from the Edge Tools toolbar.

Command: JoinEdge

2 Select the edges indicated in figure 7-19.

Fig. 7-19. Edges being joined.

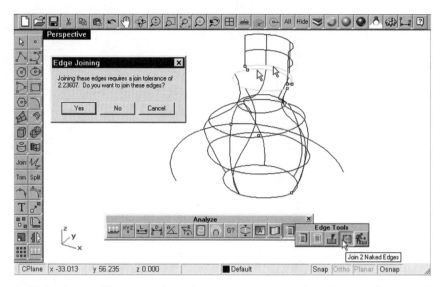

A dialog box telling you the tolerances between the edges is displayed. Continue with the following.

3 In the Edge Joining dialog box, click on the OK button.

The display of the edges shows them closed. This works for rendering purposes but not for manufacturing purposes, as the edges are not really joined.

Rebuilding Edges

To restore the original edge after you have joined edges, you explode the joined surface into individual surfaces and rebuild the edges, as follows.

1 Select Edit > Explode, or the Explode button from the Main toolbar.
Command: Explode

2 Select the polysurface and press the Enter key.

3 Select Analyze > Edge Tools > Rebuild Edges, or the Rebuild Edges button from the Edge Tools toolbar.

4 *Command:* RebuildEdges

5 Select the surface indicated in figure 7-20 and press the Enter key.

Show Broken Edges

To check for edge errors, perform the following steps.

1 Select Analyze > Diagnostics > Show Broken Edges, or the Show Broken Edges button from the Diagnostics toolbar.
Command: ShowBrokenEdges

2 Select the surfaces and press the Enter key.

Fig. 7-20. Polysurface exploded and edges restored.

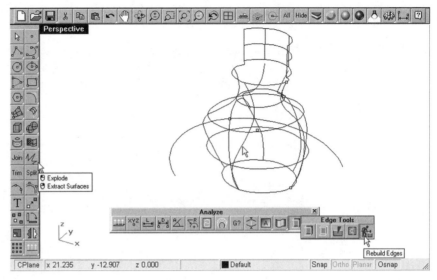

The diagnostics result is displayed in the command line area.

Direction

To check the normal direction of a surface, perform the following steps.

1 Select Analyze > Direction, or the Direction button from the Analyze toolbar.
   Command: Dir
2 Select the surface and press the Enter key.

Surface direction arrows are displayed. (See figure 7-21.)

Fig. 7-21. Direction arrows.

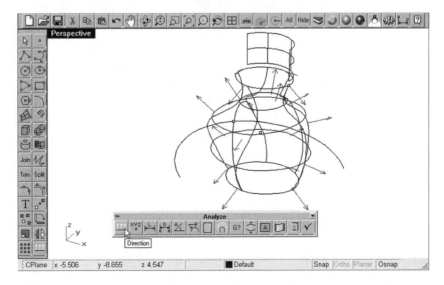

▪▪ Surface Modeling Projects

In this section you will work on a number surface modeling projects in which you will use the curves you constructed in Chapter 4. These projects include the joypad, joystick, mobile phone, and car body.

Joypad

The joypad shown in figure 7-22 has a number of surface types: extruded, planar trimmed, lofted, offset, and filleted.

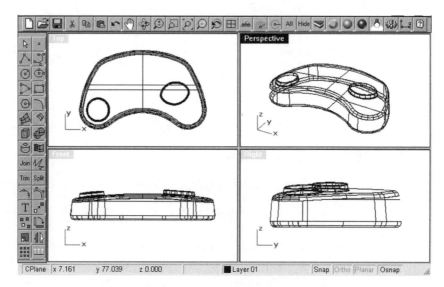

Fig. 7-22. Surface model of a joypad.

To construct a new layer, set the current layer to the new layer, and construct surfaces on the new layer, perform the following steps.

1 Open the file *Joypad.3dm* you constructed in Chapter 4.

2 Select Edit > Layers > Edit Layers, or the Edit Layers button from the Layer toolbar.

Command: Layer

3 In the Edit Layers dialog box, click on the New button to construct a new layer, check the *Layer 01* box in the layer list to set it as the current layer, and click on the OK button.

4 Select Surface > From Planar Curves, or the Surface from Planar Curves button from the Surface toolbar.

Command: PlanarSrf

5 Select the curve indicated in figure 7-23 and press the Enter key.

Surface Modeling Projects

Fig. 7-23. Planar surface constructed.

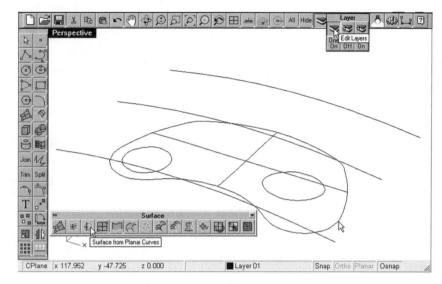

A trimmed planar surface is constructed. To construct an extruded surface, continue with the following steps.

1 Select Surface > Extrude > Straight, or the Extrude Straight button from the Extrude toolbar.

Command: Extrude

2 Select the curve indicated in figure 7-24.

Fig. 7-24. Extruded surface being constructed.

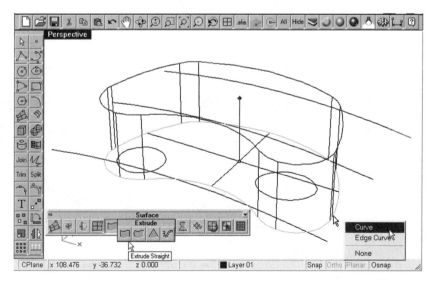

3 Because the location where you select has two edges (the curve and the edge curve of the planar surface), select Curve in the pop-up menu.

4 Press the Enter key.

5 Type *T* at the command line area to use the Taper option.

6 Type *-5* at the command line area to specify a taper angle of -5 degrees.

7 Type *50* at the command line area to specify the extrusion height.

An extruded surface is constructed. To construct a lofted surface, continue with the following steps.

8 Select Surface > Loft, or the Loft button from the Surface toolbar.
   Command: Loft

9 Select the curves indicated in figure 7-25. (You may select the curves from right to left or from left to right. The order of selection affects how the surface interpolates among the curves.)

10 In the Loft Options dialog box, click on the OK button.

A lofted surface is constructed

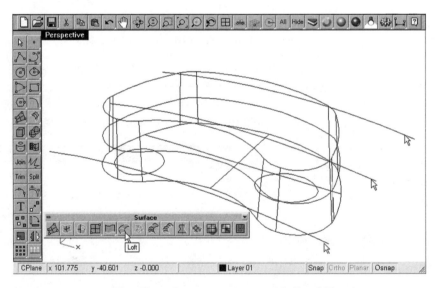

Fig. 7-25. Lofted surface being constructed.

To construct two filleted surfaces, continue with the following steps.

11 Select Surface > Fillet, or the Fillet Surface button from the Surface Tools toolbar.
   Command: FilletSrf

12 Type *R* at the command line area to use the Radius option.

13 Type *5* at the command line area to specify a filleted radius of 5 units.

14 With reference to figure 7-26, select the central part of the lofted surface and the lower part of the extruded surface.

15 Repeat the FilletSrf command.

Surface Modeling Projects 323

Fig. 7-26. Filleted surface between the extruded surface and the lofted surface being constructed.

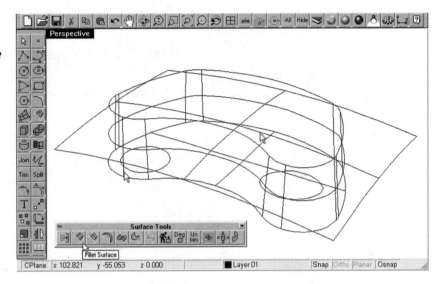

16 With reference to figure 7-27, select the upper part of the extruded surface and the central part of the planar surface.

Two filleted surfaces are constructed.

Fig. 7-27. Filleted surface between the extruded surface and the planar surface being constructed.

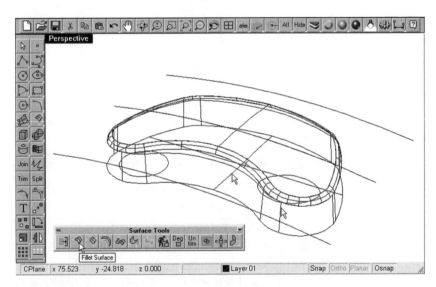

To construct two raised features on the joypad, continue with the following steps.

17 Select Surface > Extrude > Straight, or the Extrude Straight button from the Extrude Toolbar.
Command: Extrude

Chapter 7: Surface Modeling: Part 3

18 Select the circle and ellipse indicated in figure 7-28 and press the Enter key.

19 Type *50* at the command line area to specify the extrusion height.

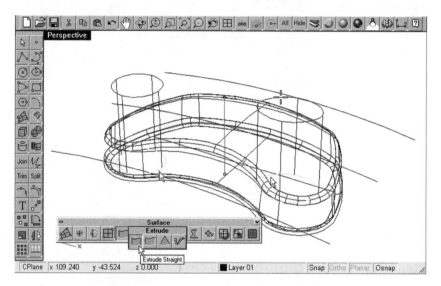

Fig. 7-28. Extruded surfaces being constructed.

The curves are not required. To turn off the default layer, continue with the following steps.

20 Select Edit > Layers > Edit Layers, or the Edit Layers button from the Layer toolbar.

Command: Layer

21 In the Edit Layers dialog box, turn off the default layer and click on the OK button.

To construct two filleted surfaces, continue with the following steps.

22 Select Surface > Fillet, or the Fillet Surface button from the Surface Tools toolbar.

Command: FilletSrf

23 Type *R* at the command line area to use the Radius option.

24 Type *1* at the command line area to set the fillet radius to 1 unit.

25 Select the surfaces indicated in figure 7-29.

26 Repeat the FilletSrf command.

27 Select the surfaces indicated in figure 7-30.

Surface Modeling Projects

Fig. 7-29. First filleted surface constructed.

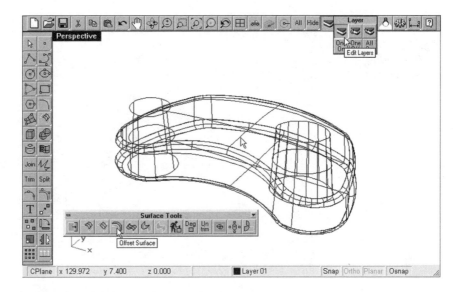

Fig. 7-30. Second filleted surface constructed.

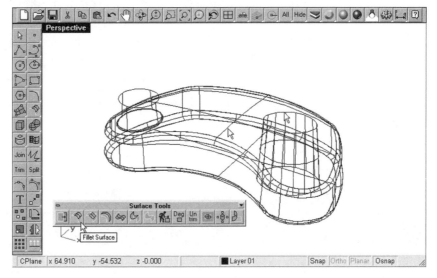

To construct an offset surface from the lofted surface, continue with the following steps.

28 Select Surface > Offset, or the Offset Surface button from the Surface Tools toolbar.

Command: OffsetSrf

29 Select the surface indicated in figure 7-31 and press the Enter key.

Now you need to specify an offset value. It is important to understand that a surface has a normal direction. A positive offset value constructs an offset surface in the normal direction, and a negative offset value

Fig. 7-31. Offset surface being constructed.

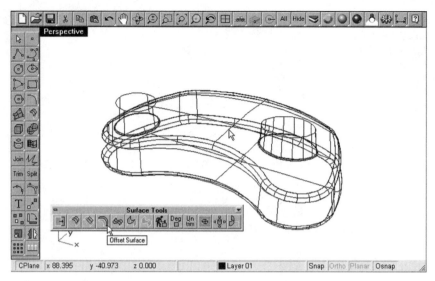

constructs an offset surface in the opposite direction. For a lofted surface, the normal direction depends on the sequence and location of selection of the input curves. To try a positive value of 6 units, perform the following.

30 Type *6* at the command line area.

An offset surface is constructed. (See figure 7-32.) The offset surface should be 6 units above the original surface. If, however, you constructed an offset surface in the opposite direction, undo the last command and construct the surface again by changing the positive offset value to a negative offset value.

Fig. 7-32. Offset surface constructed.

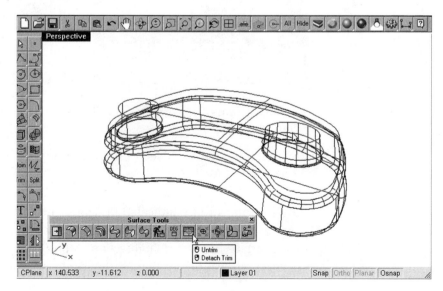

Surface Modeling Projects

To untrim the offset surface, continue with the following steps.

31 Select Surface > Edit Tools > Untrim, or the Untrim button from the Surface Tools toolbar.
Command: Untrim

32 Select the trim boundary indicated in figure 7-32.

Isoparm curves on a surface serve at least two purposes: visualization of the profile of the surface and selection of the surface. To modify the isoparm of the offset surface to a higher density to facilitate selection, continue with the following steps.

33 Select Edit > Object Properties, or the Object Properties button from the Properties toolbar.

34 Select the offset surface indicated in figure 7-33 and press the Enter key.

Fig. 7-33. Offset surface untrimmed.

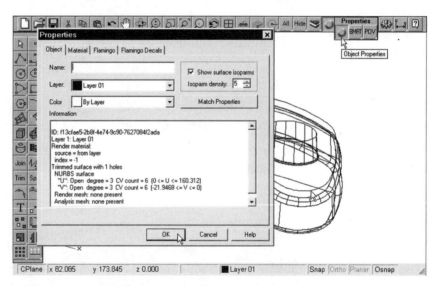

35 In the Properties dialog box, select the Object tab, set the Isoparm density to 5, and click on the OK button.

To construct a filleted surface between the offset surface and the extruded surface from the ellipse, continue with the following steps.

36 Select Surface > Fillet, or the Fillet Surface button from the Surface Tools toolbar.

37 *Command:* FilletSrf

38 Select the surfaces indicated in figure 7-34.

A filleted surface is constructed.

Fig. 7-34. Isoparm modified and filleted surface being constructed.

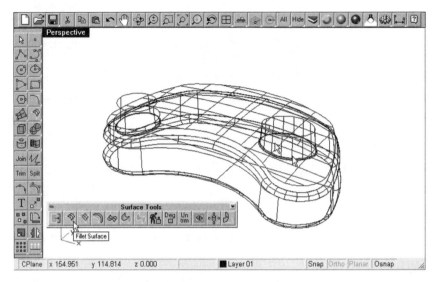

As mentioned previously, a surface has direction. To reveal the direction of a surface, continue with the following steps.

39 Select Analyze > Direction, or the Direction button from the Analyze toolbar.

Command: Dir

40 Select the lofted surface indicated in figure 7-35.

Fig. 7-35. Filleted surface constructed.

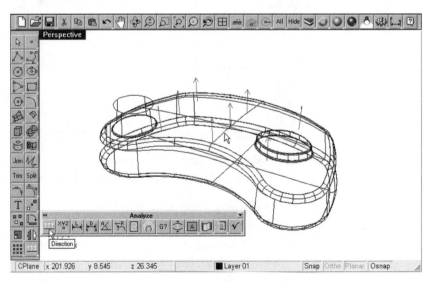

You can follow a similar course of action to construct another offset surface from the lofted surface, and to construct a filleted surface between the offset surface and the extruded surface of the circle. (See figure 7-36.)

Surface Modeling Projects   329

Fig. 7-36. Offset surface and filleted surface constructed.

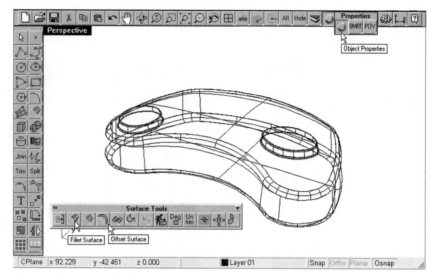

The surface model of the joypad is complete. Save your file.

Joystick

This section deals with the joystick you developed in Chapter 4. To construct the surface model of a joystick, perform the following steps.

1 Open the file *JoyStick.3dm*.
2 Select Surface > From Curve Network, or the Surface from Curve Network button from the Surface toolbar.
   Command: NetworkSrf
3 Select all curves. (See figure 7-37.)

Fig. 7-37. Surface from network curves being constructed.

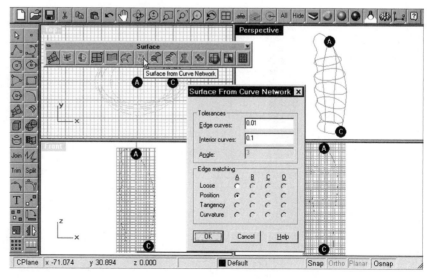

330 CHAPTER 7: Surface Modeling: Part 3

4   In the Surface From Curve Network dialog box, click on OK.

A surface is constructed. Save your file. (See figure 7-38.)

Fig. 7-38. Completed surface.

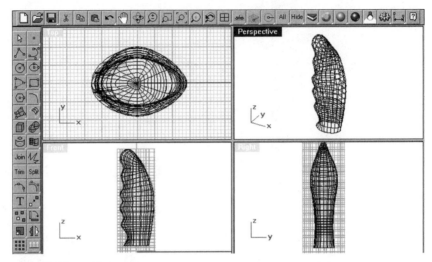

Mobile Phone

To construct the surface model of a mobile phone, perform the following.

1   Open the file *Mphone.3dm*.
2   Set the current layer to *Layer 01*.
3   Select Surface > Sweep 1 Rail, or the Sweep Along 1 Rail button from the Surface toolbar.
    Command: Sweep1
4   Select the curve indicated in figure 7-39 as the rail curve.

Fig. 7-39. Rail curve selected.

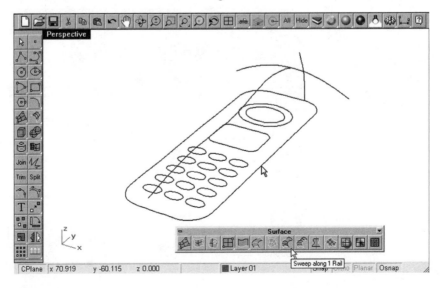

Surface Modeling Projects

5 Select the curve indicated in figure 7-40 as the cross-section curve and press the Enter key.

6 In the Sweep 1 Rail Options dialog box, click on the OK button.

Fig. 7-40. Cross-section curve selected.

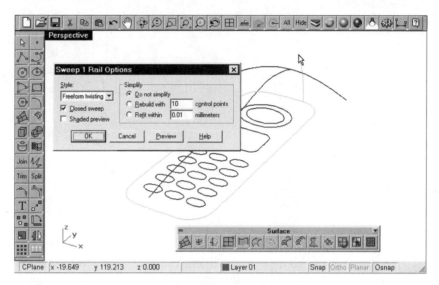

A swept surface is constructed. To construct another swept surface, continue with the following steps.

7 Repeat the Sweep1 command.

8 Select the curve indicated in figure 7-41 as the rail curve.

Fig. 7-41. Rail curve selected.

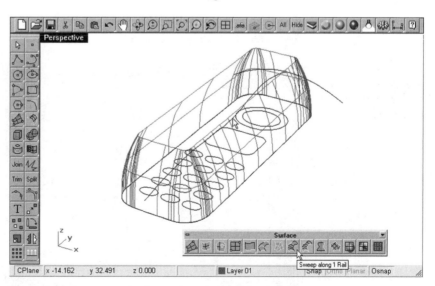

9 Select the curve indicated in figure 7-42 as the cross-section curve and press the Enter key.

10 Click on the OK button in the Sweep 1 Rail Options dialog box.

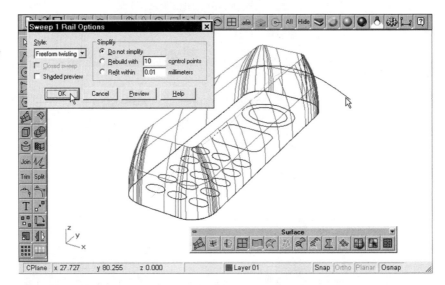

Fig. 7-42. Second swept surface being constructed.

To construct a filleted surface between the two swept surfaces, continue with the following steps.

11 Select Surface > Fillet, or the Fillet Surface button from the Surface Tools toolbar.

Command: FilletSrf

12 Type *R* at the command line area to use the Radius option.

13 Type *4* at the command line area to specify the fillet radius.

14 Select the surfaces indicated in figure 7-43.

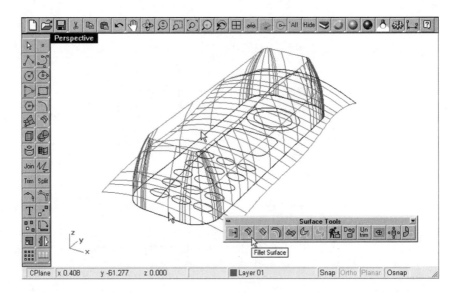

Fig. 7-43. Filleted surface being constructed.

Surface Modeling Projects

To trim the top surface of the mobile phone, continue with the following steps.

15 Maximize the Top viewport.

16 With reference to figure 7-44, trim the surface.

Fig. 7-44. Surface trimmed.

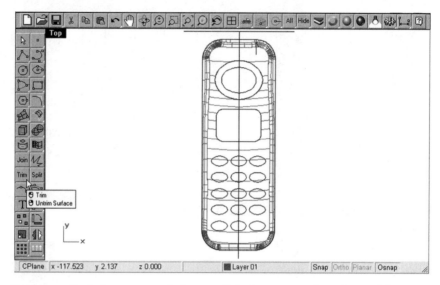

You may find that some of the small holes cannot be trimmed because the isoparm lines of the surface are too coarse. To modify the isoparm of the top surface and trim it again, continue with the following steps.

17 Select Edit > Object Properties, or the Object Properties button from the Properties toolbar.

Command: Properties

18 Select the top surface indicated in figure 7-45 and press the Enter key.

19 In the Properties dialog box, select the Object tab, set isoparm density to 5, and click on the OK button. Trim the remaining holes.

To construct an offset surface from the top surface, continue with the following steps.

20 Maximize the Perspective viewport.

21 Select Analyze > Direction, or the Direction button from the Analyze toolbar.

Command: Dir

22 Select the top surface indicated in figure 7-46.

23 If the normal direction differs from figure 7-46, type *F* at the command line area. Otherwise, press the Enter key to exit.

334 CHAPTER 7: Surface Modeling: Part 3

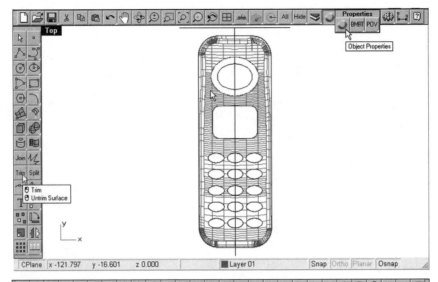

Fig. 7-45. Isoparm density modified and surface trimmed.

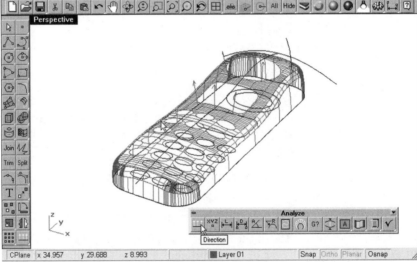

Fig. 7-46. Normal direction.

24 Select Surface > Offset, or the Offset Surface button from the Surface Tools toolbar.

Command: OffsetSrf

25 Select the top surface and press the Enter key.

26 Type -3 at the command line area to specify an offset distance of -3 units. An offset surface is constructed. To untrim this surface, continue with the following steps.

27 Select Surface > Edit Tools > Untrim, or the Untrim button from the Surface Tools toolbar.

Command: Untrim

Surface Modeling Projects 335

28 Select the boundary edge indicated in figure 7-47 and press the Enter key.

Fig. 7-47. Offset surface constructed and boundary edge being untrimmed.

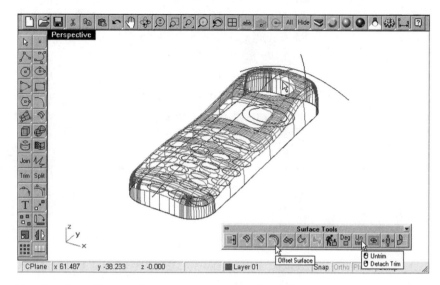

To trim the offset surface, continue with the following steps.

29 Maximize the Top viewport.

30 Select Edit > Trim, or the Trim button from the Main toolbar.
Command: Trim

31 With reference to figure 7-48, select the ellipse as the cutting object and press the Enter key. Then select the offset surface and press the Enter key.

To construct a lofted surface, continue with the following steps.

Fig. 7-48. Offset surface being trimmed.

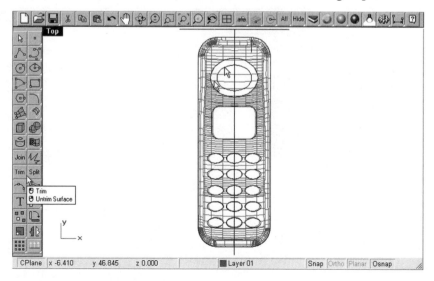

32 Maximize the Perspective viewport.
33 Turn off the Default layer.
34 Zoom the display in accordance with figure 7-49.
35 Select Surface > Loft, or the Loft button from the Surface toolbar.
Command: Loft
36 Select the edges indicated in figure 7-49 and press the Enter key.
37 If the seam point is congruent with figure 7-49, press the Enter key.

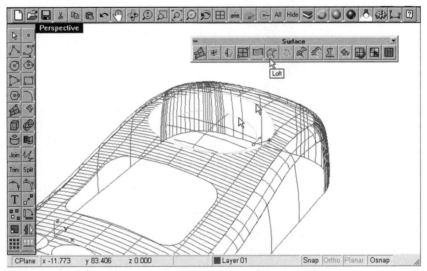

Fig. 7-49. Loft surface being constructed.

38 In the Loft Options dialog box, click on the OK button.

The surface model of the mobile phone is complete. (See figure 7-50.) Save your file.

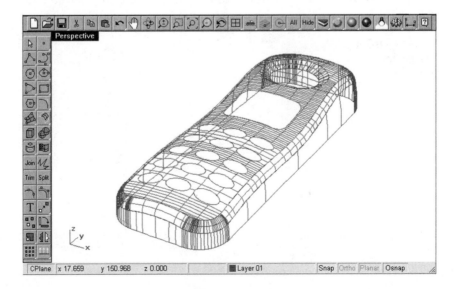

Fig. 7-50. Completed mobile phone model.

Scale Model Car Body

To construct the surface of a model car using the curves you constructed in Chapter 4, perform the following steps.

1 Open the file *Beetle.3dm*. (See figure 7-51.)

2 Select Edit > Layers > Edit Layers, or the Edit Layers button from the Layers toolbar.
Command: Layer

3 Set the current layer to *Layer 05*.

4 Set the color for *Layer 05* to red.

5 Turn off *Layer 02*, *Layer 03*, and *Layer 04*.

6 Click on the OK button.

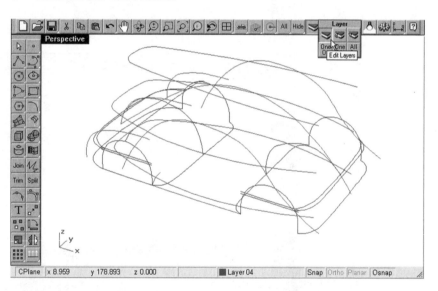

Fig. 7-51. Curves for the model car.

To construct the surfaces for the main body of the car, continue with the following steps.

7 Select Surface > Loft, or the Loft button from the Surface toolbar.
Command: Loft

8 Select the curves indicated in figure 7-52 sequentially, either from right to left or left to right, and press the Enter key.

9 In the Loft Options dialog box, click on the OK button.

10 Repeat the Loft command.

11 Select the curves indicated in figure 7-53 sequentially, either from right to left or left to right, and press the Enter key.

12 Click on the OK button in the Loft Options dialog box.

Fig. 7-52. Loft surface being constructed.

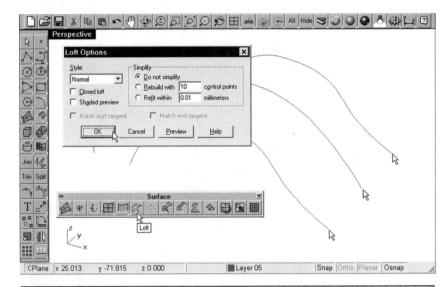

Fig. 7-53. Second loft surface being constructed.

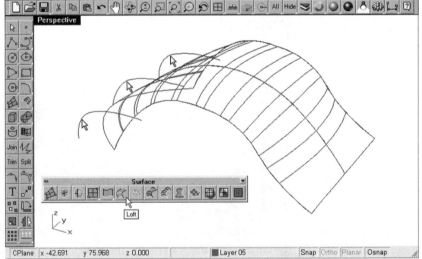

13 Turn off *Layer 02* and turn on *Layer 03*.

14 Select Surface > Loft, or the Loft button from the Surface toolbar.
Command: Loft

15 Select the curves indicated in figure 7-54 sequentially, either from bottom to top or top to bottom, and press the Enter key.

16 In the Loft Options dialog box, click on the OK button.

Because a cutting object has to be wider than the object to be trimmed, you will construct a blended surface for use as a cutting object. To trim the surfaces, continue with the following steps.

Surface Modeling Projects

Fig. 7-54. Third loft surface being constructed.

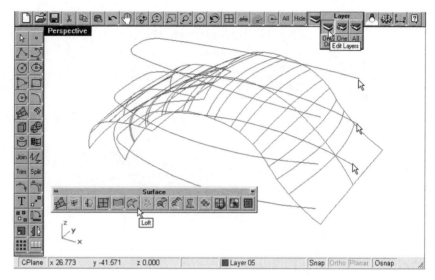

17 Turn off *Layer 03*.

18 Select Surface > Blend, or the Blend button from the Surface toolbar.

Command: BlendSrf

19 Select the edges indicated in figure 7-55. (Select either the lower or the upper part of both edges.)

20 In the Blend Bulge dialog box, click on the OK button.

To trim the surfaces, continue with the following steps.

21 Select Edit > Trim, or the Trim button from the Main toolbar.
Command: Trim

Fig. 7-55. Blend surface being constructed.

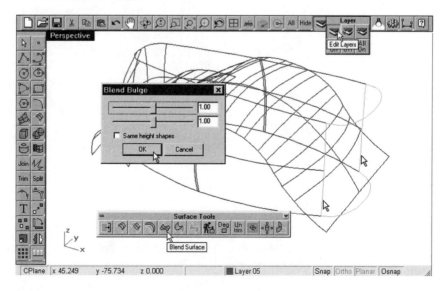

22 Select the surfaces indicated in figure 7-56 and press the Enter key.
23 Select the portion of the surfaces indicated in figure 7-56 and press the Enter key.

Fig. 7-56. Surfaces being trimmed.

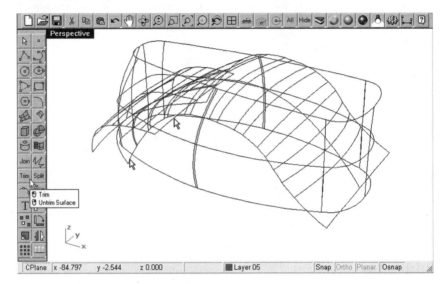

24 Repeat the Trim command.
25 Select the surfaces A, B, C, and D indicated in figure 7-57 and press the Enter key.
26 Select the portion of the surfaces at A, B, D, and E indicated in figure 7-57 and press the Enter key.

Fig. 7-57. Surfaces being trimmed.

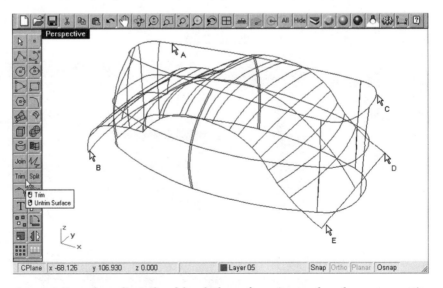

As mentioned earlier, the blended surface is used only as a cutting object. To erase the blended surface, continue with the following.

Surface Modeling Projects 341

27 Select the blended surface indicated in figure 7-58 and press the Delete key.

Fig. 7-58. Blended surface being deleted.

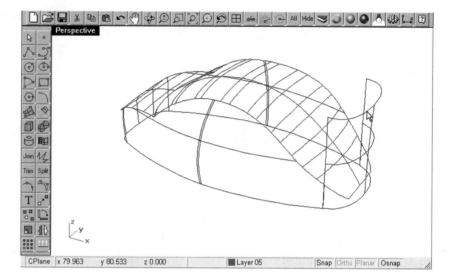

The surfaces for the main body of the car are complete. To construct the surfaces for the front and rear parts of the car, continue with the following steps.

28 Turn on *Layer 03*.

29 Select Surface > From Curve Network, or the Surface from Curve Network button from the Surface toolbar.

Command: NetworkSrf

30 Select the curves indicated in figure 7-59 and press the Enter key.

Fig. 7-59. Surface from curve network being constructed.

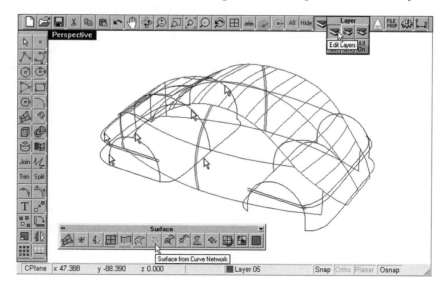

31 In the Surface From Curve Network dialog box, click on the OK button.

A surface derived from a curve network is constructed. To trim two surfaces, continue with the following steps.

32 Select Edit > Trim, or the Trim button from the Main toolbar.
Command: Trim

33 Select the surfaces indicated in figure 7-60 and press the Enter key.

Fig. 7-60. Surfaces being trimmed.

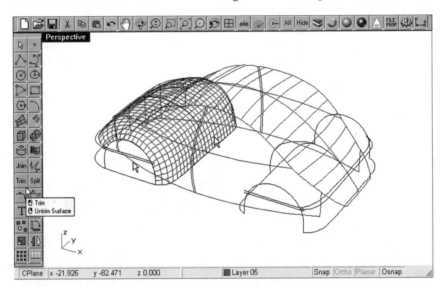

34 Select the portion of the surfaces indicated in figure 7-60 and press the Enter key.

35 Select Surface > From Curve Network, or the Surface from Curve Network button from the Surface toolbar.
Command: NetworkSrf

36 Select the curves indicated in figure 7-61 and press the Enter key.

37 In the Surface From Curve Network dialog box, click on the OK button.

The surface from curve network is complete. To trim three surfaces, continue with the following steps.

38 Select Edit > Trim, or the Trim button from the Main toolbar.
Command: Trim

39 Select the surfaces A, B, and C indicated in figure 7-62 and press the Enter key.

40 Select the portions of the surfaces A, B, C, and D indicated in figure 7-62 and press the Enter key.

Surface Modeling Projects

Fig. 7-61. Surface from curve network being constructed.

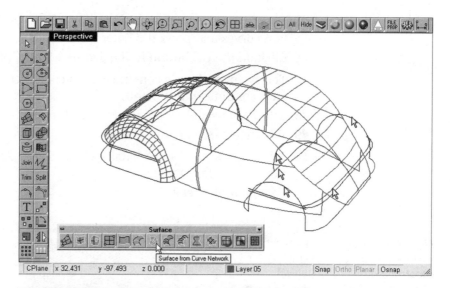

Fig. 7-62. Surfaces being trimmed.

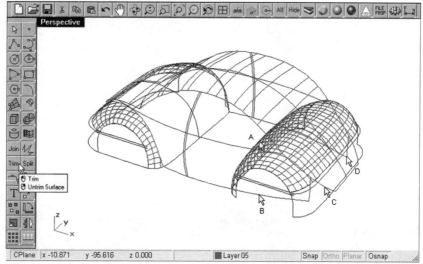

To construct a swept surface, continue with the following steps.

41 Select Rotate View from the Standard toolbar.

42 With reference to figure 7-63, click and drag the viewport to rotate it.

43 Select Surface > Sweep 2 Rails, or the Sweep Along 2 Rails button from the Surface toolbar.
Command: Sweep2

44 Select curves A and B indicated in figure 7-63 as the rail curves.

Chapter 7: Surface Modeling: Part 3

45 Select curves C, D, and E indicated in figure 7-63 as the cross-section curves and press the Enter key.

46 Click on the OK button in the Sweep 2 Rails Options dialog box.

47 The swept surface is constructed. To construct another swept surface, continue with the following steps.

48 Select Rotate View from the Standard toolbar.

49 With reference to figure 7-64, click and drag the viewport to rotate it.

50 Select Surface > Sweep 2 Rails, or the Sweep Along 2 Rails button from the Surface toolbar.

Command: Sweep2

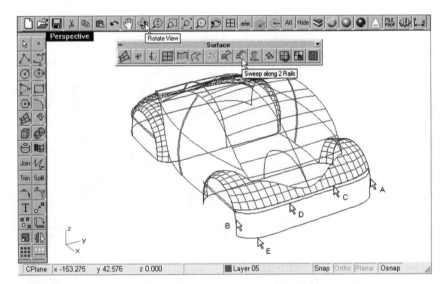

Fig. 7-63. Swept surface being constructed.

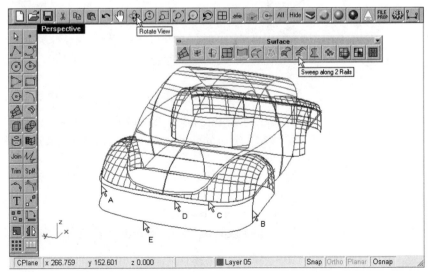

Fig. 7-64. Second swept surface being constructed.

Surface Modeling Projects

51 Select curves A and B indicated in figure 7-64 as the rail curves.

52 Select curves C, D, and E indicated in figure 7-64 as the cross-section curves and press the Enter key.

53 Click on the OK button in the Sweep 2 Rails Options dialog box.

The surfaces for the front and rear parts of the car are complete. To construct the surfaces for the skirts of the car, continue with the following steps.

54 Right-click on the Perspective viewport title.

55 Select Set View > Perspective to set the viewport display to the default perspective view.

56 Turn off *Layer 03* and turn on *Layer 04*.

57 Select Surface > Sweep 1 Rail, or the Sweep Along 1 Rail button from the Surface toolbar.

Command: Sweep1

58 Select curve A indicated in figure 7-65.

59 Select curve B indicated in figure 7-65 and press the Enter key.

60 Click on the OK button in the Sweep 1 Rail Options dialog box.

To mirror the swept surface, continue with the following steps. Because the surface model is placed symmetrically along the X axis of the construction plane, the mirror line will be along the X axis.

61 Turn off *Layer 04*.

62 Select Transform > Mirror, or the Mirror button from the Transform toolbar.

Command: Mirror

Fig. 7-65. Swept surface being constructed.

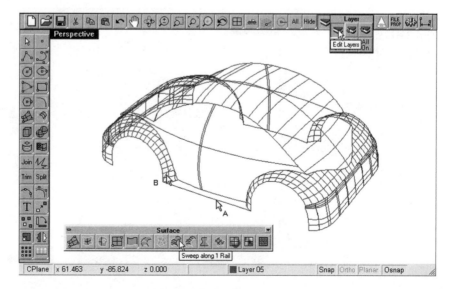

63 Select the surface indicated in figure 7-66 and press the Enter key.
64 Type *0,0* at the command line area to specify the start of the mirror plane.
65 Type *1,0* at the command line area to specify the end of the mirror plane.

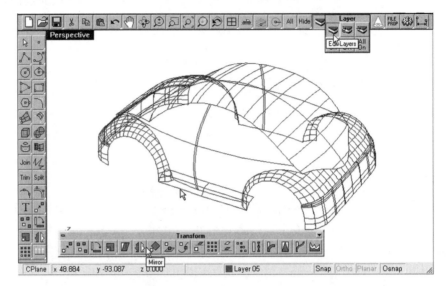

Fig. 7-66. Surface being mirrored.

To trim the surfaces, continue with the following steps.

66 Select Edit > Trim, or the Trim button from the Main toolbar.
Command: Trim
67 Select the surfaces indicated in figure 7-67 and press the Enter key.

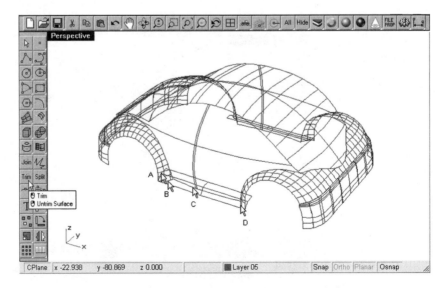

Fig. 7-67. Surfaces being trimmed.

Surface Modeling Projects

68 Select the portions of the surfaces A, B, C, and D indicated in figure 7-67 and press the Enter key.

The surfaces are trimmed. (See figure 7-68.)

69 With reference to figure 7-69, rotate the viewport and trim the surfaces.

The surface model for the car is complete. To inspect the surfaces, continue with the following steps.

Fig. 7-68. Surfaces trimmed.

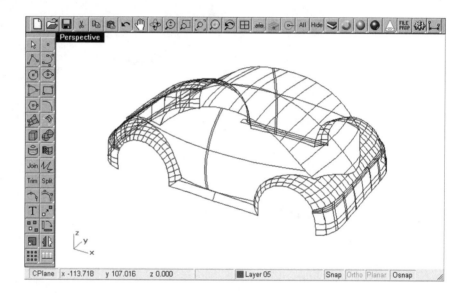

Fig. 7-69. Completed model.

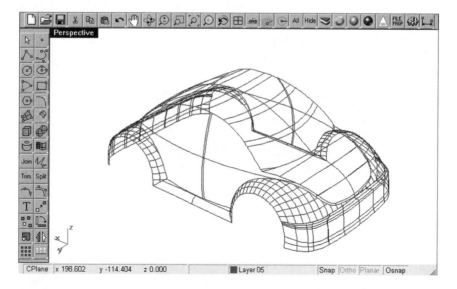

70 Select Render > Shade, or the Shade button from the Main toolbar. (See figure 7-70.)

Command: Shade

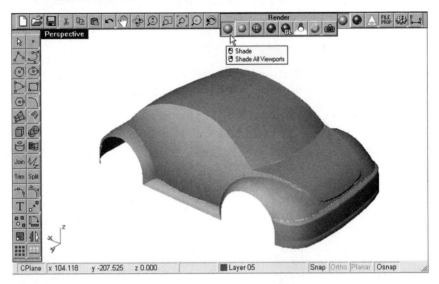

Fig. 7-70. Shaded view of the model car.

Note in the shaded view that there are some minute gaps between contiguous surfaces. These gaps are the result of mathematical approximation when contiguous surfaces are generated by the computer. Potentially, these gaps may cause problems in downstream computerized operations. To heal the gap, you join contiguous naked edges, as follows.

71 Select Analyze > Edge Tools > Join 2 Naked Edges, or the Join 2 Naked Edges button from the Edge Tools toolbar.

Command: JoinEdge

72 Select the edge indicated in figure 7-71 twice. (Because there are two edges at any two contiguous surfaces, you need to select an edge twice.)

73 In the Edge Joining dialog box, click on the Yes button.

Two edges are joined. Continue with the following steps.

74 Apply the JoinEdge command on all remaining edges.

75 Shade your viewport again. (See figure 7-72.)

The model is complete. Save your file.

Summary

Fig. 7-71. Edges being joined.

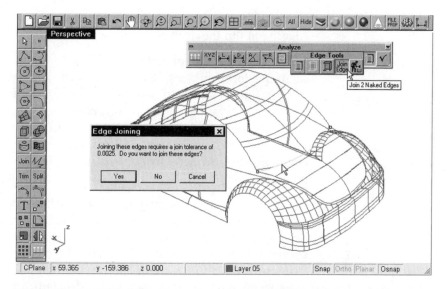

Fig. 7-72. All naked edges joined.

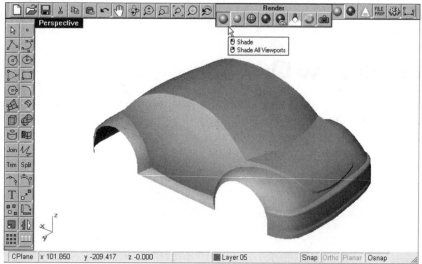

▪▪ Summary

In this chapter, you learned how to analyze surfaces. You learned how to construct a bounding box; how to evaluate area, area centroid, and area moments; how to render curvature, draft angle, environment map, and zebra stripes; and how to construct curvature arc on a surface and points on a surface. You also learned how to show the edges of a surface, show naked edges, create points at the end points of naked edges, split an edge, merge a split edge, join edges, rebuild edges, show broken edges, and flip the normal direction.

To construct a model of a 3D free-form object in the computer, you use a set of free-form surfaces. Before making the surfaces, you first study and analyze the 3D object to determine what types of surfaces are needed. You consider various types of primitive surfaces, basic free-form surfaces, and derived surfaces. Among the surfaces, free-form surfaces are those most commonly used. All free-form surfaces have one thing in common: they all need to be constructed from smooth curves and/or point objects.

It is natural to start thinking about the surfaces but not the points and curves while you are designing and making a surface model. However, because the computer constructs basic free-form surfaces from defined curves and/or point objects, the first task you need to tackle in making free-form surfaces is to think about what types of curves and/or points are needed and how they can be constructed. After making the curves and/or points, you then let the computer generate the required surfaces.

■■ Review Questions

1 List the methods you might use to analyze a NURBS surface.
2 Write a brief account of the general surface modeling approach.

Chapter 8

Solid Modeling

■■ Objectives

The goals of this chapter are to introduce the concepts of solid modeling, to outline various solid modeling methods, to explain Rhino solid modeling methods, and to let you practice making 3D solids using Rhino. After studying this chapter, you should be able to:

❒ Describe the key concepts of solid modeling

❒ State various types of solid modeling methods

❒ Construct 3D solid objects

❒ Use Rhino as a design tool to construct solid models

■■ Overview

Among the three types of 3D models, a solid model provides the most comprehensive information about an object. A solid model has volume data and face, edge, and vertex information associated with it. There are many ways to represent a solid in the computer: pure primitive instancing, generalized sweeping, spatial occupancy enumeration, cellular deconstruction, constructive solid geometry, and boundary representation.

Most contemporary solid modeling applications use a hybrid approach. Basically, Rhino represents a solid using a polysurface that encloses a volume without any gap or opening. In this chapter, you will learn solid modeling concepts and methods of representing solids in the computer. You will also construct solid objects using Rhino. In the next chapter, you will consolidate your learning on curves, surfaces, and solids.

Solid Modeling Concepts

Unlike a wireframe model (which is a set of unassociated curves that depict the edges of a 3D object) and a surface model (which is a set of unassociated surfaces that delineate the boundary faces of a 3D object), a solid model integrates the mathematical data of a 3D object in the computer. A solid model has associated with it the edge, vertex, face, and volume data of the object it represents. Solid models are used in various downstream computerized operating processes, including finite element analysis, rapid prototyping, and CNC machining.

Because any individual 3D object is unique in shape, it would be impossible to derive a general mathematical formula to represent all types of shapes. Several of the mathematical methods used to derive various types of shapes are explored in the sections that follow.

Primitive Instancing

Primitive instancing predefines a library of primitive solid objects. To make a solid object, you select a primitive from the library and specify the value of the parameters of the primitive. Figure 8-1 shows a family of solid objects derived from a primitive. Using this method to construct a solid model, you construct model shapes defined by the primitives contained in the library.

Fig. 8-1. A family of instances derived from a pure primitive.

Generalized Sweeping

Generalized sweeping creates a solid object by sweeping a 2D or 3D closed-loop curve or lamina in 3D space. You construct a closed-loop curve or lamina and sweep it in a linear direction, about an axis, or along a 3D curve. (See figures 8-2 through 8-4.) This method provides a flexible way of constructing solids of various shapes.

Fig. 8-2. Solid represented by sweeping linearly.

(Left): Fig. 8-3. Solid represented by sweeping about an axis.

(Right): Fig. 8-4. Solid represented by sweeping along a 3D curve.

Spatial Occupancy Enumeration

Spatial occupancy enumeration divides the entire 3D space into a number of identical cubical cells. You construct a solid by listing the cells the object occupies in 3D space. The accuracy of the representation of the solid is a function of the size of the cubical cells.

Fig. 8-5. Solids represented as a measure of the cubical cells they occupy in 3D space.

To represent an object more accurately, you use cells of smaller size. As a result of a decrease in cell size, file size will increase tremendously. Figure 8-5 shows a solid object represented at two resolution settings. This method is not widely used in geometry-based modeling systems. However, it is a basis for finite element analysis.

Cellular Deconstruction

Cellular deconstruction is similar to the spatial occupancy enumeration method in that you construct a solid by listing the cells it occupies in 3D space. However, unlike the spatial occupancy enumeration method, the cells are not always identical or cubical in shape. As a result, this method requires less memory to represent objects in a more accurate way. (See figure 8-6.) Again, this method is not widely used in geometric modeling.

Fig. 8-6. Solid represented by cells of different shapes and sizes.

Boundary Representation

Boundary representation (B-rep) derives from surface modeling techniques, with the addition of information on the connectivity of surfaces and specification of which side of the surface is solid and which is void. This method defines a solid using a set of surfaces that must not intersect with each other except at their common vertices or edges and that must form a closed loop encompassing a volume.

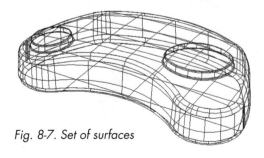

This method of constructing a solid is tedious because you have to construct all surfaces one at a time, as if you were making a surface model. Nevertheless, it enables you to construct a solid object with free-form surfaces. (See figure 8-7.)

Fig. 8-7. Set of surfaces

Constructive Solid Geometry

Constructive solid geometry (CSG), also known as set-theoretic modeling, provides a range of primitive solids in a manner similar to the pure primitive instancing method. In addition, it provides an extra facility for you to combine primitive solids to obtain a solid of complex shape using Boolean operations.

To construct a solid of complex shape, you combine solid objects using the union, intersection, and difference operators of the set theory. A union of two solids, A and B, forms a complex solid representing the volume encompassed by solid A or solid B. An intersection of two solids, A and B, forms a complex solid representing the volume encompassed by solid A and solid B. A difference of solid A and solid B forms a complex solid representing the volume encompassed by solid A but not solid B. (See figures 8-8 through 8-10.)

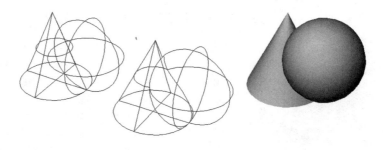

Fig. 8-8. Union of two solids: cone and sphere.

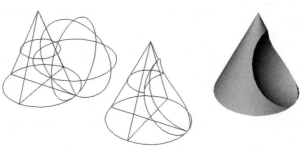

Fig. 8-9. Intersection of two solids: cone and sphere.

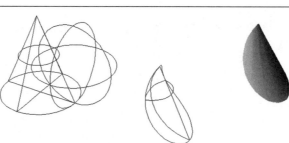

Fig. 8-10. Difference of two solids: cone and sphere.

Hybrid System

Most contemporary computer-aided design applications use a hybrid approach to the representation of solids. In general, they provide the following.

- ❒ A set of primitive solids of basic geometric shapes
- ❒ A facility for constructing solids by sweeping curves or laminas
- ❒ A facility for constructing a solid from a set of surfaces enclosing a volume
- ❒ A facility for combining solids using Boolean operations

■■ Rhino Solid Modeling Tools

Rhino uses the B-rep method to represent a solid object. A Rhino solid is a polysurface that encloses a volume without any gap or opening. Individual surfaces of the polysurface must not intersect except at their edges and vertices.

Rhino enables you to construct solid primitives ranging from simple 3D boxes to 3D text objects, to construct solids by extruding planar curves and surfaces, to convert a set of surfaces to a solid by adding planar surfaces to define an enclosed volume, to modify edges of solids by filleting, and to combine solid objects using Boolean operations. You perform these operations from the Solid pull-down menu, shown in figure 8-11.

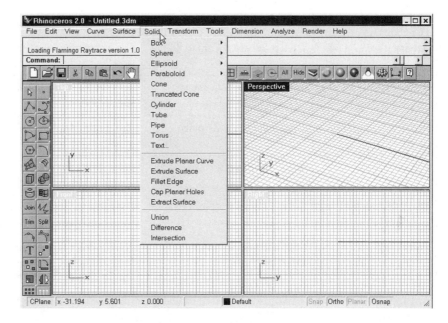

Fig. 8-11. Solid modeling pull-down menu items.

Making Primitive Solids

Rhino primitive solids are closed polysurfaces of basic geometric shapes. There are eleven types: box, sphere, ellipsoid, paraboloid, cone, truncated cone, cylinder, tube, pipe, torus, and text. Like primitive surfaces, curves and/or points are not required. To make a primitive solid, you select a type and specify the parameters. Figure 8-12 shows primitive solids commands on the Solid and Main toolbars.

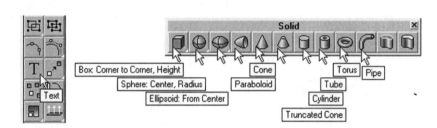

Fig. 8-12. Primitive solids commands on the Solid and Main toolbars.

Table 8-1 outlines the functions of primitive solids commands and their location in the pull-down menu.

Table 8-1: Primitive Solids Commands and Their Functions

Command Name	Pull-down Menu	Function
Box	Solid > Box > Corner to Corner, Height	For constructing a box by specifying the diagonal corners of its base and height
Box3pt	Solid > Box > 3 Points, Height	For constructing a box by specifying three corner points of its base and height
Sphere	Solid > Sphere > Center, Radius	For constructing a sphere by specifying its center and the radius
SphereD	Solid > Sphere > Diameter	For constructing a sphere by using two points to specify its diameter
Sphere3Pt	Solid > Sphere > 3 Points	For constructing a sphere by specifying three points on the surface of the sphere
Ellipsoid	Solid > Ellipsoid > From Center	For constructing an ellipsoid by specifying its center, first axis, second axis, and third axis
Ellipsoid FromFoci	Solid > Ellipsoid > From Foci	For constructing an ellipsoid by specifying the foci and a point on the ellipsoid
Paraboloid	Solid > Paraboloid > Focus, Direction	For constructing a paraboloid by specifying the focus, direction, and end point of the paraboloid

Command Name	Pull-down Menu	Function
Paraboloid Vertex	Solid > Paraboloid > Vertex, Focus	For constructing a paraboloid by specifying the vertex, direction, and end point of the paraboloid
Cone	Solid > Cone	For constructing a cone by specifying the center, radius of the base, and end point of the cone
TCone	Solid > Truncated Cone	For constructing a truncated cone by specifying the center, two radii, and end point of the truncated cone
Cylinder	Solid > Cylinder	For constructing a cylinder by specifying its center, radius, and height
Tube	Solid > Tube	For constructing a tube by specifying its center, inner and outer radii, and height
Pipe	Solid > Pipe	For constructing a pipe by specifying a curve, and inner and outer radii at both ends of the pipe
Torus	Solid > Torus	For constructing a torus by specifying the center, radius of the torus, and radius of the cross section
TextObject	Solid > Text	For constructing text objects

Box

A box is a polysurface with six planar surfaces. There are two ways to construct a box, as follows.

1. Start a new file. Use the metric (millimeter) template.
2. Select Solid > Box > Corner to Corner, Height, or the Box/Corner to Corner, Height button from the Box toolbar.
   Command: Box
3. Select two points in the Top viewport to indicate two diagonal points of the base of the box, and select a point in the Front viewport to indicate the height of the box. (See figure 8-13.)

A box is constructed.

Fig. 8-13. Box being constructed.

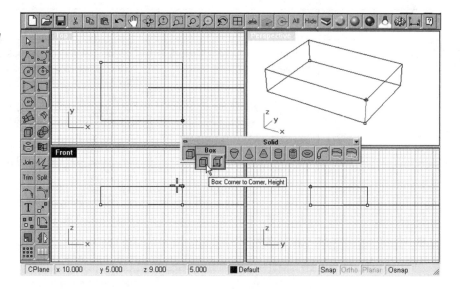

To construct another box, perform the following steps.

4 Select Solid > Box > 3 Points, Height, or the Box/3 Points, Height button from the Box toolbar.

Command: Box3Pt

5 Select three points in the Front viewport to describe a rectangle and select a point in the Right viewport to indicate the height. (See figure 8-14.)

The boxes are complete. Save your file as *Box.3dm*.

Fig. 8-14. Second box being constructed.

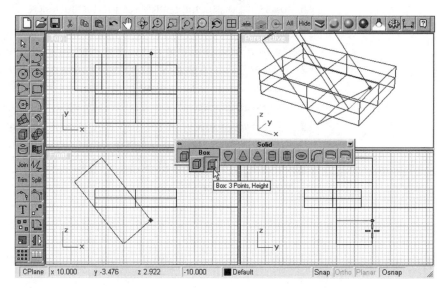

Sphere

A sphere is a single closed-loop surface. There are three ways to construct a sphere, as follows.

1. Start a new file. Use the metric (millimeter) template.
2. Select Solid > Sphere > Center, Radius, or the Sphere/Center, Radius button from the Sphere toolbar.

   Command: Sphere

3. Select two points in the Front viewport to indicate the center and radius. (See figure 8-15.)

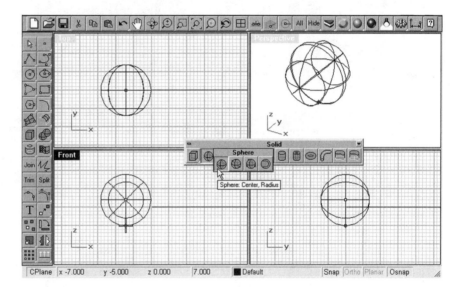

Fig. 8-15. Sphere being constructed by specifying the center and radius.

4. Select Solid > Sphere > Diameter, or the Sphere/Diameter button from the Sphere toolbar.

   Command: SphereD

5. Select two points in the Top viewport to indicate the diameter of the sphere. (See figure 8-16.)

6. Select Solid > Sphere > 3 Points, or the Sphere/3 Points button from the Sphere toolbar.

   Command: Sphere3Pt

7. Select two points in the Top viewport and a point in the Front viewport to indicate three points on the surface of the sphere. (See figure 8-17.)

8. Select the Baseball/Sphere button from the Sphere toolbar.

   Command: Baseball

360 CHAPTER 8: Solid Modeling

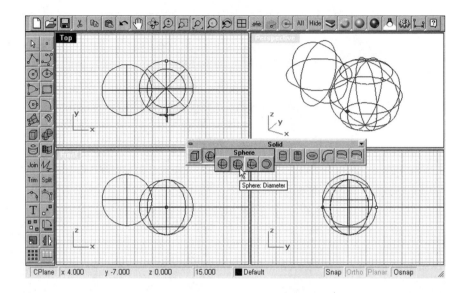

Fig. 8-16. Sphere being constructed by specifying the diameter.

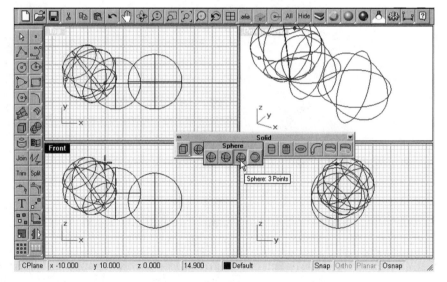

Fig. 8-17. Sphere being constructed by specifying three points on the surface.

9 Select two points in the Right viewport. (See figure 8-18.)

The spheres are complete. Save your file as *Sphere.3dm*.

Ellipsoid

An ellipsoid is also a single closed-loop surface. To construct an ellipsoid, perform the following steps.

1 Start a new file. Use the metric (millimeter) template.

Fig. 8-18. Baseball sphere being constructed.

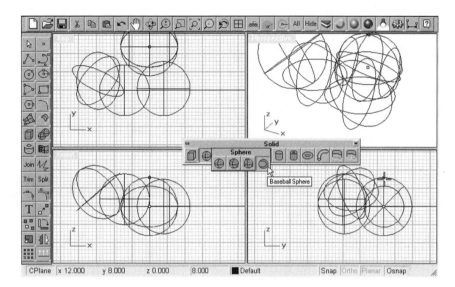

2 Select Solid > Ellipsoid > From Center, or the Ellipsoid/From Center button from the Ellipsoid toolbar.

Command: Ellipsoid

3 Select three points in the Top viewport to indicate the center, first axis, and second axis, and select a point in the Front viewport to indicate the third axis. (See figure 8-19.)

Fig. 8-19. Ellipsoid being constructed.

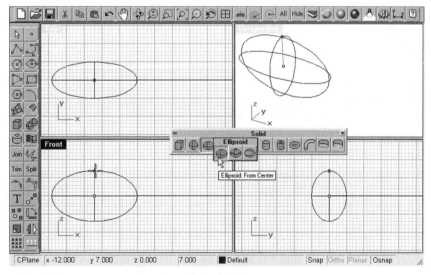

4 Select Solid > Ellipsoid > From Foci or the Ellipsoid/From Foci button from the Solid Tools toolbar.

Command: Ellipsoid FromFoci

362   CHAPTER 8: Solid Modeling

5   Select three points in the Top viewport to indicate the foci, and select a point on the ellipsoid. (See figure 8-20.)

6   Select the Baseball Ellipsoid button from the Ellipsoid toolbar.
Command: BaseballEllipsoid

7   Select three points in the Top viewport to indicate the center, first axis, and second axis, and select a point in the Front viewport to indicate the third axis. (See figure 8-21)

The ellipsoids are complete. Save your file as *Ellipsoid.3dm*.

Fig. 8-20. Ellipsoid being constructed by specifying the foci.

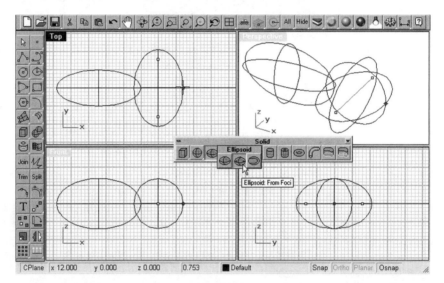

Fig. 8-21. Baseball ellipsoid being constructed.

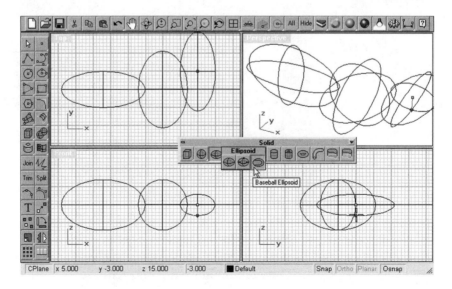

Rhino Solid Modeling Tools

Paraboloid

A paraboloid is a single closed-loop surface. To construct a paraboloid, perform the following steps.

1. Start a new file. Use the metric (millimeter) template.
2. Select Solid > Paraboloid > Focus, Direction, or the Paraboloid button from the Solid toolbar.
   Command: Paraboloid
3. Type C at the command line area if the prompt there indicates "Cap = No."
4. Select three points to indicate the focus, direction, and end point. (See figure 8-22.)
5. Select Solid > Paraboloid > Vertex, Focus, or select the Paraboloid button of the Solid toolbar and use the Vertex option.
   Command: Paraboloid/Vertex
6. Select three points to indicate the vertex, focus, and end point. (See figure 8-23.)

Two paraboloids are constructed. Save your file as *Paraboloid.3dm*.

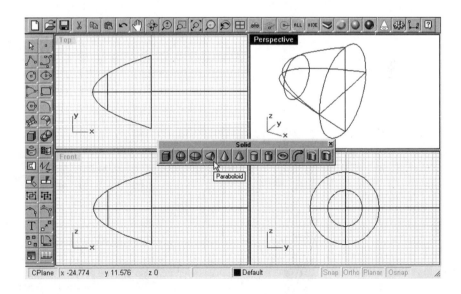

Fig. 8-22. Paraboloid being constructed by specifying the focus, direction, and end point.

Cone

A cone is a polysurface consisting of two joined surfaces. The slant surface is a closed surface and the base is a circular planar surface. To construct a cone, perform the following steps.

1. Start a new file. Use the metric (millimeter) template.

364 CHAPTER 8: Solid Modeling

Fig. 8-23. Paraboloid being constructed by specifying the vertex, focus, and end point.

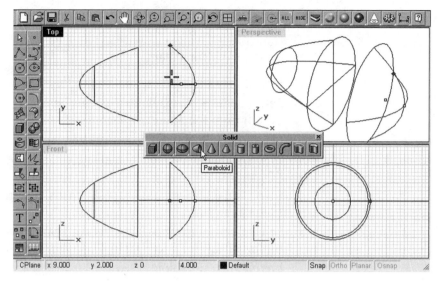

2 Select Solid > Cone, or the Cone button from the Solid toolbar.
Command: Cone

3 Select two points in the Top viewport to indicate the center and radius, and select a point in the Front viewport to indicate the height. (See figure 8-24.)

Fig. 8-24. Cone being constructed.

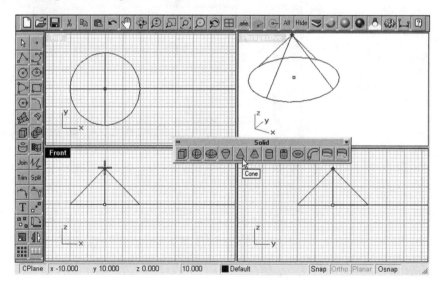

Truncated Cone

A truncated cone is a polysurface consisting of three joined surfaces. The slant surface is a closed surface and the top and bottom surfaces are circular planar surfaces. To construct a truncated cone, perform the following steps.

1 Select Solid > Truncated Cone, or the Truncated Cone button from the Solid toolbar.
Command: TCone

2 Select three points in the Top viewport to indicate the center and radii at both ends, and select a point in the Front viewport to indicate the height. (See figure 8-25.)

The cone is constructed. Save your file as *Cone.3dm*.

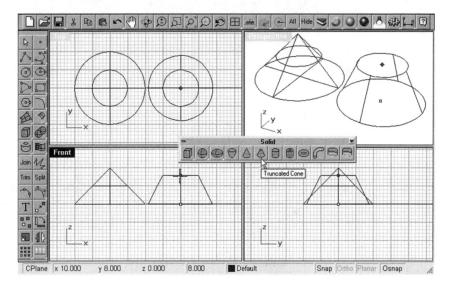

Fig. 8-25. Truncated cone being constructed.

Cylinder

A cylinder is a polysurface consisting of three surfaces. The body is a cylindrical surface and the top and bottom surfaces are circular planar surfaces. To construct a cylinder, perform the following steps.

1 Start a new file. Use the metric (millimeter) template.

2 Select Solid > Cylinder, or the Cylinder button from the Solid toolbar.
Command: Cylinder

3 Select two points in the Top viewport to indicate the center and radius, and select a point in the Front viewport to indicate the height. (See figure 8-26.)

Fig. 8-26. Cylinder being constructed.

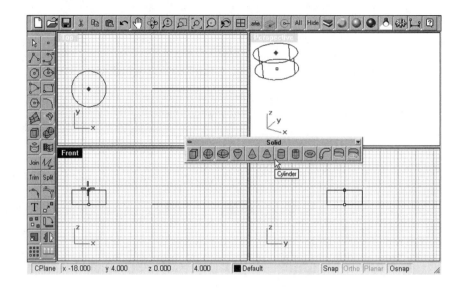

Tube

A tube is a polysurface consisting of two cylindrical surfaces and two planar surfaces. To construct a tube, perform the following steps.

1   Select Solid > Tube, or the Tube button from the Solid toolbar.
    Command: Tube

2   Select three points in the Top viewport to indicate the center, inner radius, and outer radius, and select a point in the Front viewport to indicate the height. (See figure 8-27.)

Fig. 8-27. Tube being constructed.

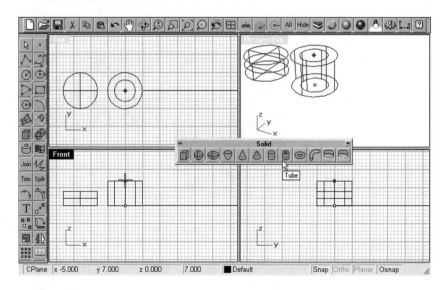

Torus

A torus is single closed-loop surface. To construct a torus, perform the following steps.

1   Select Solid > Torus, or the Torus button from the Solid toolbar.
    Command: Torus

2   Select three points in the Top viewport to indicate the center, radius of the torus, and radius of the tube. (See figure 8-28.)

Save your file as *Cylinder.3dm*.

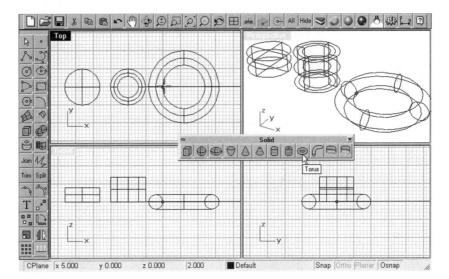

Fig. 8-28. Torus being constructed.

Pipe

A pipe is polysurface consisting of four surfaces: inner pipe, outer pipe, and two planar surfaces. To construct a pipe, perform the following steps.

1   Start a new file. Use the metric (millimeter) template.
2   Construct a free-form curve in the Top viewport. (See figure 8-29.)
3   Select Solid > Pipe, or the Pipe button from the Solid toolbar.
    Command: Pipe
4   Select the free-form curve.
5   Select a point in the Top viewport to indicate the starting radius.
6   Type *C* if the prompt is "Cap = No."
7   Type *T* if the prompt is "Thickness = No."
8   Select two points to indicate the starting inner and outer radii. (See figure 8-30.)

Fig. 8-29. Free-form curve constructed.

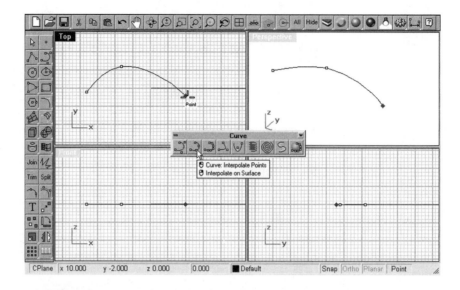

Fig. 8-30. Starting inner and outer radii being specified.

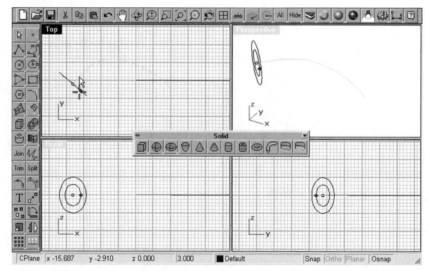

9 Press the Enter key (if the inner and outer radii at the other end of the pipe are the same as the starting end).

A pipe is constructed. (See figure 8-31.) Save your file as *Pipe.3dm*.

Text

A text object is a polysurface consisting of three elements: top and bottom planar surfaces that resemble the text, and surfaces joining the top and bottom surfaces. To construct a text object, perform the following steps.

Rhino Solid Modeling Tools

Fig. 8-31. Pipe constructed.

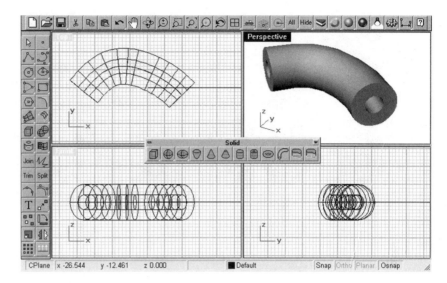

1. Start a new file. Use the metric (millimeter) template.

   Select Solid > Text, or the Text button of the Solid toolbar. (See figure 8-32.)

   Command: TextObject

Fig. 8-32. Text object constructed.

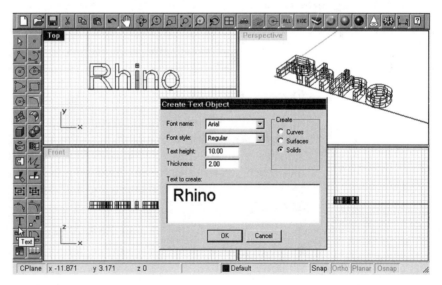

2. In the Create Text Object dialog box, type a text string, select a font and font style, specify text height and thickness, check the Solids box, and click on the OK button.

A text object is constructed. Save your file as *Text.3dm*.

Extruding and Filleting

There are two ways to construct an extruded solid: extruding a planar curve or extruding a surface. To modify the edge of a solid, you round it off by adding a fillet. Figure 8-33 shows the extrude and fillet commands in the Solid and Solid Tools toolbars.

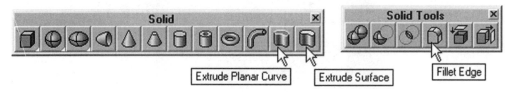

Fig. 8-33. Solid and Solid Tools toolbar extrude and fillet commands.

Table 8-2 outlines the functions of extrude and fillet commands and their location in the pull-down menu.

Table 8-2: Extrude and Fillet Commands and Their Functions

Command Name	Pull-down Menu	Function
Extrude	Solid > Extrude Planar Curve	For constructing a solid by extruding a planar curve
ExtrudeSrf	Solid > Extrude Surface	For constructing a solid by extruding a surface
FilletEdge	Solid > Fillet Edge	For rounding off the edges of a solid

Extrude Planar Curve

By extruding a closed planar curve, you construct a closed polysurface consisting of a surface extruded from the planar curve and two planar surfaces enclosing the extruded surface. If you extrude an open-loop planar curve, you get only an extruded surface. To see the difference between extruding a closed curve and an open planar curve, perform the following steps.

1   Start a new file. Use the metric (millimeter) template.
2   With reference to figure 8-34, construct an open-loop and a closed-loop planar curve.
3   Select Solid > Extrude Planar Curve, or the Extrude Planar Curve button from the Solid toolbar.
    Command: Extrude
4   Select the curves and press the Enter key.
5   Type C to cap the ends if the prompt says "Cap = No."

Fig. 8-34. Curves constructed.

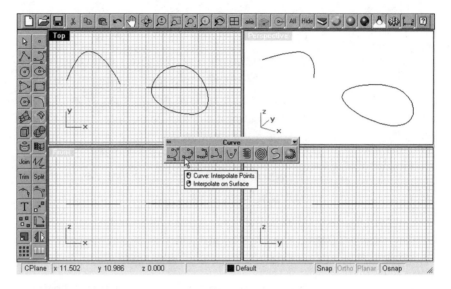

6 Select a point in the Front viewport to specify the extrusion height. (See figure 8-35.)

Fig. 8-35. Curves being extruded.

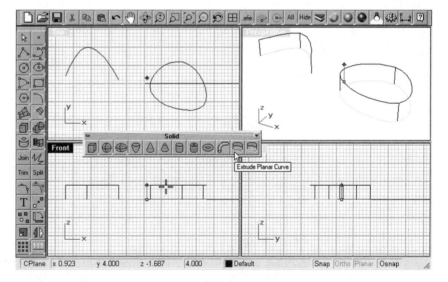

Note that extruding an open-loop curve creates a single surface. Save your file as *Extrude.3dm*.

Extrude Surface

By extruding a surface, you construct a polysurface consisting of two copies of the original surface at a distance apart and a set of surfaces joining their edges. To construct a solid by extruding a surface, perform the following steps.

1. Open the file *SrfPtGrid.3dm* you constructed in Chapter 7.
2. Set the display to a four-viewport layout.
3. Select File > Save As and specify a file name of *ExtrudeSrf.3dm*.
4. Select Solid > Extrude Surface, or the Extrude Surface button from the Solid toolbar.
   Command: ExtrudeSrf
5. Select the surface in the Front viewport and press the Enter key.

The surface will be extruded in a direction perpendicular to the construction plane.

6. Select a point in the Right viewport to indicate an extruded distance. (See figure 8-36.)

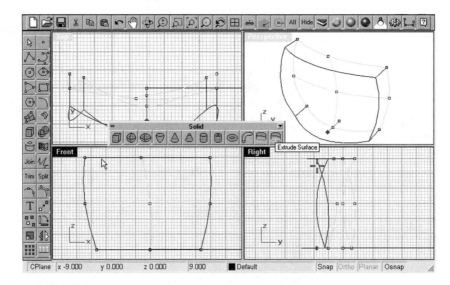

Fig. 8-36. Surface being extruded.

Fillet Edge

Adding a fillet to edges of a closed polysurface produces filleted surfaces at the selected edges of the polysurface. To construct filleted edges, perform the following steps.

1. Select Solid > Fillet Edge, or the Fillet Edge button from the Solid Tools toolbar.
   Command: FilletEdge
2. Select two points in the Front viewport to indicate the fillet radius.
3. Select the edges indicated in figure 8-37 and press the Enter key.

Rhino Solid Modeling Tools

The edges are filleted. Save your file.

Fig. 8-37. Edges being filleted.

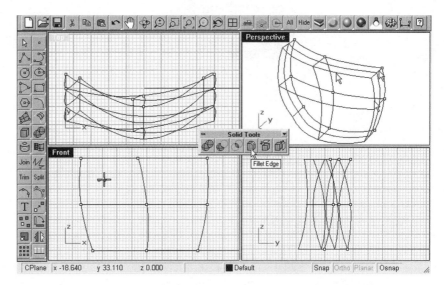

Capping Planar Holes of a Polysurface

Fig. 8-38. Cap Planar Holes button of the Solid Tools toolbar.

As mentioned, a Rhino solid is a closed polysurface. An existing polysurface that does not form a closed volume because of planar holes can be capped to close the holes. This is done with the Cap Planar Holes button of the Solid Tools toolbar, shown in figure 8-38. Table 8-3 outlines the function of the Cap command and its location in the pull-down menu.

Table 8-3: Cap Command and Its Function

Command Name	Pull-down Menu	Function
Cap	Solid > Cap Planar Holes	For constructing a solid by capping planar holes of a polysurface.

To construct a solid by capping planar holes, perform the following steps.

1 Open the file *Extrude.3dm*.

2 Select File > Save As and specify a file name of *Cap.3dm*.

In this file, there are two objects: a surface and a polysurface. To add a planar surface, continue with the following steps.

3 Select Surface > Corner Points, or the Surface from 3 or 4 Corner Points button from the Surface toolbar.
Command: **SrfPt**

4 Select (clockwise or counterclockwise) the four corner points indicated in figure 8-39.

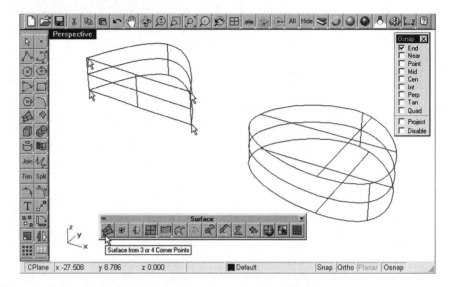

Fig. 8-39. Planar surface constructed.

To join the two surfaces, continue with the following steps.

5 Select Edit > Join, or the Join button from the Main toolbar.
Command: Join

6 Select the surfaces indicated in figure 8-40 and press the Enter key.

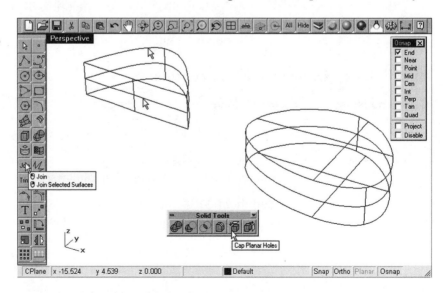

Fig. 8-40. Polysurface being capped.

To convert the polysurface to a solid, you will cap the planar holes, as follows.

Rhino Solid Modeling Tools

7 Select Solid > Cap Planar Holes, or the Cap Planar Holes button from the Solid Tools.
Command: Cap

8 Select the polysurface and press the Enter key.

The polysurface is capped. Save your file. Note that only planar holes can be capped this way. If your polysurface has a lot of openings that are not planar holes, you have to construct surfaces and join them in order to form a solid. Save your file.

Combining Solid Objects

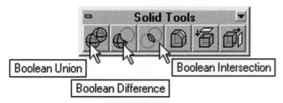

Fig. 8-41. Boolean commands on the Solid Tools toolbar.

There are not many objects that can be represented simply by one of the primitives previously delineated. By combining solids in one of the three ways (union, difference, and intersection), you get a more complex object. Figure 8-41 shows Boolean operation commands of the Solid Tools toolbar.

Table 8-4 outlines the functions of Boolean operation commands and their location in the pull-down menu.

Table 8-4: Boolean Operation Commands and Their Functions

Command Name	Pull-down Menu	Function
BooleanUnion	Solid > Union	For constructing a solid having the volume of all selected solids
BooleanDifference	Solid > Difference	For constructing a solid that has a volume enclosed by the first set of selected solids but not the second set of selected solids
BooleanIntersection	Solid > Intersection	For constructing a solid that has a volume enclosed by the first and second sets of selected solids

Union

A union of a set of solids produces a solid that has the volume of all solids in the set. To construct a unioned solid, perform the following steps.

1 Open the file *Ellipsoid.3dm*.

2 Select File > Save As and specify a file name of *Union.3dm*.

3 Maximize the Perspective viewport.

4 Select Solid > Union, or the Boolean Union button from the Solid Tools toolbar.

Command: BooleanUnion

5 Select the ellipsoids and press the Enter key. (See figure 8-42.)

The solids are united. Save your file.

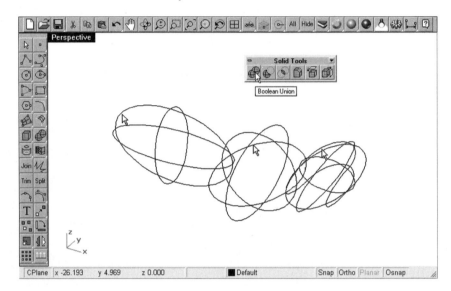

Fig. 8-42. Solids being united.

Difference

A difference of two sets of solids produces a solid that has the volume contained in the first set of solids but not the second set of solids. To construct a differenced solid, perform the following steps.

1 Open the file *Sphere.3dm*.
2 Select File > Save As and specify a file name of *Difference.3dm*.
3 Maximize the Perspective viewport.
4 Select Solid > Difference, or the Boolean Difference button from the Solid Tools toolbar.

Command: BooleanDifference

5 Select the sphere indicated in figure 8-43 and press the Enter key.
6 Select the sphere indicated in figure 8-44 and press the Enter key.

The second selected sphere is subtracted from the first selected sphere.

Rhino Solid Modeling Tools

Fig. 8-43. First set of objects selected.

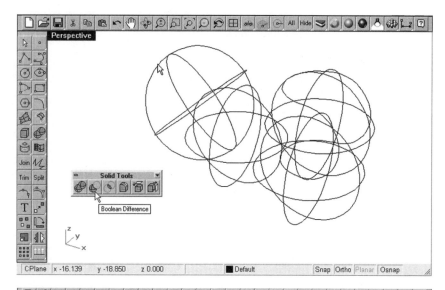

Fig. 8-44. Second set of objects selected.

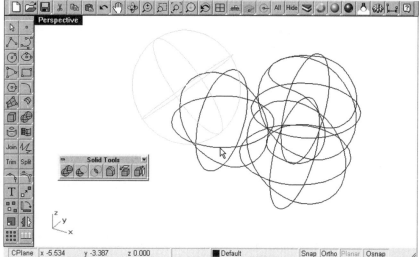

Intersection

An intersection of two sets of solids produces a solid that has the volume contained in the first set of solids and the second set of solids. To construct an intersected solid, perform the following steps.

1 Select Solid > Intersection, or the Boolean Intersection button from the Solid Tools toolbar.

Command: BooleanIntersection

2 Select the left sphere indicated in figure 8-45 and press the Enter key.

Fig. 8-45. Objects to be selected.

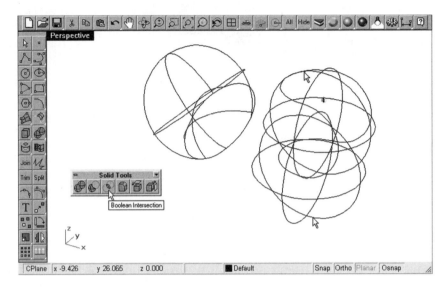

3 Select the right sphere indicated in figure 8-45 and press the Enter key.

An intersection of the two selected spheres is constructed. Save your file.

Editing Polysurfaces

There are several ways to edit polysurfaces. By joining surfaces to a closed polysurface, a solid is implied because a Rhino solid is a closed polysurface. To revert a solid or polysurface to individual surfaces, you explode the object. To separate a surface from a solid or polysurface without exploding it, you extract a surface.

In addition, you trim a solid or polysurface to remove a portion of the surfaces or split a solid or polysurface to split it into two objects. Because a solid is a polysurface, extracting a surface from a solid and trimming and splitting a solid create open polysurfaces.

Joining and Exploding

In Chapter 6, you learned how to join surfaces to a polysurface. If the polysurface encloses a volume without any gap or opening, a solid is implied. Contrary to joining, you explode a polysurface into individual surfaces.

Extracting a Surface from a Solid or Polysurface

Unlike exploding, which breaks down a joined polysurface into individual surfaces, extracting separates the selected surfaces from the polysurface, leaving the remaining surfaces joined. Figure 8-46 shows the Extract Surface command on the Solid Tools toolbar.

Rhino Solid Modeling Tools

Fig. 8-46. Extract Surface command on the Solid Tools toolbar.

Table 8-5 outlines the functions of the Extract Surface command and its location in the pull-down menu.

Table 8-5: Extract Surface Command and Its Function

Command Name	Pull-down Menu	Function
ExtractSrf	Solid > Extract Surface	For constructing a surface by duplicating a surface of a polysurface

To extract a surface from a solid, perform the following steps.

1. Start a new file. Use the metric (millimeter) template.
2. With reference to figure 8-47, construct a solid box.
3. Select Solid > Extract Surface, or the Extract Surface button from the Solid Tools toolbar.
   Command: ExtractSrf
4. Select the surface indicated in figure 8-47 and press the Enter key.

A surface is extracted. To view the extracted surface clearly, you will hide the box. Continue with the following steps.

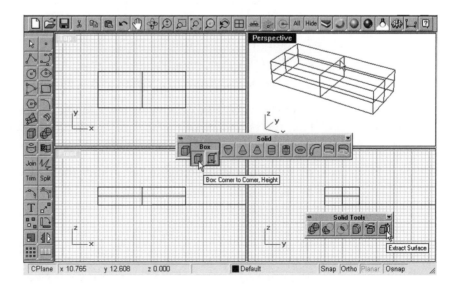

Fig. 8-47. Box constructed and current layer set.

5 Select Edit > Visibility > Hide, or the Hide Objects button of the Visibility toolbar.

Command: Hide

6 Select the extracted surface, if it is not already selected.

The extracted surface is hidden. (See figure 8-48.) Save your file as *ExtractSrf.3dm*.

Fig. 8-48. Extracted surface hidden.

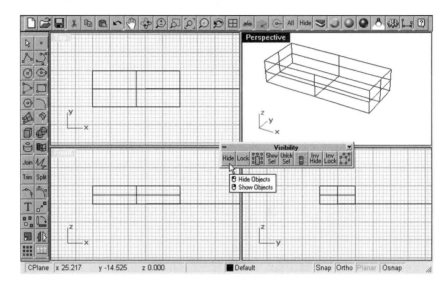

Trimming

To construct a box and use a surface to trim a portion of it, perform the following steps.

1 Start a new file. Use the metric (millimeter) template.
2 With reference to figure 8-49, construct a box and a free-form curve.
3 Extrude the curve to a surface. (See figure 8-50.)

Now you have a box and a surface. To use the surface to cut the solid box, the boundary of the surface has to lie outside the box. Continue with the following steps.

4 With reference to figure 8-51, move the surface.
5 Select Edit > Trim, or the Trim button from the Main toolbar.

Command: Trim

6 Select the surface and press the Enter key.
7 Select the box and press the Enter key. (See figure 8-51.)

The selected portion of the box is trimmed. Note that the trimmed box becomes an open polysurface. Save your file as *TrimSolid.3dm*.

Rhino Solid Modeling Tools 381

Fig. 8-49. Box and free-form curve constructed.

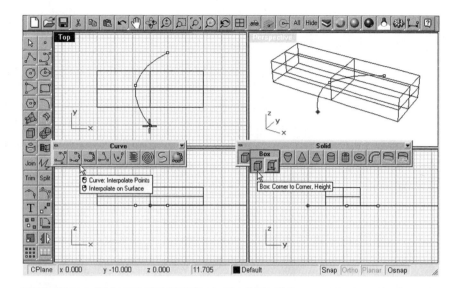

Fig. 8-50. Curve extruded.

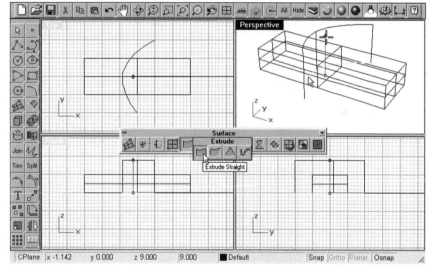

To turn the trimmed box (now a polysurface) into a solid, continue with the following steps.

8 Select Edit > Trim, or the Trim button from the Main toolbar.
   Command: Trim

9 Select the polysurface and press the Enter key.

10 With reference to figure 8-52, select the surface and press the Enter key.

11 Select Edit > Join, or the Join button of the Main toolbar.
   Command: Join

Fig. 8-51. Box being trimmed by the surface.

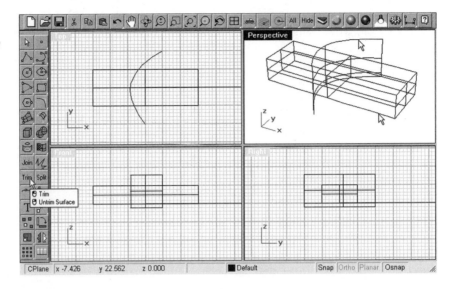

Fig. 8-52. Surface being trimmed.

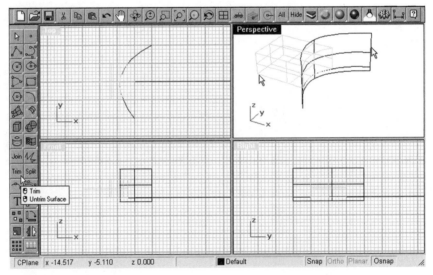

12 Select the surface and polysurface and press the Enter key. (See figure 8-53.)

A solid is formed. Save your file.

Splitting

To construct a box and use a surface to split it into two polysurfaces, perform the following steps.

1 Start a new file. Use the metric (millimeter) template.

Fig. 8-53. Surface and polysurface being joined.

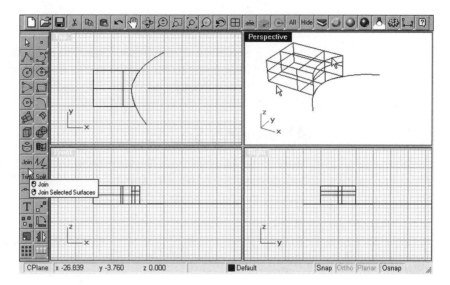

2 With reference to figures 8-49 through 8-51, construct a box and a curve, extrude the curve to a surface, and move the extruded surface so that its boundary lies completely outside the box boundary.

3 Select Edit > Split, or the Split button from the Main toolbar.
Command: Split

4 Select the box.

5 Select the surface and press the Enter key. (See figure 8-54.)

The box is split into two polysurfaces. After splitting, you get two open polysurfaces. If you want to get two solids from the polysurfaces, refer to figures 8-52 and 8-53. Save your file as *SplitSolid.3dm*.

Fig. 8-54. Box being split.

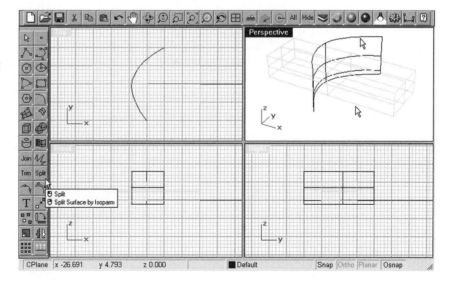

Transforming Polysurfaces

With the exception of the Twist, Bend, Taper, Flow, and Smooth commands (which apply to individual surfaces only), you use the transform commands on polysurfaces to translate and deform them. See chapters 3 and 6 on using the transform commands.

Analyzing Solids

Most of the analysis tools you learned in Chapter 7 that apply to surfaces and polysurfaces are also applicable to solids. Figure 8-55 shows the analysis commands specific to solids.

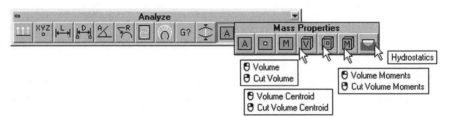

Fig. 8-55. Analysis commands specific to solids.

Table 8-6 outlines the functions of the analysis commands and their location in the pull-down menu.

Table 8-6: Analysis Commands and Their Functions

Command Name	Pull-down Menu	Function
Volume	Analyze > Mass Properties > Volume	For evaluating the volume of a solid
VolumeCentroid	Analyze > Mass Properties > Volume Centroid	For evaluating the volume centroid of a solid
VolumeMoments	Analyze > Mass Properties > Volume Moments	For evaluating a solid's volume, volume centroid, volume moments (first, second, and product), volume moments of inertia about world coordinate axes, volume radii of gyration about world coordinate axes, volume moments of inertia about centroid coordinate axes, and volume radii of gyration about centroid coordinate axes
Hydrostatics	Analyze > Mass Properties > Hydrostatics	For evaluating the hydrostatics values of surfaces, with the waterline set at the horizontal plane of the world coordinate system

Rhino Solid Modeling Tools

Volume, Volume Centroid, and Volume Moments

To construct a solid box and evaluate its volume, volume centroid, and volume moments, perform the following steps.

1. Start a new file. Use the metric (millimeter) template.
2. With reference to figure 8-56, construct a box.
3. Select Analyze > Mass Properties > Volume, or the Volume button from the Mass Properties toolbar.
   Command: Volume
4. Select the box and press the Enter key.

Volume information is displayed in the command line area.

Fig. 8-56. Volume information.

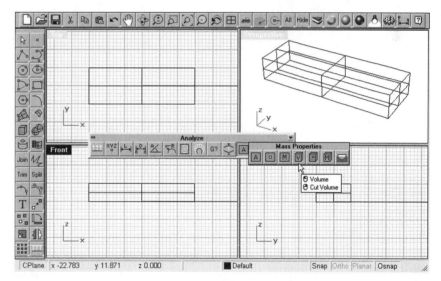

5. Select Analyze > Mass Properties > Volume Centroid, or the Volume Centroid button from the Mass Properties toolbar.
   Command: VolumeCentroid
6. Select the box and press the Enter key.

Volume centroid information is displayed. Continue with the following steps.

7. Select Analyze > Mass Properties > Volume Moments, or the Volume Moments button from the Mass Properties toolbar.
   Command: VolumeMoments
8. Select the box and press the Enter key.

Volume moment information is displayed.

Hydrostatics

Presuming the waterline is set at the world coordinate plane, you can evaluate the hydrostatic value of a solid, as follows.

1 Start a new file. Use the metric (millimeter) template.

2 With reference to figure 8-57, construct an ellipsoid.

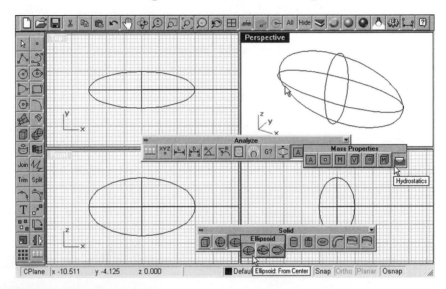

Fig. 8-57. Ellipsoid constructed.

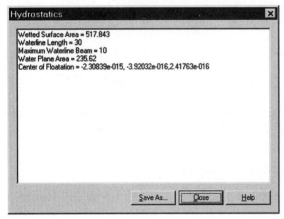

3 Select Analyze > Mass Properties > Hydrostatic, or the Hydrostatics button from the Mass Properties toolbar

Command: Hydrostatics

4 Select the ellipsoid and press the Enter key.

Hydrostatics information is displayed in the Hydrostatics dialog box. (See figure 8-58.)

Fig. 8-58. Hydrostatics dialog box.

■■ Solid Modeling Projects

This section offers solid modeling projects that will deepen your knowledge and enhance your skills. These projects include model car parts, a joypad, a joystick, and a medal.

Model Car Parts

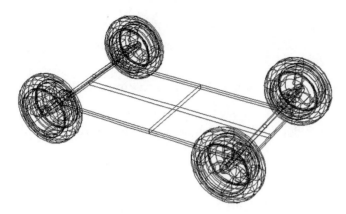

The model car designed in this section has a number of component parts: tire, wheel, base, and shaft. (See figure 8-59.) You will construct solid models of the component parts in this chapter and in the next chapter merge them to form a single file representing the assembled model car.

Fig. 8-59. Tire, wheel, shaft, and base.

Tire

To construct the solid model of a tire, perform the following steps. You will construct a solid tube and fillet its edges.

1 Start a new part file. Use the metric (millimeter) template.

2 Check the Snap pane of the status bar.

3 Select Solid > Tube, or the Tube button from the Solid Tools toolbar.
Command: Tube

4 Select the origin point in the Front viewport (figure 8-60) to specify the center.

Fig. 8-60. Tube being constructed.

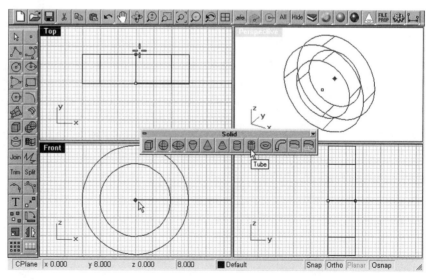

5 Type *10* at the command line area to specify the radius of the tube.

388   CHAPTER 8: **Solid Modeling**

6 Type *15* at the command line area to specify the second radius of the tube.

7 Type *8* at the command line area to specify the height of the tube.

8 Select the point indicated in the Top viewport (figure 8-60) to specify the direction.

A solid tube is constructed. To fillet its edges, continue with the following steps.

9 Select Solid > Fillet Edge, or the Fillet Edge button from the Solid Tools toolbar.
Command: FilletEdge

10 Type *R* at the command line area to use the Radius option.

11 Type *2.5* at the command line area to set the fillet radius to 2.5 units.

12 Select the edges indicated in figure 8-61 and press the Enter key.

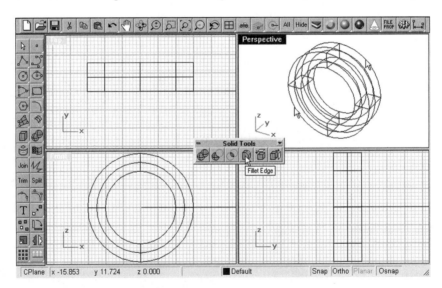

Fig. 8-61. First edges being filleted.

13 Repeat the FilletEdge command.

14 Type *R* and then *1* at the command line area to set the fillet radius to 1 unit.

15 Select the edges indicated in figure 8-62 and press the Enter key.

The edges are filleted (figure 8-63) and the solid model for the tire is complete. Save your file as *Tire.3dm*.

Solid Modeling Projects

Fig. 8-62. Second edges being filleted.

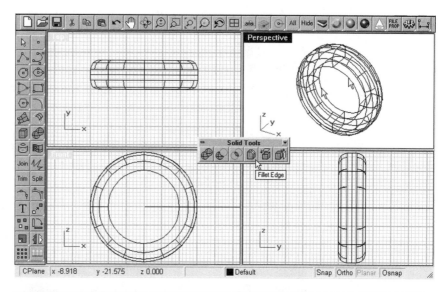

Fig. 8-63. Edges filleted.

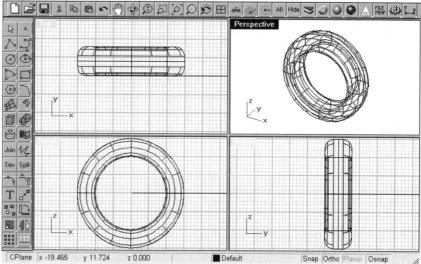

Wheel

In the following, you will construct the solid model of a wheel. This is a complex solid model consisting of cylinders, ellipsoids, paraboloid, and truncated cone combined using Boolean union and Boolean difference operations.

1 Start a new file. Use the metric (millimeter) template.

2 Select Solid > Cylinder, or the Cylinder button from the Solid Tools toolbar.

Command: Cylinder

3   Select the origin point in the Front viewport (figure 8-64) to specify the center.
4   Type *10* at the command line area to specify the radius.
5   Type *8* at the command line area to specify the height of the cylinder.
6   Select a point in the Top viewport indicated in figure 8-64 to specify the direction.

Fig. 8-64. Cylinder being constructed.

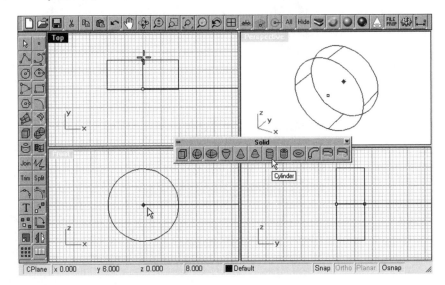

7   Repeat the Cylinder command.
8   Select the origin point in the Front viewport (concentric to the last cylinder) to specify the center.
9   Type *11* at the command line area to specify the radius.
10 Type *1* at the command line area to specify the height.
11 Select the point indicated in figure 8-65 to specify the direction.
12 With reference to figure 8-66, copy a cylinder.

To unite the three solid cylinders into a single solid, continue with the following steps. (See figure 8-67.)

13 Select Solid > Union, or the Boolean Union button from the Solid Tools toolbar.
    Command: Boolean
14 Select the cylinders and press the Enter key.

To construct a cylinder and subtract it from the union of the three cylinders, continue with the following steps.

Solid Modeling Projects

Fig. 8-65. Second cylinder being constructed.

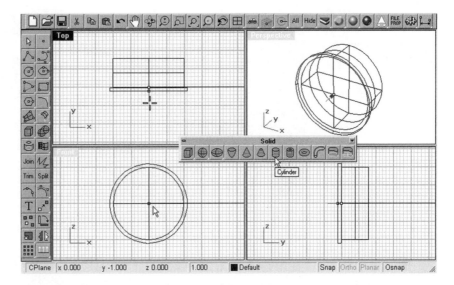

Fig. 8-66. Cylinder being copied.

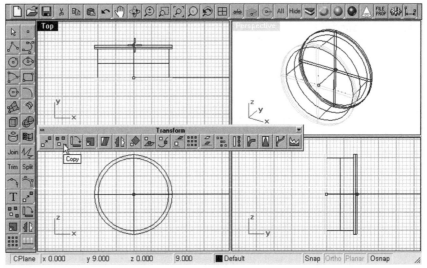

15 Select Solid > Cylinder, or the Cylinder button from the Solid Tools toolbar.

Command: Cylinder

16 Select the point indicated in the Top viewport of figure 8-68 to specify the center.

17 Type *8* at the command line area to specify the radius.

18 Type *10* at the command line area to specify the height.

19 Select the point indicated in the Top viewport shown in figure 8-68 to specify the direction.

Fig. 8-67. Cylinders being united.

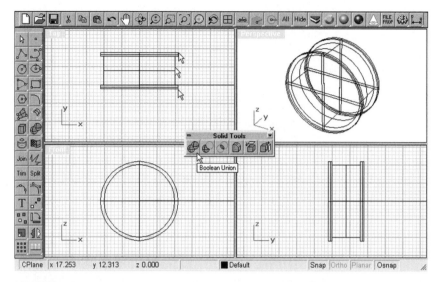

Fig. 8-68. Cylinder being constructed.

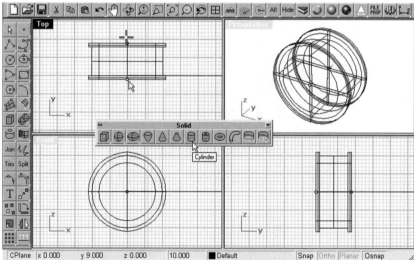

20 Select Solid > Difference, or the Boolean Difference button from the Solid Tools toolbar.

Command: BooleanDifference

21 Select solid A in figure 8-69 and press the Enter key.

22 Select solid B in figure 8-69 and press the Enter key.

To construct an ellipsoid and an array of the ellipsoid, continue with the following steps.

23 Maximize the Front viewport.

Solid Modeling Projects

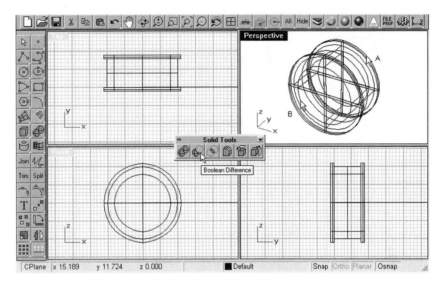

Fig. 8-69. Cylinder being subtracted.

24 Select Solid > Ellipsoid > From Center, or the Ellipsoid/From Center button from the Ellipsoid toolbar.

Command: Ellipsoid

25 Select the point 0,5 in the Front viewport to specify the center.

26 Type *r4 < 270* at the command line area to specify the end of the first axis.

27 Type *r2 < 0* at the command line area to specify the end of the second axis.

28 Type *4* at the command line area to specify the distance of the third axis. (See figure 8-70.)

29 Return to a four-viewport display.

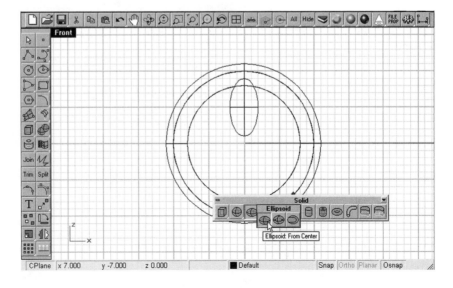

Fig. 8-70. Ellipsoid constructed.

Fig. 8-71. Ellipsoid moved.

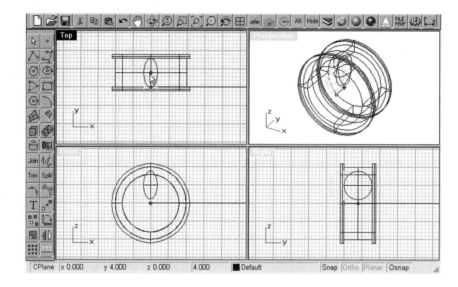

30 Select and drag the ellipsoid to move it to a new location in accordance with figure 8-71.

31 Select Transform > Array > Polar, or the Polar Array button from the Array toolbar.
Command: ArrayPolar

32 Select the ellipsoid and press the Enter key.

33 Select the origin point in the Front viewport to specify the center of the polar array. (See figure 8-72.)

34 Type 3 at the command line area to specify the number of elements.

Fig. 8-72. Ellipsoid arrayed.

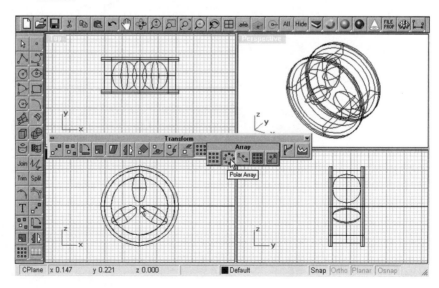

Solid Modeling Projects

35 Type *360* at the command line area to specify the angle to be filled. (If the default angle is 360, press the Enter key to accept.) To construct a paraboloid and a truncated cone, continue with the following steps.

36 Maximize the Top viewport.

37 Select Solid > Paraboloid > Focus, Direction, or the Paraboloid button from the Solid Tools toolbar.
Command: Paraboloid

38 Select any point in the Top viewport to specify the paraboloid focus.

39 Type *r1 < 90* at the command line area to specify the paraboloid direction.

40 Type *r2,2* at the command line area to specify the paraboloid end. (See figure 8-73.)

Fig. 8-73. Paraboloid constructed.

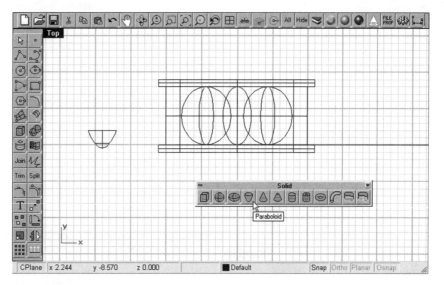

41 Select Solid > Truncated Cone, or the Truncated Cone button from the Solid Tools toolbar.
Command: TCone

42 Select the point indicated in figure 8-74 to indicate the base of the truncated cone.

43 Type *2* at the command line area to specify a radius.

44 Type *3* at the command line area to specify the second radius.

45 Type *r8 < 90* at the command line area to specify the end location of the cone.

*Fig. 8-74.
Truncated cone
constructed.*

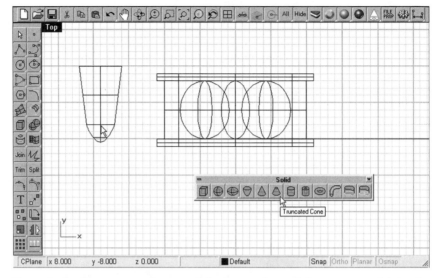

To explode the truncated cone, delete a surface of the cone, and join the remaining surfaces of the truncated cone with the paraboloid to form a solid, continue with the following steps.

46 Maximize the Perspective viewport.

47 Select Edit > Explode, or the Explode button from the Main toolbar.
Command: Explode

48 Select the truncated cone and press the Enter key.

49 Select the surface indicated in figure 8-75 and press the Enter key.

50 Select Edit > Join, or the Join button from the Main toolbar.
Command: Join

*Fig. 8-75.
Truncated cone
being exploded.*

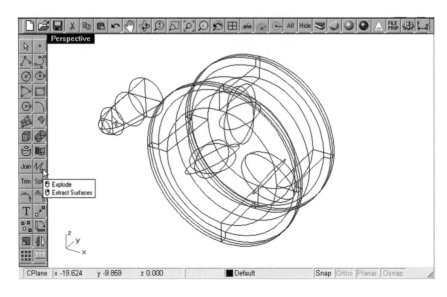

Solid Modeling Projects

51 Select the surfaces indicated in figure 8-76 and press the Enter key.

Fig. 8-76. Paraboloid and truncated cone surfaces being joined.

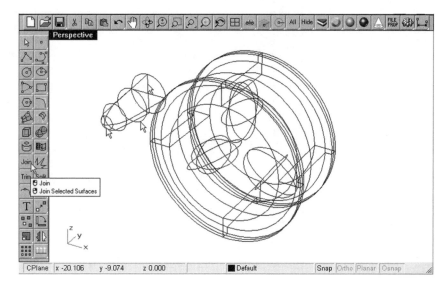

To move the combined paraboloid and truncated cone and unite it with the other solids to form a single solid, continue with the following steps.

52 Set the display to 4-viewport.

53 Select the joined paraboloid and the truncated cone and drag them to the origin location in accordance with figure 8-77.

Fig. 8-77. Paraboloid and truncated cone moved.

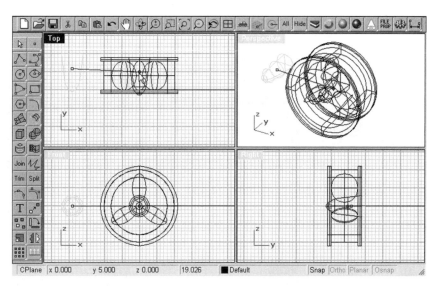

54 Select Solid > Union, or the Boolean Union button from the Solid Tools toolbar.

Command: BooleanUnion

55 Select all solid objects and press the Enter key. (See figure 8-78.)

Fig. 8-78. Solids being united.

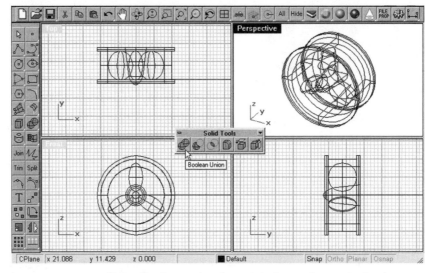

To construct a solid cylinder and subtract it from the main body, continue with the following steps.

56 Maximize the Top viewport.

57 Select Solid > Cylinder, or the Cylinder button from the Solid Tools toolbar.

Command: Cylinder

58 Type *0,4* at the command line area to specify the center.

59 Type *1* at the command line area to specify the radius.

60 Type *r10<90* at the command line area to specify the end of the cylinder. (See figure 8-79.)

Fig. 8-79. Cylinder constructed.

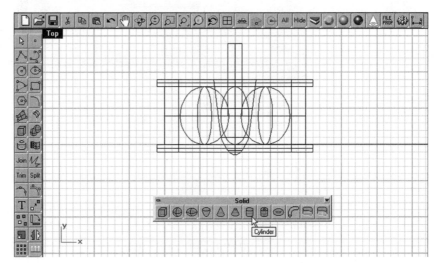

Solid Modeling Projects 399

61 Return to a four-viewport display.

62 Select Solid > Difference, or the Boolean Difference button from the Solid Tools toolbar.

Command: BooleanDifference

63 Select solid A indicated in figure 8-80 and press the Enter key.

64 Select solid B indicated in figure 8-80 and press the Enter key.

Fig. 8-80. Cylinder being subtracted.

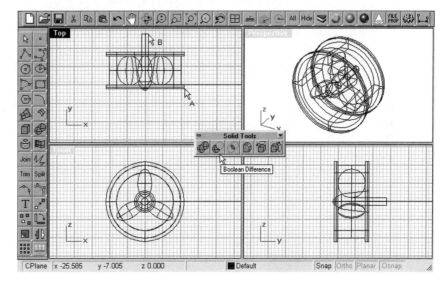

The solid model for the wheel is complete. (See figure 8-81.) Save your file as *Wheel.3dm*.

Fig. 8-81. Wheel constructed.

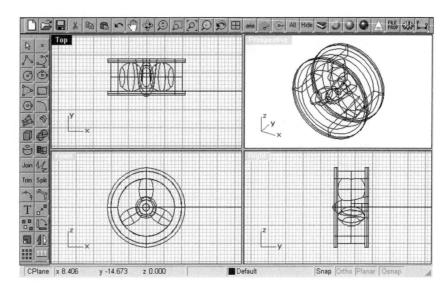

Base

To construct the solid model of the base of the car, perform the following steps.

1 Start a new file. Use the metric (millimeter) template.

2 With reference to the dimensions shown in figure 8-82, construct a polyline in the Right viewport and a circle and polyline in the Front viewport. (Do not add the dimensions to your curves. You will learn dimensioning in Chapter 10.)

Fig. 8-82. Polylines constructed in the Front and Right viewports.

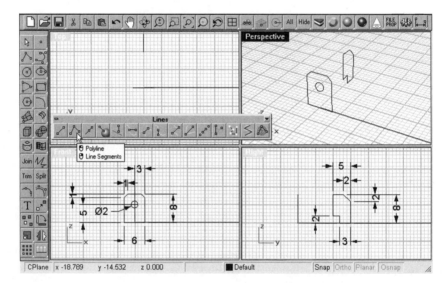

To construct two extruded solids, continue with the following steps.

3 Maximize the Perspective viewport

4 Select Solid > Extrude Planar Curve, or the Extrude Planar Curve button from the Solid Tools toolbar.
Command: Extrude

5 Select the curves indicated in figure 8-83 and press the Enter key.

6 Type *-10* at the command line area to specify the extrusion distance. (A distance of -10 here means in the negative Y direction.)

7 Repeat the Extrude command.

8 Select the polyline indicated in figure 8-84 and press the Enter key.

9 Type *10* at the command line area to specify the extrusion distance.

10 Select and drag the solid indicated in figure 8-85 to a new location.

Combine the two solids using a Boolean intersection, as follows.

Solid Modeling Projects

Fig. 8-83. Curves being extruded.

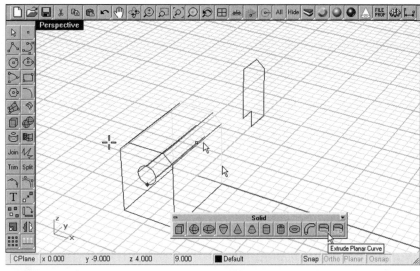

Fig. 8-84. Polyline being extruded.

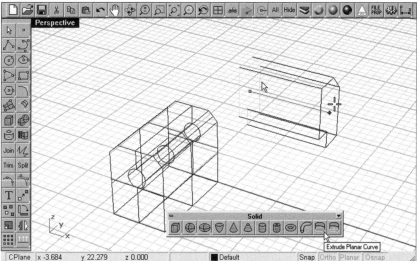

11 Select Solid > Intersection, or the Boolean Intersection button from the Solid Tools toolbar.

Command: BooleanIntersection

12 Select solid A indicated in figure 8-86 and press the Enter key.

13 Select solid B indicated in figure 8-86 and press the Enter key.

The curves for making the extruded solids are not required. You will hide them, as follows.

14 Select Edit > Visibility > Hide, or the Hide Objects button from the Visibility toolbar.

Command: Hide

Fig. 8-85. Solid being dragged to a new position.

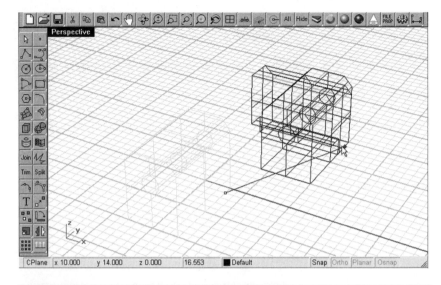

Fig. 8-86. Solids being combined by Boolean intersection.

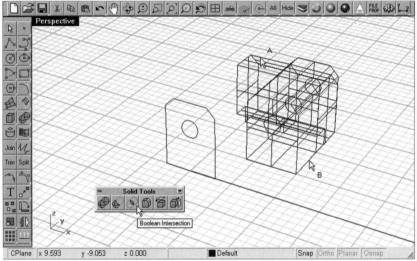

15 Select curves A, B, and C indicated in figure 8-87.

16 Select Transform > Copy, or the Copy button from the Transform toolbar.

Command: Copy

17 Select solid D indicated in figure 8-87 and press the Enter key.

18 Select any point in the viewport to specify a point to copy from.

19 Type *r84<0* at the command line area to specify the distance and direction of the copy.

20 Press the Enter key.

To construct a box, continue with the following steps.

Solid Modeling Projects

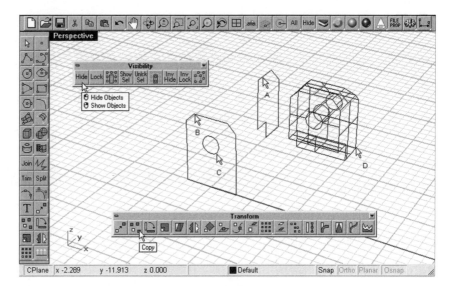

Fig. 8-87. Curves being hidden and solid being copied.

21 Turn off grid and grid axes display.
22 Zoom the display in accordance with figure 8-88.
23 Select Solid > Box > Corner to Corner, Height, or the Box/Corner to Corner, Height button from the Box toolbar.
Command: Box
24 Select the point indicated in figure 8-88 to specify the first corner of the box.
25 Type *r90,44* at the command line area to specify the other corner.
26 Type *2* at the command line area to specify the height.

To move the solid box, continue with the following steps.

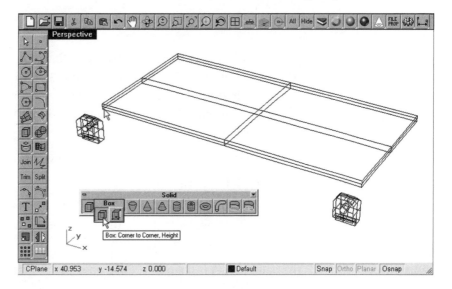

Fig. 8-88. Solid box constructed.

27 Select Transform > Move, or the Move button from the Transform toolbar.

Command: Move

28 Select the box and press the Enter key.

29 Select end point A indicated in figure 8-89 to specify the point to move from.

30 Select end point B indicated in figure 8-89 to specify the point to move to.

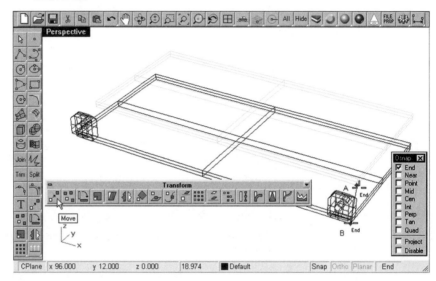

Fig. 8-89. Solid box being moved.

31 Select Transform > Mirror, or the Mirror button from the Transform toolbar.

Command: Mirror

32 Select solids A and B indicated in figure 8-90 and press the Enter key.

33 Check the Mid box in the Osnap toolbar.

34 Select mid point C indicated in figure 8-90 to specify the start of mirror plane.

35 Select mid point D indicated in figure 8-90 to specify the end of mirror plane.

To unite the solids, continue with the following steps.

36 Select Solid > Union.

Command: Boolean

37 Select the solids and press the Enter key. The completed solid model is shown in figure 8-91.

The solid model is complete. Save your file as *Base.3dm*.

Solid Modeling Projects   405

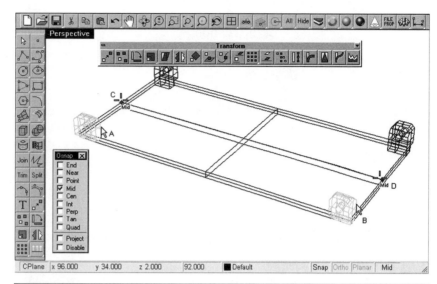

Fig. 8-90. Solids being mirrored.

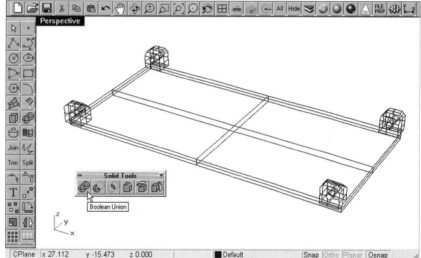

Fig. 8-91. Completed model.

Shaft

In the following, you will construct the solid model of the shaft of the car. It is a cylinder.

1   Start a new file. Use the metric (millimeter) template.

2   Select Solid > Cylinder, or the Cylinder button from the Solid toolbar.
      Command: Cylinder

3   Select any point in the Front viewport to specify the base.

4   Type *1* at the command line area to specify the radius.

5 Type *r70 < 90* at the command line area to specify the end position of the cylinder.

The solid model is complete. (See figure 8-92.) Save your file as *Shaft.3dm*.

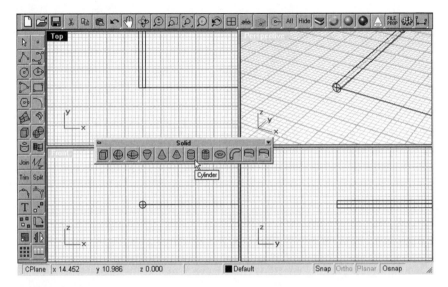

Fig. 8-92. Solid cylinder constructed.

Joypad

The set of surfaces representing the joypad model you constructed in the last chapter already encloses a volume. Therefore, you can join the individual surfaces to form a solid. However, there may be some minute gaps between contiguous surfaces that prevent the surfaces from joining properly. To rectify any such problem, you use the edge tools to join the naked edges of those problematic contiguous surfaces, as follows.

1 Open the file *Joypad.3dm*.
2 Select Edit > Join, or the Join button from the Main toolbar.
Command: Join
3 With reference to figure 8-93, select the contiguous surfaces at their edges.

As can be seen in the command line area, the program is unable to join this object. To join the naked edges, continue with the following steps.

4 Select Analyze > Edge Tools > Join 2 Naked Edges, or the Join 2 Naked Edges button from the Edge Tools dialog box.
Command: JoinEdge
5 Select the edge twice.
6 In the Edge Joining dialog box, select the Yes button.

Solid Modeling Projects

Fig. 8-93. Joining naked edges of contiguous surfaces.

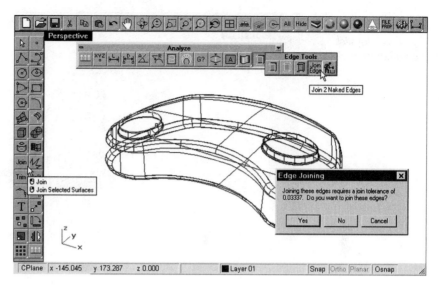

To join all of the surfaces, continue with the following steps.

7 Select Edit > Join, or the Join button from the Main toolbar.
Command: Join

8 Select a surface and then its contiguous surfaces one by one until all surfaces are selected. (If you encounter any problem during joining, use the JoinEdge command.)

A solid is constructed. Save your file.

Joystick

In the following, you will cap the planar hole in the joystick model to convert it to a solid.

1 Open the file *Joystick.3dm*.
2 Maximize the Perspective viewport.
3 Select Solid > Cap Planar Holes, or the Cap Planar Holes button from the Solid Tools toolbar.
Command: Cap
4 Select the surface indicated in figure 8-94 and press the Enter key.

The planar hole in the joystick model is capped. A solid is formed. Save your file.

Medal

In the following, you will construct a solid model of a medal.

1 Open the file *Heightfield.3dm* that you constructed in Chapter 5.

Fig. 8-94. Joystick being capped.

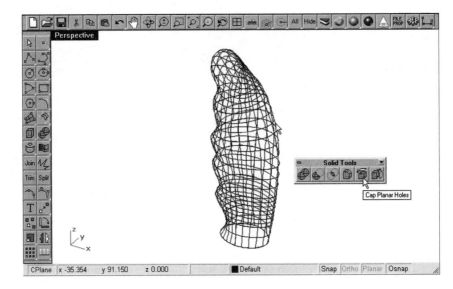

2   Select Solid > Cylinder, or the Cylinder button from the Solid toolbar.

Command: Cylinder

3   Select a point in the Top viewport to specify the center.
4   Type *V* at the command line area to use the Vertical option.
5   Select the point indicated in the Top viewport shown in figure 8-95 to specify the radius.
6   Select the point indicated in the Front viewport shown in figure 8-95 to specify the height.

Fig. 8-95. Cylinder being constructed.

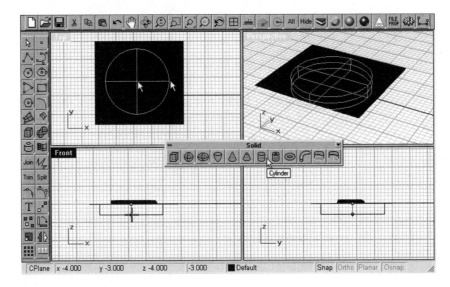

Solid Modeling Projects

A cylinder is constructed. To trim the cylinder and the heightfield surface, continue with the following steps.

7 With reference to figure 8-96, select and drag the cylinder in the Front viewport to move it to a new location so that it overlaps with the heightfield surface.

8 Select Edit > Trim, or the Trim button from the Main toolbar.
Command: Trim

9 Select the cylinder and the surface and press the Enter key.

10 Select the portion of the surface and the cylinder indicated in figure 8-97 and press the Enter key.

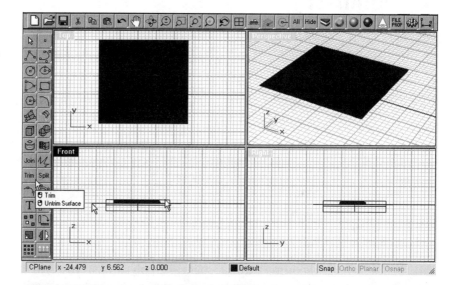

Fig. 8-96. Cylinder moved, and surface and cylinder being trimmed.

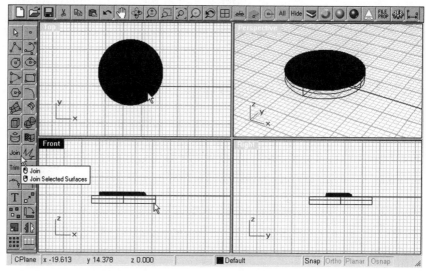

Fig. 8-97. Cylinder and surface being joined.

After trimming, the solid cylinder becomes an open-ended cylinder. To join the heightfield surface and the cylinder, continue with the following steps.

11 Select Edit > Join, or the Join button from the Main toolbar.
Command: Join

12 Select the surface and the cylinder and press the Enter key.

The medal is complete. Save your file.

Summary

A solid model represents integrated data (vertices, edges, surfaces, and volume) of the 3D object the model represents. There are many ways to represent a solid in the computer: primitive instancing, generalized sweeping, spatial occupancy enumeration, cellular deconstruction, boundary representation, and constructive solid geometry. A Rhino solid is a polysurface that encloses a volume with no gap or opening.

There are several ways to construct Rhino solids: using primitive polysurfaces, extruding closed-loop curves, extruding surfaces, joining surfaces to form a closed polysurface, and capping planar holes in polysurfaces to form a closed polysurface. To construct solids of more complex shape, you combine solid objects using Boolean operations, add filleted edges, and transform the solid (polysurface) in different ways, which you learned about in Chapter 6.

Because a Rhino solid is a polysurface, you join surfaces that enclose a volume into a solid. On the other hand, you explode a solid to get individual surfaces, and you trim, split, or extract a solid to get an open-loop polysurface.

Review Questions

1 Illustrate, with the aid of simple sketches, various ways of representing a solid in the computer.

2 Explain in detail how solids are constructed in Rhino.

Chapter 9

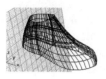

Consolidation

■■ Objectives

The goals of this chapter are to consolidate the curve, surface, and solid manipulation methods introduced in chapters 2 through 4, to elaborate on the techniques involved in setting up construction planes in 3D space, to explore the import and export utilities, and to introduce the use of polygon meshes. After studying this chapter, you should be able to:

- ❐ Use various methods to construct points, NURBS curves, and NURBS surfaces
- ❐ Set up construction planes
- ❐ Describe methods of importing and exporting data
- ❐ Use polygon meshes

■■ Overview

You use points and NURBS curves as frameworks for making NURBS surfaces. Joining contiguous NURBS surfaces, you get a polysurface. If the polysurface encloses a volume with no gap or opening, a solid is implied.

Logically, you start making points and curves before you construct freeform NURBS surfaces. To detail your models and to be able to improvise during the design process, you derive points and curves from existing surfaces and construct more surfaces using the derived points and curves.

One way to construct points and curves is to use the mouse as an input device. However, you are restricted to selecting locations on the default construction planes unless you set up construction planes in various orientations. The ability to manipulate construction planes helps you construct points and curves in 3D space.

Modeling an object in the computer is not the end of the design process. You use the computer model in downstream processes by exporting it to various file formats. To reuse computer models constructed using other computer applications, you import them.

As mentioned in Chapter 3, there are two ways to represent a surface in the computer: using NURBS surfaces to exactly represent surfaces or using polygon meshes to approximate surfaces. To cope with other upstream and downstream computer applications that use polygon meshes to represent surfaces, you need to know how to manipulate polygon meshes.

NURBS Surface Modeling

The ultimate product of using Rhino as a design tool is a computer model (surface model or solid model) that represents a 3D free-form object. Because Rhino solids are closed polysurfaces, both surface and solid modeling enter into the 3D NURBS surface creation process. Hence, you need a very thorough understanding of the various methods of constructing points, NURBS curves, and NURBS surfaces.

Design development using the computer is a two-stage iterative process. With an idea or concept in mind, you start deconstructing the 3D object into discrete surface elements by identifying and matching various portions of the object with various types of surfaces. Unless the individual surfaces are primitive surfaces, you need to think about the shapes and locations of curves and points required to build the surface, as well as the surface construction commands to be applied on the curves and points. This is a top-down thinking process.

After this thinking process, you create the curves and points and apply appropriate surface construction commands. By putting the surfaces together properly in 3D space, you get the 3D object. This is the bottom-up construction process. If the surfaces you construct do not conform to your concept, you think about the curves and points again. With practice you gain the experience that makes this process more efficient.

Logically, you construct points and curves and from these construct surfaces and solids. In detailing your design, you add surface features. To construct surface features that conform in shape to existing surfaces, you create points and curves from existing objects.

Points Derived from Surface Objects

On a surface, you select a region and drape a set of points. Draping points is similar to draping surfaces, which you learned about in Chapter 3. The difference is that draping points creates a set of point meshes instead of a set of surfaces. You specify a region of a surface and the

NURBS Surface Modeling

computer constructs a set of points on the surface within the region. In the following you will construct an ellipsoid and drape a set of point objects on the surface of the ellipsoid.

1 Start a new file. Use the metric (millimeter) template.

2 With reference to figure 9-1, construct an ellipsoid.

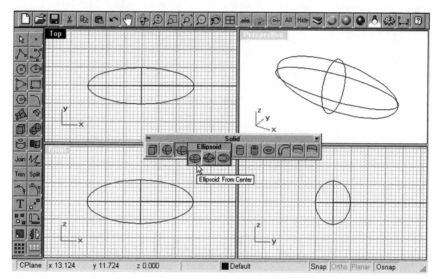

Fig. 9-1. Ellipsoid constructed.

3 Select Curve > Point Object > Drape Points, or the Drape Points on Shaded Preview button from the Point toolbar.
Command: DrapePt

4 Select two points in the Top viewport to specify a rectangular region. (See figure 9-2.)

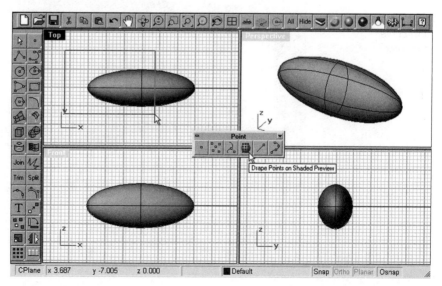

Fig. 9-2. Rectangular region specified.

414 CHAPTER 9: Consolidation

A set of point objects is constructed within the specified region on the surface of the ellipsoid. (See figure 9-3.) Save your file as *Drapt-Points.3dm*.

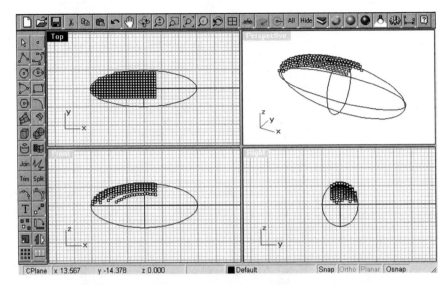

Fig. 9-3. Set of points constructed.

Curve Extended to a Surface

In Chapter 3, you learned various ways of extending a curve. To extend a curve to a surface, perform the following steps.

1 Start a new file. Use the metric (millimeter) template.
2 With reference to figure 9-4, construct a sphere and a curve.

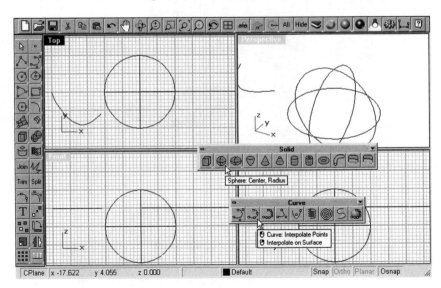

Fig. 9-4. Sphere and curve constructed.

NURBS Surface Modeling

3 Select Curve > Extend > Curve on Surface, or the Extend Curve on Surface button from the Extend toolbar.
Command: ExtendCrvOnSrf

4 Select the curve and select the surface. (See figure 9-5.)

The curve is extended to meet the surface. (See figure 9-6.) Save your file as *CurveExtSrf.3dm*.

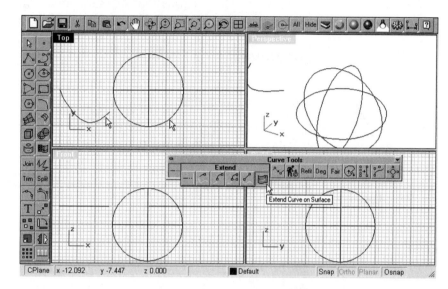

Fig. 9-5. Curve and surface selected.

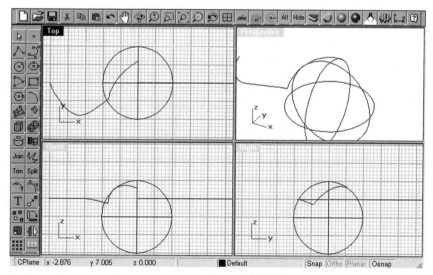

Fig. 9-6. Curve extended to the surface.

Constructing Basic Curves on Existing Objects

There are many ways of constructing a basic curve. In previous chapters you used the mouse and selected points on the construction plane. You can construct curves and polylines on existing NURBS surfaces or polygon meshes. The sections that follow take you through the process of creating these types of derived curves and polylines.

Line Normal to a Surface

To construct an ellipsoid and a line normal to the ellipsoid surface, perform the following steps.

1. Start a new file. Use the metric (millimeter) template.
2. With reference to figure 9-7, construct an ellipsoid.

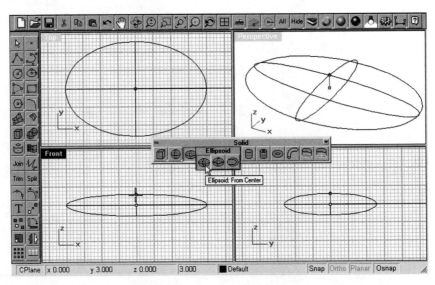

Fig. 9-7. Ellipsoid constructed.

3. Select Curve > Line > Normal to Surface, or the Surface Normal button from the Lines toolbar.

   Command: Normal

4. Select the ellipsoid.
5. Select the point in the Top viewport indicated in figure 9-8 to specify a point on the surface.
6. Select the point in the Right viewport indicated in figure 9-8 to specify the length of the normal line.

A line normal to the ellipsoid surface is constructed.

NURBS Surface Modeling

Fig. 9-8. Line normal to the surface being constructed.

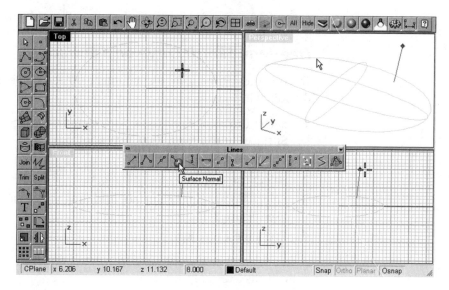

Free-form Curve on a Surface

To construct a free-form curve on the ellipsoid surface, perform the following steps.

1 Select Curve > Free-form > Interpolate on Surface, or the Interpolate on Surface button from the Curve toolbar.

   Command: InterpCrvOnSrf

2 Select the ellipsoid.

3 With reference to figure 9-9, select points on the surface and press the Enter key.

Fig. 9-9. Free-form curve being constructed on the surface.

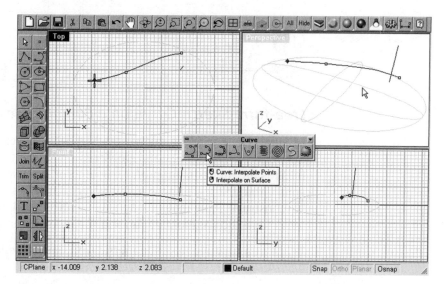

A free-form curve is constructed on the ellipsoid surface.

Free-form Sketch Curve on a Surface

To construct a free-form sketch curve on the ellipsoid surface, perform the following steps.

1 Select Curve > Free-form > Sketch on Surface, or the Sketch on Surface button from the Curve toolbar.
Command: SketchOnSrf

2 Select the ellipsoid.

3 Select a point on the surface and drag the cursor along the surface. (See figure 9-10.)

4 Press the Enter key.

A free-form sketch curve is constructed on the ellipsoid surface. Save your file as *SketchSrf.3dm*.

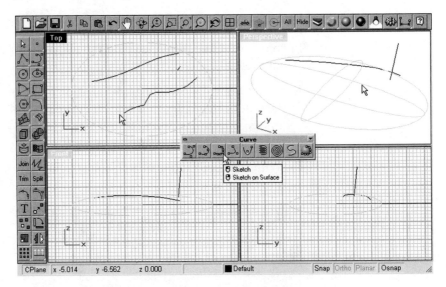

Fig. 9-10. Free-form sketch curve being constructed on the surface.

Free-form Sketch Curve on a Polygon Mesh

In addition to creating curves on NURBS surfaces, you can construct free-form sketch curves and polylines on a polygon mesh. A polygon mesh is an approximation of a smooth surface via a set of planar polygons. (You will learn more about polygon meshes later in this chapter.) To construct a paraboloid (NURBS surface) and construct a polygon mesh from the paraboloid, perform the following steps.

1 Start a new file. Use the metric (millimeter) template.

NURBS Surface Modeling 419

2 Select Solid > Paraboloid > Vertex, Focus, or the Paraboloid button from the Solid toolbar and type *V* to use the Vertex option.
   Command: Paraboloid Vertex

3 Select two points in the Front viewport to indicate the paraboloid vertex and focus point, and select a point in the Top viewport to indicate the end of the paraboloid. (See figure 9-11.)

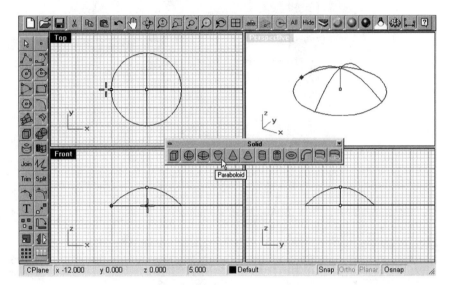

Fig. 9-11. Paraboloid constructed.

A NURBS paraboloid is constructed. Continue with the following steps.

4 Select Tools > Polygon Mesh > From NURBS Object, or the Mesh from NURBS Object button from the Mesh toolbar.
   Command: Mesh

5 Select the paraboloid and press the Enter key.

6 In the Polygon Mesh Options dialog box, move the slider bar to the left to have fewer polygons, and click on the OK button. (See figure 9-12.)

You now have a polygon mesh paraboloid. Continue with the following.

7 Select Curve > Free-form > Sketch on Polygon Mesh, or the Sketch on Polygon Mesh button from the Curve toolbar.
   Command: SketchOnMesh

The polygon mesh is shaded automatically. Continue with the following steps.

8 Select a point on the surface, hold down the left mouse button, and drag along the surface. (See figure 9-13.)

9 Release the left mouse button when you finish.

A free-form sketch curve is constructed on the polygon mesh.

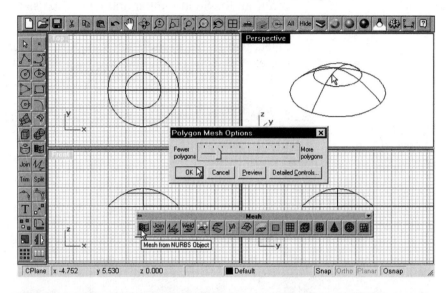

Fig. 9-12. Polygon mesh being constructed from a NURBS surface.

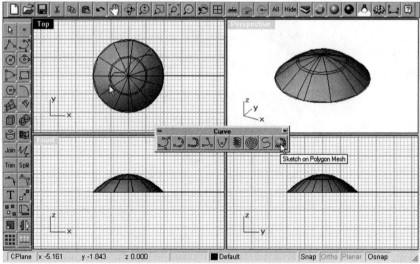

Fig. 9-13. Free-form polygon mesh being constructed on the polygon mesh.

Polyline on a Polygon Mesh

To construct a polyline on the polygon mesh, perform the following steps.

1. Select Curve > Line > Polyline on Mesh, or the Polyline On Mesh button from the Lines toolbar.

   Command: PolylineOnMesh

2. Select the polygon mesh. (See figure 9-14.)

NURBS Surface Modeling

3 Select a number of points on the polygon mesh to indicate the end points of the polyline segments, and press the Enter key.

A polyline is constructed on the polygon mesh. Save your file as *SketchMesh.3dm*.

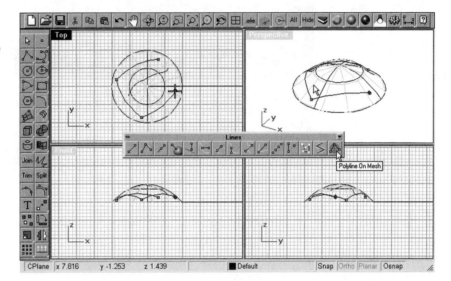

Fig. 9-14. Polyline being constructed on the polygon mesh.

Deriving Curves from Existing Objects

Chapter 3 stated that there are thirteen ways of deriving curves from existing objects. In the sections that follow, you have the opportunity to work with an example of each of these thirteen methods.

Projecting and Pulling a Curve to a Surface

One method of better controlling the shape of a curve constructed on a surface is to construct a curve and project or pull it on the surface. In this method, the direction of projection is perpendicular to the construction plane and the direction of pull is normal to the surface. To construct a circle and an ellipsoid and project the circle on the ellipsoid surface, perform the following steps.

1 Start a new file. Use the metric (millimeter) template.

2 With reference to figure 9-15, construct an ellipsoid.

3 Select Curve > Circle > Center, Radius, or the Circle/Center, Radius button from the Circle toolbar.
Command: Circle

4 Select a point in the Front viewport to indicate the center, and select a point in the Top viewport to specify the radius.

Fig. 9-15. Circle being constructed.

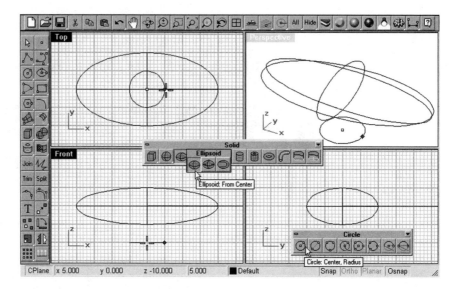

5 Select Curve > From Objects > Project, or the Project Curve to Surface button from the Curve From Object toolbar.

6 *Command:* Project

7 Select the circle and press the Enter key. (See figure 9-16.)

Fig. 9-16. Circle being projected.

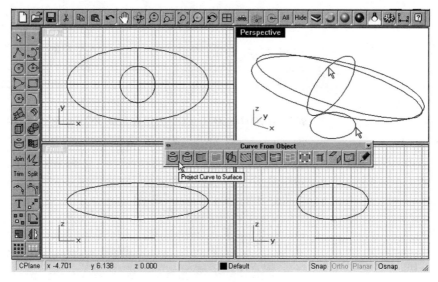

8 Select the ellipsoid and press the Enter key.

The circle is projected onto the ellipsoid surface. Note that two curves are constructed. To pull the circle to the ellipsoid surface, continue with the following steps.

NURBS Surface Modeling

9 Select Curve > From Objects > Pullback, or the Pull Curve to Surface by Closest Points button from the Curve From Object toolbar.
Command: Pull

10 Select the circle and press the Enter key. (See figure 9-17.)

11 Select the ellipsoid.

12 The circle is pulled to the ellipsoid surface. Compare the results of pulling and projection.

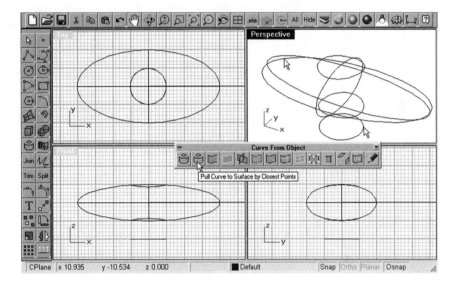

Fig. 9-17. Circle being pulled.

Contour Lines and Sections

Contour lines typically form a set of section lines spaced at a regular interval. To construct a set of contour lines across the ellipsoid surface, perform the following steps.

1 Set the current layer to *Layer 01*.

2 Select Curve > From Objects > Contour, or the Contour button from the Curve From Object toolbar.
Command: Contour

3 Select the ellipsoid and press the Enter key.

4 Select two points in the Top viewport to indicate the contour plane base point and contour plane direction.

5 Type *3* at the command line area to specify the distance between contours. (See figure 9-18.)

Contour lines at 3-mm spacing are constructed along the ellipsoid surface. To construct a section across the surface, continue with the following steps.

Fig. 9-18. Contour lines constructed.

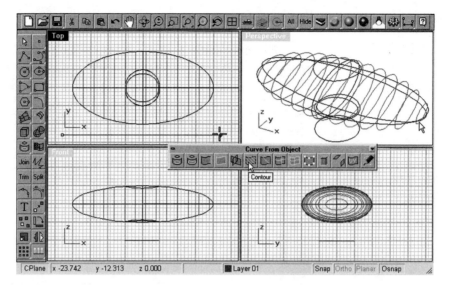

6 Turn off *Layer 01* and set the current layer to *Layer 02*.

7 Select Curve > From Objects > Section, or the Section button from the Curve From Object toolbar.

Command: Section

8 Select the ellipsoid and press the Enter key.

9 Select two points in the Top viewport to define a section plane and press the Enter key. (See figure 9-19.)

A section curve is constructed. Note that these commands also work on polygon meshes, explored later in this chapter. You can construct a NURBS surface from a set of contour lines or sections through a polygon mesh.

Fig. 9-19. Section curve being constructed.

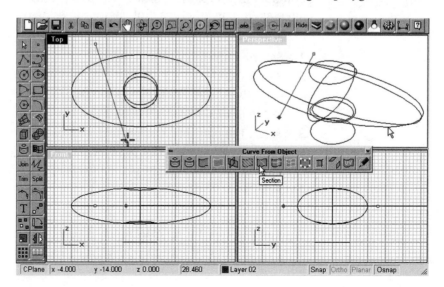

NURBS Surface Modeling

Extracting Isoparms from a Surface

Isoparm lines are directionally U and V lines along a surface. You construct these U and V curves at designated locations on the surface. To construct U and V isoparm lines on the ellipsoid surface, perform the following steps.

1   Select Curve > From Objects > Extract Isoparm, or the Extract Isoparm button from Curve From Object toolbar.

    Command: ExtractIsoparm

2   Select the ellipsoid.

3   Type *D* and then *B* at the command line area to extract both U- and V-direction isoparms.

4   Select a point on the surface and press the Enter key. (See figure 9-20.)

U and V isoparm lines are constructed on the ellipsoid surface.

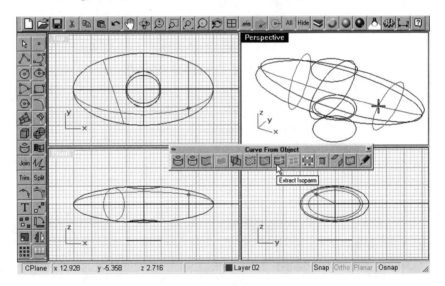

Fig. 9-20. Isoparm lines being constructed.

Silhouette of a Surface in a Viewport

The shape of the silhouette of a free-form surface resembles the edge of the shadow of the object projected in the viewing direction of the viewport. You see a different silhouette of the same object at different viewing angles. To construct a silhouette of the ellipsoid surface in a selected viewport, perform the following steps.

1   Select Curve > From Objects > Silhouette, or the Silhouette button from the Curve From Object toolbar.

    Command: Silhouette

2   Select the ellipsoid in the Perspective viewport and press the Enter key.

A silhouette of the ellipsoid in the Perspective viewport is constructed. (See figure 9-21.) Save your file as *CurveFromObjects1.3dm*.

Fig. 9-21. Silhouette in the Perspective viewport constructed.

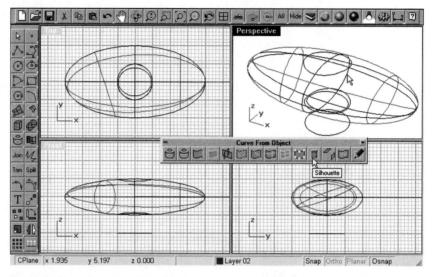

Extracting Control Points

NURBS surfaces and curves have control points that govern their shape. To construct a surface and two curves and extract their control points, perform the following steps.

1 Start a new file. Use the metric (millimeter) template.

2 With reference to figure 9-22, construct an ellipsoid and two free-form curves.

Fig. 9-22. Ellipsoid and free-form curves.

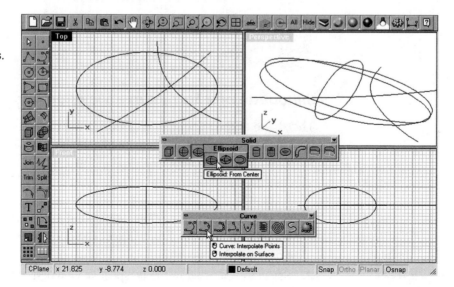

NURBS Surface Modeling

3 Set the current layer to *Layer 01*.

4 Select Curve > From Objects > Extract Points, or the Extract Points button from the Curve From Object toolbar.
Command: ExtractPt

5 Select the ellipsoid and the free-form curves and press the Enter key. (See figure 9-23.)

Fig. 9-23. Point objects constructed.

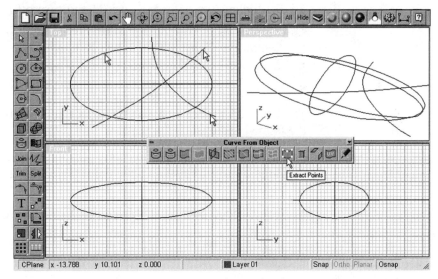

Points are constructed at the control point locations of the surface and curves. (See figure 9-24.)

Fig. 9-24. Points constructed.

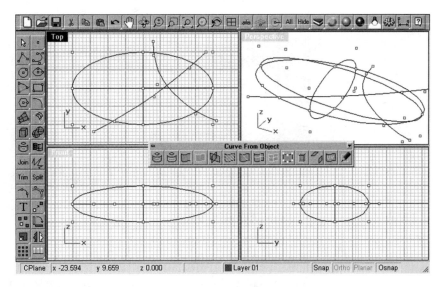

Duplication of Surface Edge and Border

An open surface can have multiple edges but only a single border. For example, a rectangular surface has four edges but just one continuous border. The ellipsoid you constructed is a closed surface. In the following you will use the curves to trim the ellipsoid surface to make it an open surface. Then you will duplicate an edge of the trimmed surface.

1 Set the current layer to *Layer 02* and turn off *Layer 01*.
2 Select Edit > Trim, or the Trim button from the Main toolbar.
   Command: Trim
3 Select the curves and press the Enter key.
4 Select the surface at the locations indicated in figure 9-25.

The trimmed surface now has two edges and one border. Continue with the following steps.

5 Select Curve > From Objects > Duplicate Edge, or the Duplicate Edge button from the Curve From Object toolbar.
   Command: DupEdge
6 Select the edge indicated in figure 9-26 and press the Enter key.

Fig. 9-25. Surface being trimmed by the curves.

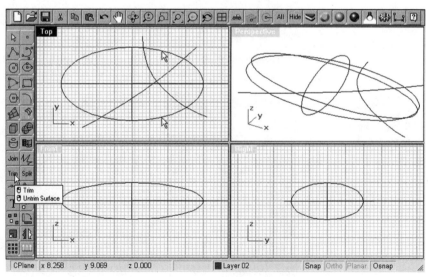

7 Select Edit > Visibility > Hide, or the Hide Objects button from the Visibility toolbar.
   Command: Hide
8 Select the surface and press the Enter key. (See figure 9-27.)

An edge is duplicated from the surface and the surface is hidden.

NURBS Surface Modeling

Fig. 9-26. Surface trimmed and edge being duplicated.

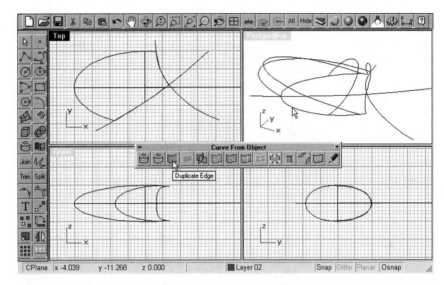

Fig. 9-27. Surface being hidden.

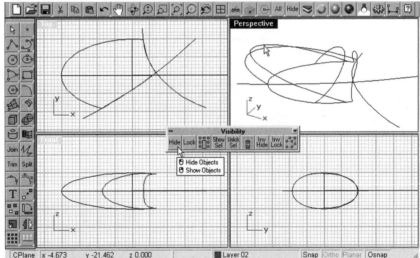

To duplicate the border of the surface, continue with the following steps.

9 Select Edit > Visibility > Show, or the Show Objects button from the Visibility toolbar.

Command: Show

10 Set the current layer to *Layer 03* and turn off *Layer 02*.

11 Select Curve > From Objects > Duplicate Border, or the Duplicate Border button from the Curve From Object toolbar.

Command: DupBorder

12 Select the border indicated in figure 9-28 and press the Enter key.

430 CHAPTER 9: Consolidation

Fig. 9-28. Border being duplicated.

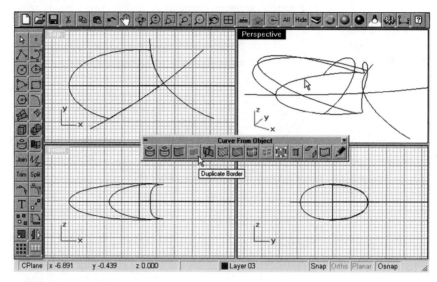

13 Hide the surface.

The border is duplicated. (See figure 9-29.) Save your file as *Border-Edge.3dm*.

Fig. 9-29. Border duplicated.

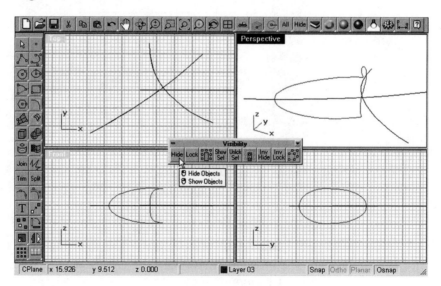

Intersection of Surfaces

At the intersection of two surfaces, you can construct an intersection curve, as follows. Using this command, you can also find the points where a curve intersects a surface.

1 Start a new file. Use the metric (millimeter) template.

NURBS Surface Modeling

2 With reference to figure 9-30, construct a cone and a cylinder.

Fig. 9-30. Cone and cylinder constructed.

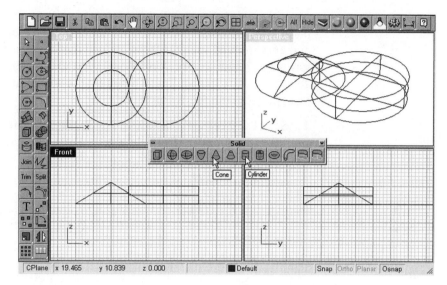

3 Select Curve > From Objects > Intersection, or the Intersection button from the Curve From Object toolbar.

Command: Intersect

4 Select the cone and cylinder and press the Enter key.

A curve at the intersection of the cone and cylinder is constructed. (See figure 9-31.)

Fig. 9-31. Intersection curve constructed.

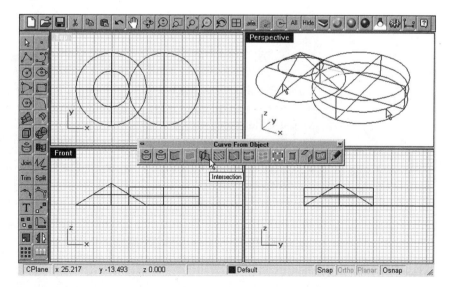

Extracting a Wireframe

As explained in Chapter 5, NURBS surfaces are smooth surfaces and isoparm lines are displayed in the viewport to help you visualize the profile and curvature of such surfaces. To use these isoparm lines as curves for further development of your model, you extract them from the surface. To extract the isoparm lines (wireframe representation) of a conical surface, perform the following steps.

1 Select Curve > From Objects > Extract Wireframe, or the Extract Wireframe button from the Curve From Object toolbar.

Command: ExtractWireframe

2 Select the cone and press the Enter key. (See figure 9-32.)

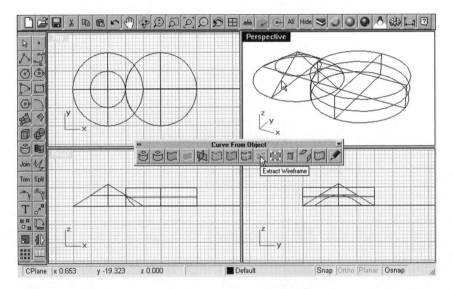

Fig. 9-32. Wireframe being extracted from the cone.

The wireframe is extracted from the cone. Because the extracted curves and the isoparm lines lie in the same location, you will not find any visual difference after you construct the curves. To see the difference, you can hide the cone and shade the viewport, as follows.

3 Select Edit > Visibility > Hide, or the Hide Objects button from the Visibility toolbar.

Command: Hide

4 Select the cone. In the pop-up menu, select Polysurface. (See figure 9-33.)

5 Press the Enter key.

After hiding the surface, you will not see any difference because the curves are still there. Continue with the following steps.

NURBS Surface Modeling

Fig. 9-33. Cone (polysurface) being hidden.

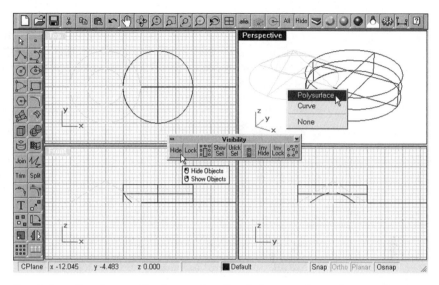

6 Select Render > Shade, or the Shade button from the standard toolbar. (See figure 9-34.)

Command: Shade

Note that the cylinder is shaded but the curves are not. Save your file as *CrOnSrf.3dm*.

Fig. 9-34. Cone hidden and viewport shaded.

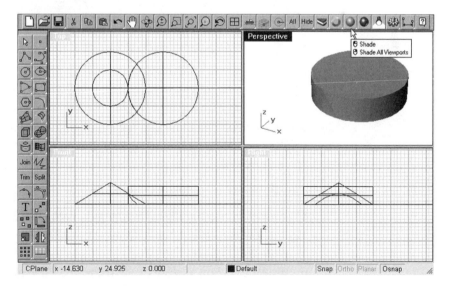

Creating Reference U and V Curves on the X-Y Plane and Applying Planar Curves on Surfaces

Sometimes you need to place a set of planar curves on a surface with reference to the U and V orientation of the surface. To properly locate the planar curve on the surface, you map the U and V curves of the surface onto the X-Y plane. Using these curves as a reference, you construct a planar curve and apply the curve to the surface. To construct a truncated cone and map the U and V curves of the slant surface on the X-Y plane, perform the following steps.

1. Start a new file. Use the metric (millimeter) template.
2. Select Solid > Truncated Cone, or the Truncated Cone button from the Solid Tools toolbar.

   Command: TCone

3. Select three points in the Top viewport indicated in figure 9-35 to specify the center, radius at the base, and radius at the top.

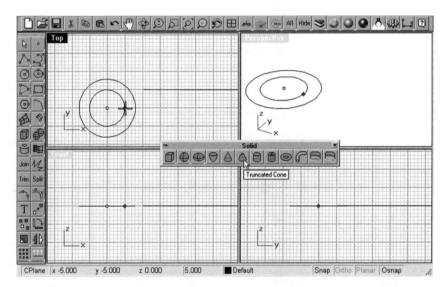

Fig. 9-35. Truncated cone being constructed.

4. Hold down the Shift key and select a point in the Front viewport to specify the height in a direction perpendicular to the Top viewport. (See figure 9-36.)

A truncated cone is constructed. Continue with the following steps.

5. Select Curve > From Objects > Create UV Curves, or the Create UV Curves button from the Curve From Object toolbar.

   Command: CreateUVCrv

6. Select the slant surface of the truncated cone and press the Enter key.

The U and V curves of the selected surface are mapped onto the X-Y plane. (See figure 9-37.) Note that the curves constructed do not represent the true lengths of the edges. They serve only as reference for applying curves to the surface.

NURBS Surface Modeling

Fig. 9-36. Truncated cone constructed.

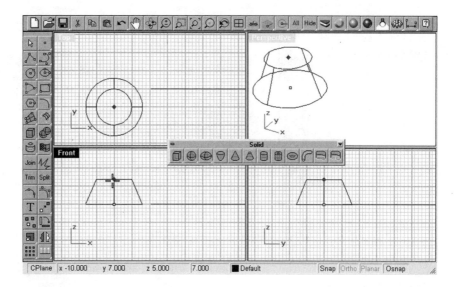

Fig. 9-37. Surface mapped to the X-Y plane.

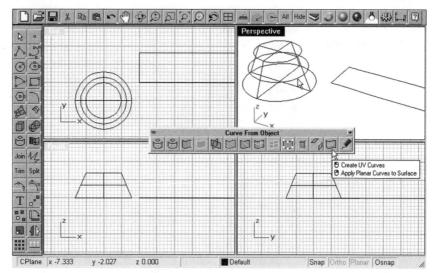

To construct a curve and apply it on the slant surface of the truncated cone, continue with the following steps.

7 Maximize the Perspective viewport.

8 With reference to figure 9-38, construct a free-form curve.

9 Select Curve > From Objects > Apply UV Curves, or the Apply Planar Curves to Surface button from the Curve From Objects toolbar.
Command: ApplyCrv

10 Select the curves indicated in figure 9-39 and press the Enter key.

11 Select the slant surface indicated in figure 9-39.

Fig. 9-38. Free-form curve constructed.

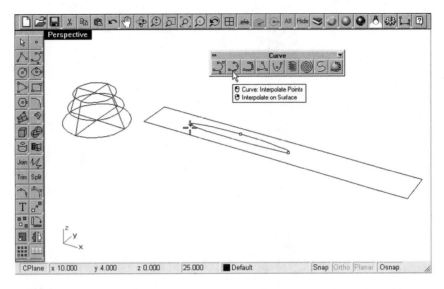

Fig. 9-39. Curve being applied to the surface.

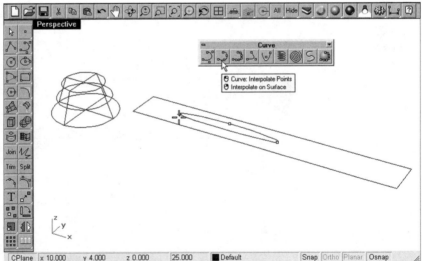

The curves are applied to the surface. (See figure 9-40.) Save your file as *ApplyCrv.3dm*.

■■ The Construction Plane

In previous chapters you learned how to construct basic curves by selecting a command and inputting parameters of the curve. You input data by entering the coordinates at the command line area, or by using the pointing device or digitizer to select a set of points. Using the digitizer, you select a point on the surface of a physical object.

The Construction Plane

Fig. 9-40. Curve applied to the surface.

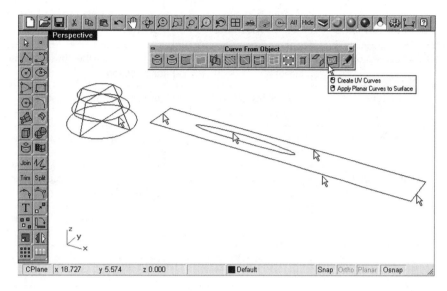

Using the pointing device, you select points on the construction plane. The construction plane is an imaginary plane. As explained in Chapter 1, there is a construction plane corresponding to each viewport. To facilitate point and curve construction using the pointing device, you set up construction planes in various ways using the commands on the Set CPlane toolbar, shown in figure 9-41.

Fig. 9-41. Set CPlane toolbar.

Table 9-1 outlines the functions of the construction plane commands and their locations in the pull-down menu.

Table 9-1: Construction Plane Commands and Their Functions

Command Name	Pull-down Menu	Function
CPlanePrev	View > Set CPlane > Previous	For enabling use of the previous construction plane
CPlaneNext	View > Set CPlane > Next	For enabling use of the original construction plane after you use the previous construction plane
CPlaneOrigin	View > Set CPlane > Origin	For repositioning the origin at a new location

Command Name	Pull-down Menu	Function
CPlaneElevation	View > Set CPlane > Elevation	For translating the construction plane in a perpendicular direction
RotateCPlane	View > Set CPlane > Rotate	For rotating the construction plane
CPlane3Pt	View > Set CPlane > 3 Points	For defining a construction plane by specifying three points
CPlaneToObject	View > Set CPlane > To Object	For setting the construction plane in relation to a selected object
CPlaneToView	View > Set CPlane > To View	For setting the construction plane in relation to the display view
CPlaneV	View > Set CPlane > Vertical	For orienting the construction plane vertical to a line defined by two selected points
CPlanePerpToCrv	View > Set CPlane > Perpendicular to Curve	For orienting the construction plane perpendicular to a curve at a selected point
CPlaneX	View > Set CPlane > X Axis	For orienting the construction plane by specifying the origin and the X axis
CPlaneZ	View > Set CPlane > Z Axis	For orienting the construction plane by specifying the origin and the Z axis
CPlaneTop	View > Set CPlane > World Top	For enabling use of the Top viewport construction plane
CPlaneRight	View > Set CPlane > World Right	For enabling use of the Right viewport construction plane
CPlaneFront	View > Set CPlane > World Front	For enabling use of the Front viewport construction plane
SaveCPlane	Named CPlane > Save	For saving the current construction plane setting
RestoreCPlane	Named CPlane > Restore	For retrieving a saved construction plane setting
NamedCPlane	Named CPlane > Edit	For renaming or deleting saved construction planes
ReadNamedCPlanesFromFile	Named CPlane > Read from File	For enabling use of a saved construction plane from a Rhino file

The Construction Plane

Setting the Construction Plane by Specifying Origin, X Axis, and Orientation

To construct a box and set the construction plane to the diagonal corners of the box, perform the following steps.

1. Start a new file. Use the metric (millimeter) template and a four-viewport configuration.
2. With reference to figure 9-42, construct a box.

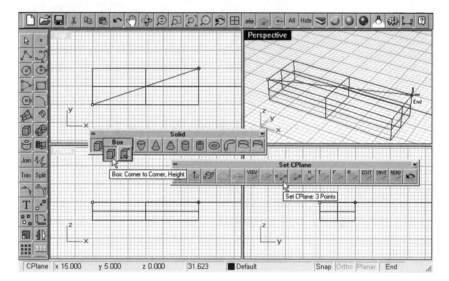

Fig. 9-42. Box constructed and construction plane being set up.

By default, the construction plane in the Perspective viewport is the same as that of the Top viewport. To set the construction plane by specifying three points, continue with the following steps.

3. Select the Perspective viewport to use it as the current viewport.
4. Select View > Set CPlane > 3 Points, or the Set CPlane/3 Points button from the Set CPlane toolbar.
   Command: CPlane3Pt
5. Select the end point indicated by the arrow in figure 9-42 to indicate the origin.
6. Select the end point indicated in figure 9-42 to indicate the X axis direction.
7. Select the end point indicated in figure 9-43 to indicate the orientation.

The construction plane in the Perspective viewport is set. Note that the construction planes in the Top, Front, and Right viewports remain unchanged.

Fig. 9-43. Construction plane orientation being specified.

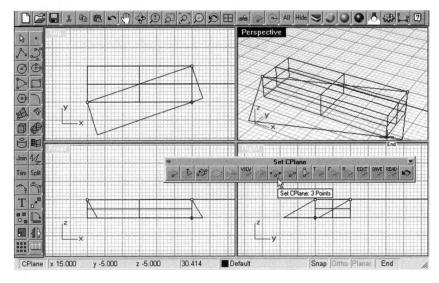

Construction Plane Origin

To set the origin of the construction plane without changing its orientation, perform the following steps.

1 Select View > Set CPlane > Origin, or the Set CPlane/Origin button from the Set CPlane toolbar.

Command: CPlaneOrigin

2 Select the end point indicated in figure 9-44.

Fig. 9-44. Origin being relocated.

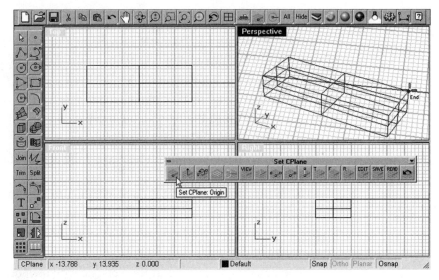

The origin of the construction plane in the Perspective viewport is relocated. (See figure 9-45.)

The Construction Plane

Fig. 9-45. Origin of the construction plane relocated.

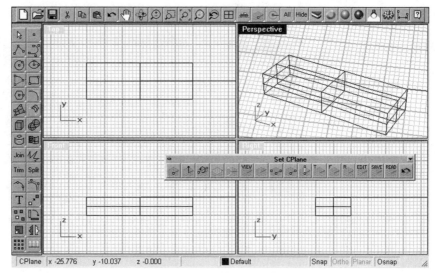

Previous Construction Plane

To return to the previous construction plane setting, perform the following steps.

1 Select View > Set CPlane > Previous, or the Previous CPlane button from the Set CPlane toolbar.
Command: CPlanePrev

The construction plane is reverted to the previous setting. (See figure 9-44.)

Next Construction Plane

After reverting to the previous construction plane, you need to change back to the *next* construction plane. *Next* refers to the next construction plane of the *previous* construction. Perform the following.

1 Select View > Set CPlane > Next, or the Next CPlane button from the Set CPlane toolbar.
Command: CPlaneNext

The construction plane setting in the Perspective viewport is changed.

Setting Construction Plane Elevation

In the following you will construct a curve in the Top viewport, change the construction plane elevation in the Top viewport, and construct another curve there.

442 CHAPTER 9: Consolidation

1 With reference to figure 9-46, construct a curve in the Top viewport.

Fig. 9-46. Curve constructed in the Top viewport.

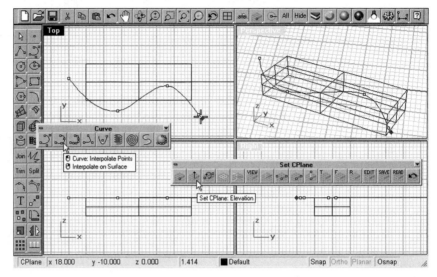

2 Select View > Set CPlane > Elevation, or the Set CPlane/Elevation button from the Set CPlane toolbar.
Command: CPlaneElevation

3 Type *10* at the command line area to change the elevation.

4 With reference to figure 9-47, construct another curve in the Top viewport.

Note in the other viewports the vertical distance between the two curves.

Fig. 9-47. Elevation changed and curve constructed.

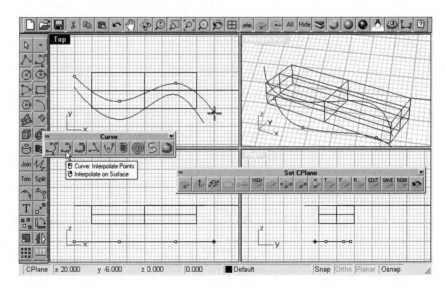

The Construction Plane

Fig. 9-48.
Construction plane rotated.

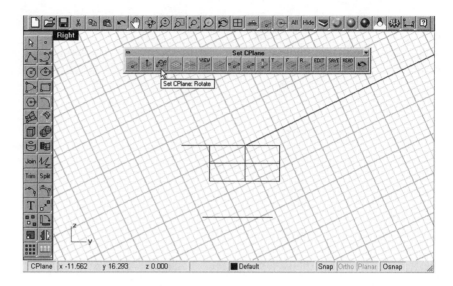

Rotating the Construction Plane

To rotate the construction plane in the Right viewport, perform the following steps.

1 Maximize the Right viewport.

2 Select View > Set CPlane > Rotate, or the Set CPlane/Rotate button from the Set CPlane toolbar.

Command: RotateCPlane

3 Type Z at the command line area to use the current construction plane Z axis as the rotation axis.

4 Type *25* to rotate the construction plane 25 degrees.

The construction plane is rotated. (See figure 9-48.)

Object Construction Plane

To set the construction plane to a selected object, perform the following steps.

1 Maximize the Front viewport.

2 Select View > Set CPlane > To Object, or the Set CPlane/To Object button from the Set CPlane toolbar.

Command: CPlaneToObject

3 Select the curve indicated in figure 9-49.

The construction plane is set to the selected object.

*Fig. 9-49.
Construction plane set to an object.*

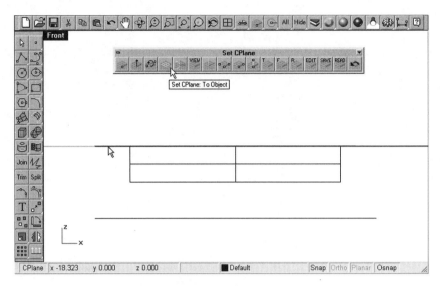

Vertical Construction Plane

To set the construction plane perpendicular to the current construction plane, perform the following steps.

1 Maximize the Top viewport.

2 Select View > Set CPlane > Vertical, or the Set CPlane/Vertical button from the Set CPlane toolbar.

Command: CPlaneV

3 Select two points indicated in figure 9-50.

*Fig. 9-50.
Construction plane being set vertical to current construction plane.*

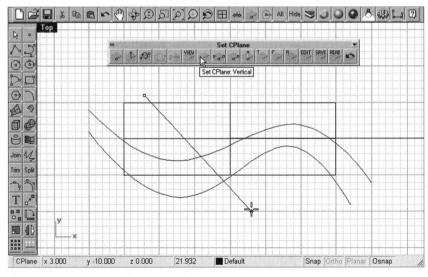

The construction plane is vertical to two selected points.

World Top Construction Plane

To align the construction plane to the top view of the world coordinate system, perform the following steps.

1. Maximize the Right viewport.
2. Select View > Set CPlane > World Top, or the Set CPlane/World Top button from the Set CPlane toolbar.
   Command: CPlaneTop

The construction plane is set to World Top. (See figure 9-51.)

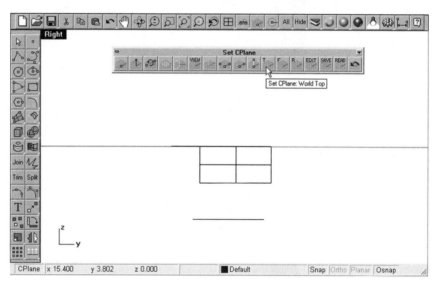

Fig. 9-51.
Construction plane set to World Top.

World Front Construction Plane

To align the construction plane to the front view of the world coordinate system, perform the following steps.

1. Select View > Set CPlane > World Front, or the Set CPlane/World Front button from the Set CPlane toolbar.
   Command: CPlaneFront

The construction plane is set to World Front. (See figure 9-52.)

World Right Construction Plane

To align the construction plane to the right view of the world coordinate system, perform the following steps.

Fig. 9-52.
Construction plane set to World Front.

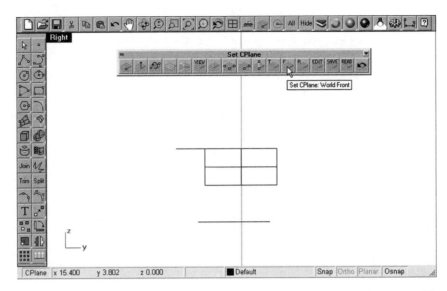

1 Select View > Set CPlane > World Right, or the Set CPlane/World Right button from the Set CPlane toolbar.

Command: CPlaneRight

The construction plane is set to World Right. (See figure 9-53.)

Fig. 9-53.
Construction plane set to World Right.

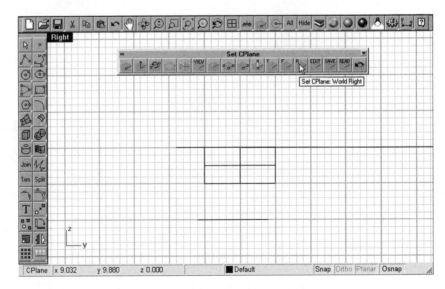

View Construction Plane

To set the construction plane parallel to the viewport, perform the following steps.

1 Set the display to a four-viewport configuration.

The Construction Plane

2 Select the Perspective viewport.

3 Select View > Set CPlane > To View, or the Set CPlane/To View button from the Set CPlane toolbar.
Command: CPlaneToView

4 Repeat the command three more times, once each for the other three viewports.

The construction planes in the viewports are set parallel to the viewport. (See figure 9-54.)

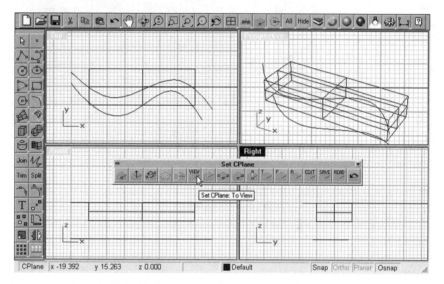

Fig. 9-54. Construction plane set.

Right-click Menu

To use the right-click menu of the Perspective viewport to set the construction plane, perform the following steps.

1 Select the Perspective viewport title and right-click.

2 In the right-click menu, select Set View > Perspective. (See figure 9-55.)

The display is set and the construction plane of the viewport follows the default setting. In this case, the default construction plane setting of the Perspective viewport is the Top view construction plane.

Construction Plane Perpendicular to a Curve

To set the construction plane perpendicular to a curve, perform the following steps.

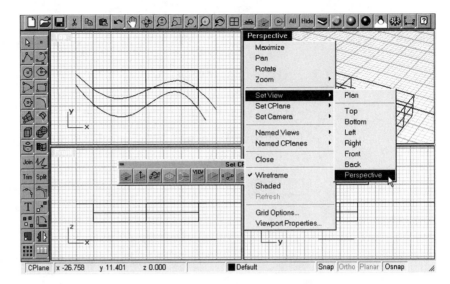

Fig. 9-55. Right-click menu.

1 Maximize the Perspective viewport.

2 Select View > Set CPlane > Perpendicular to Curve, or the Set CPlane/Perpendicular to Curve button from the Set CPlane toolbar.
Command: CPlanePerpToCrv

3 Select the curve indicated in figure 9-56.

4 Select the end point of the curve.

The construction plane is set perpendicular to the curve.

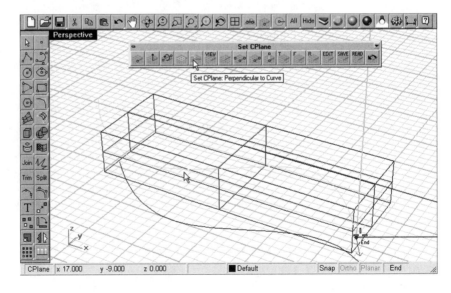

Fig. 9-56. Construction plane being set perpendicular to a curve.

The Construction Plane

Setting the Construction Plane by Specifying the X Axis

To specify the X axis of the construction plane, perform the following steps.

1. Maximize the Top viewport.
2. Select View > Set CPlane > X Axis, or the Set CPlane/X-Axis button from the Set CPlane toolbar.
   Command: CPlaneX
3. Select two points indicated in figure 9-57.

The X axis of the construction plane is specified.

Fig. 9-57. X axis being specified.

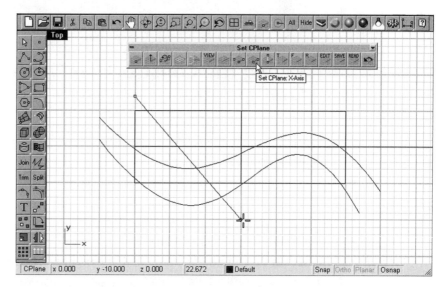

Setting the Construction Plane by Specifying the Z Axis

To specify the Z axis of the construction plane, perform the following steps.

1. Maximize the Front viewport.
2. Select View > Set CPlane > Z Axis, or the Set CPlane/Z-Axis button from the Set CPlane toolbar.
   Command: CPlaneZ
3. Select the two points indicated in figure 9-58.

The Z axis is specified.

Fig. 9-58. Z axis being specified.

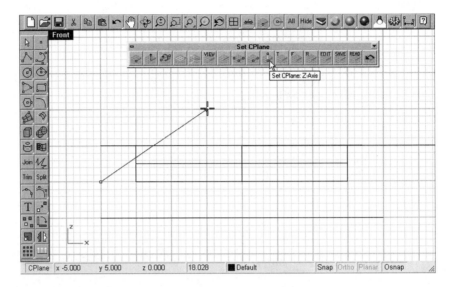

Saving the Construction Plane Configuration

To save the construction plane configuration so that you can retrieve it later, perform the following steps.

1. Return to a four-viewport display.
2. Select the Front viewport.
3. Select View > Named CPlane > Save, or the Save CPlane button from the Set CPlane toolbar.

   Command: SaveCPlane

4. In the Name of CPlane dialog box, enter a name and click on the OK button. (See figure 9-59.)

Retrieving the Construction Plane Configuration

To retrieve the saved construction plane configuration, perform the following steps.

1. Select the Perspective viewport.
2. Select View > Named CPlane > Restore, or the Restore CPlane button from the Set CPlane toolbar.

   Command: RestoreCPlane

3. In the Select CPlane to Retrieve dialog box, select the saved configuration and click on the OK button. (See figure 9-60.)

The Construction Plane

Fig. 9-59. Construction plane configuration being saved.

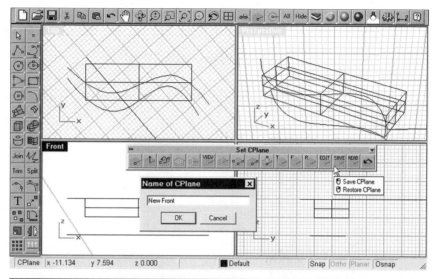

Fig. 9-60. Saved configuration being retrieved.

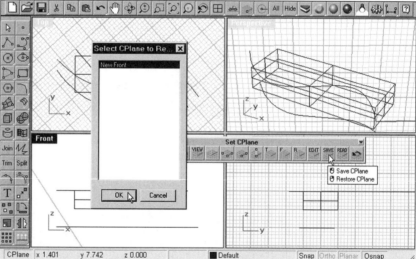

The saved construction plane configuration is applied to the Perspective viewport.

Editing the Saved Construction Plane Configuration

To edit the saved construction plane configuration, perform the following steps.

1 Select View > Named CPlane > Edit, or the Edit Named CPlanes button from the Set CPlane toolbar.

Command: NamedCPlane

2 In the Edit CPlane dialog box, you can rename or delete any configuration if you wish. Click on the Close button. (See figure 9-61.)

Save your file as *CPlane.3dm*.

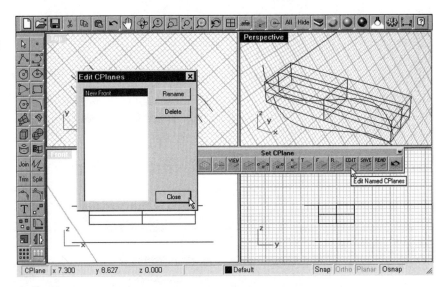

Fig. 9-61. Edit CPlane dialog box.

Reading the Construction Plane Configuration from a File

If a construction plane configuration has been saved in a file, you can retrieve the configuration from the saved file, as follows.

1 Start a new file. Use the metric (millimeter) template.

2 Select View > Named CPlane > Read from File, or the Read Named CPlanes from File button from the Set CPlane toolbar. (See figure 9-62.)

Command: ReadNamedCPlanesFromFile

3 In the File for Named CPlane dialog box, select the file that contains the saved configuration and click on the Open button.

The construction plane configuration is imported. Now you can use the saved configuration by selecting View > Named CPlane > Restore, or by clicking on the Restore CPlane button on the Set CPlane toolbar. When working with construction plane commands, the active viewport is the one that gets changed. If you are using parallel views, it is sometimes more effective to skew them a bit before changing the CPlane.

Import and Export

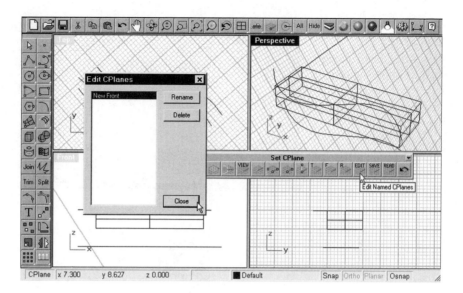

Fig. 9-62. Reading the configuration from another file.

■■ Import and Export

If your free-form models are constructed to facilitate downstream computerized operations, you can export them to various file formats for this purpose. On the other hand, you import files in order to reuse data constructed in upstream computer applications. To import a file, you open it. To export a file, you save it to an appropriate file format. To import or export files, perform the following steps.

1 To open a file, select File > Open, or the Open button from the File toolbar.

Command: Open

2 In the Open dialog box, select a file format, select a file, and click on the Open button. (See figure 9-63.) To import and merge files, select File > Import/Merge.

Command: Import

3 To export an entire file, select File > Save As, or the Save As button from the File toolbar.

Command: SaveAs

4 In the Save As dialog box, select a file format, specify a file name, and click on the Save button. (See figure 9-64.)

5 To export selected objects from a file, select File > Export Selected.

Command: Export

Basically, data you import or export can be categorized as four major types: curves and points, NURBS surfaces and polysurfaces, solids, and

Fig. 9-63. Open dialog box for importing files.

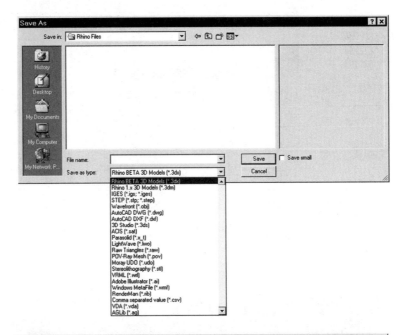

Fig. 9-64. Save As dialog box for exporting files.

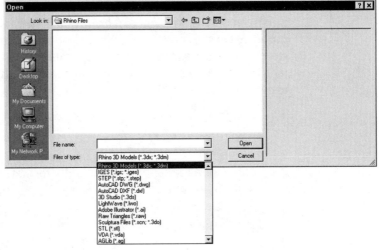

polygon meshes. The types of files you export depend on the types of objects you constructed in Rhino. For example, you can export a solid to a surface or to a solid file, but you cannot export an open polysurface or polygon mesh to a valid solid file. When you open a solid, you get a polysurface. However, you will not get a NURBS surface if the source file is a polygon mesh. Table 9-2 summarizes the file formats you can import and export in Rhino.

Table 9-2: Rhino Import (Open) and Export (Save As) File Formats

Open File Format	Save As File Format
IGES (*.igs, *.iges)	IGES (*.igs, *.iges)
STEP (*.stp, *.step)	STEP (*.stp, *.step)
—	WaveFront (*.obj)
AutoCAD DWG (*.dwg)	AutoCAD DWG (*.dwg)
AutoCAD DXF (*.dxf)	AutoCAD DXF (*.dxf)
3D Studio (*.3ds)	3D Studio (*.3ds)
—	ACIS (*.sat)
—	Parasolid (*.x_t)
LightWave (*.lwo)	LightWave (*.lwo)
Adobe Illustrator (*.ai)	Adobe Illustrator
Raw Triangles (*.raw)	Raw Triangles (*.raw)
—	POV Ray Mesh (*.pov)
—	Moray UDO (*.udo)
Sculptura Files (*.scn, *.3do)	—
STL (*.stl)	STL (*.stl)
—	VRML (*.wrl)
—	Windows MetaFile (*.wmf)
—	RenderMan (*.rib)
—	Comma separated value (*.csv)
VDA (*.vda)	VDA (*.vda)
AGLib (*.ag)	AGLib (*.ag)

▪▪ Use of Polygon Meshes

Although a polygon mesh is an approximation of a smooth surface, it is still used in many computer applications. For example, STL (a rapid prototype file format) represents a surface as a set of polygon meshes. When you import surfaces from other computerized applications, you are often importing a set of polygon meshes, which are dealt with via commands on the Mesh toolbar, shown in figure 9-65.

Fig. 9-65. Mesh toolbar.

Table 9-3 outlines the functions of the polygon mesh commands and their locations in the pull-down menu.

Table 9-3: Polygon Mesh Commands and Their Functions

Command Name	Pull-down Menu	Function
Mesh	Tools > Polygon Mesh > From NURBS Object	For constructing polygon meshes from NURBS surface objects
ExplodeMesh	Tools > Polygon Mesh > Explode	For exploding a joined polygon mesh object into individual polygon meshes
JoinMesh	Tools > Polygon Mesh > Join	For joining polygon meshes to form a single object
Weld	Tools > Polygon Mesh > Weld	For welding polygon meshes to form a single polygon mesh
UnifyMeshNormals	Tools > Polygon Mesh > Unify Normals	For unifying the normal direction of a set of joined polygon meshes
ReduceMesh	Tools > Polygon Mesh > Reduce	For simplifying a polygon mesh by reducing its number of constituent polygons
ApplyMesh	Tools > Polygon Mesh > Apply to Surface	For applying a polygon mesh to a NURBS surface
MeshPolyline	Tools > Polygon Mesh > From Closed Polyline	For constructing a polygon mesh from a closed polyline
ExtractControlPolygon	Tools > Polygon Mesh > From NURBS Control Polygon	For constructing a polygon mesh from the control points of a NURBS curve or surface
3Dface	Tools > Polygon Primitives > 3-D Face	For constructing a quadrilateral polygon mesh
MeshPlane	Tools > Polygon Primitives > Plane	For constructing a rectangular polygon mesh
MeshBox	Tools > Polygon Primitives > Box	For constructing a polygon mesh box

Use of Polygon Meshes

Command Name	Pull-down Menu	Function
MeshSphere	Tools > Polygon Primitives > Sphere	For constructing a polygon mesh sphere
MeshCone	Tools > Polygon Primitives > Cone	For constructing a polygon mesh cone
MeshCylinder	Tools > Polygon Primitives > Cylinder	For constructing a polygon mesh cylinder
MeshDensity	Tools > Polygon Primitives > Density	For setting the mesh density of polygon mesh boxes, planes, spheres, cylinders, and cones

In the sections that follow you will learn how to construct and modify polygon meshes.

Mesh Density Setting

Because a polygon mesh is a set of planar polygons that approximates a surface, the number of polygons used in the mesh has a direct impact on the accuracy of representation and the file size. The more numerous the constituent polygons, the more accurate the model.

However, file size increases quickly with a decrease in polygon size (smaller size meaning a greater number of polygons in the mesh). Hence, you need to consider using the most appropriate mesh density to provide an optimum balance between the representational accuracy of the model and the size of the file. To work with the Density setting, perform the following steps.

1. Start a new file. Use the metric (millimeter) template.
2. Select Tools > Polygon Primitives > Density, or the Densities for Mesh Primitives button from the Mesh toolbar. (See figure 9-66.)
   Command: MeshDensity
3. In the Densities for Mesh Primitives dialog box, set all values to *10* and click on the OK button.

Construction of Mesh Primitives

Mesh primitives are mesh objects of basic geometric shape. There are six types of mesh primitives: mesh cylinder, mesh cone, mesh sphere, mesh box, mesh plane, and 3D face. The sections that follow take you through the construction process for each of these primitives. You construct mesh primitives by selecting a command and specifying the parameters of the primitive.

Fig. 9-66. Densities for mesh primitives.

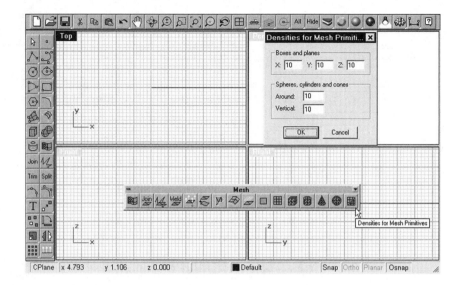

Mesh Cylinder

To construct a mesh cylinder, perform the following steps.

1 Select Tools > Polygon Primitives > Cylinder, or the Mesh Cylinder button from the Mesh toolbar.

Command: MeshCylinder

2 Select two points in the Top viewport to specify the center and radius.

3 Select a point in the Front viewport to indicate the height of the cylinder.

A mesh cylinder is constructed. (See figure 9-67.)

Mesh Cone

To construct a mesh cone, perform the following steps.

1 Select Tools > Polygon Primitives > Cone, or the Mesh Cone button from the Mesh toolbar.

Command: MeshCone

2 Select two points in the Top viewport to specify the center and radius.

3 Select a point in the Front viewport to indicate the height of the cone.

A mesh cone is constructed. (See figure 9-68.)

Use of Polygon Meshes

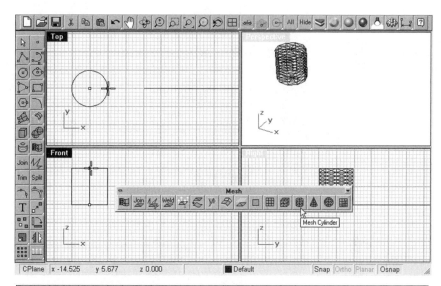

Fig. 9-67. Mesh cylinder constructed.

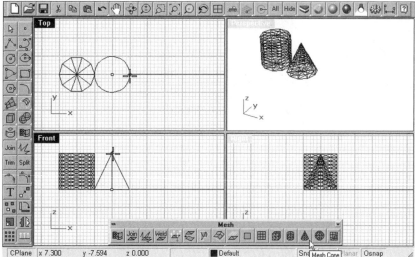

Fig. 9-68. Mesh cone constructed.

Mesh Sphere

To construct a mesh sphere, perform the following steps.

1. Select Tools > Polygon Primitives > Sphere, or the Mesh Sphere button from the Mesh toolbar.
   Command: MeshSphere

2. Select two points in the Top viewport to specify the center and radius of the sphere.

A mesh sphere is constructed. (See figure 9-69.)

Fig. 9-69. Mesh sphere constructed.

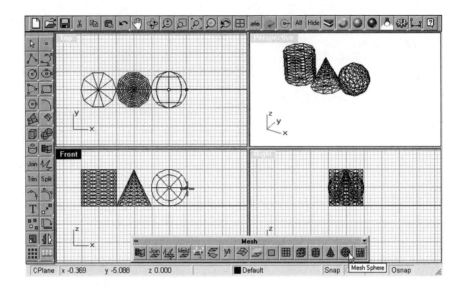

Mesh Box

To construct a mesh box, perform the following steps.

1 Select Tools > Polygon Primitives > Box, or the Mesh Box button from the Mesh toolbar.

Command: MeshBox

2 Select two points in the Top viewport to specify the diagonal corners of the base of the box.

3 Select a point in the Front viewport to indicate the height of the box.

A mesh box is constructed. (See figure 9-70.)

Mesh Rectangular Plane

To construct a mesh rectangular plane, perform the following steps.

1 Select Tools > Polygon Primitives > Plane, or the Mesh Plane button from the Mesh toolbar.

Command: MeshPlane

2 Select two points in the Top viewport to specify the diagonal end points of the rectangular plane.

A mesh rectangular plane is constructed. (See figure 9-71.)

Use of Polygon Meshes

Fig. 9-70. Mesh box constructed.

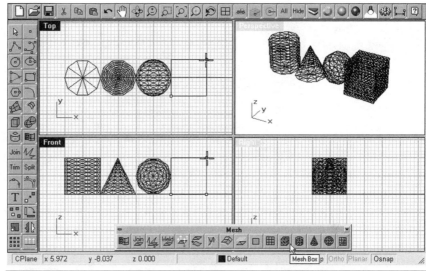

Fig. 9-71. Mesh rectangular plane constructed.

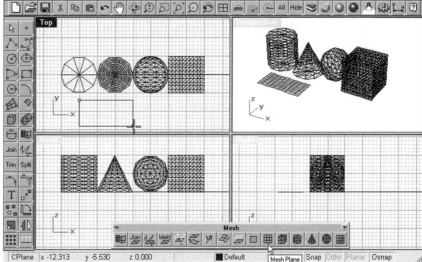

Mesh 3D Face

To construct a mesh 3D face, perform the following steps.

1 Select Tools > Polygon Primitives > 3-D Face, or the 3-D Face button from the Mesh toolbar.
   Command: 3DFace

2 Select three points in the Top viewport and a point in the Front viewport.

A 3D face is constructed. (See figure 9-72.) Save your file as *MeshPrimitives.3dm*.

Fig. 9-72. 3D face constructed.

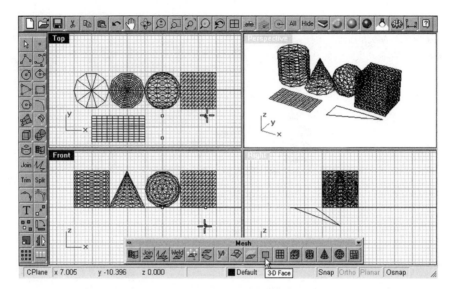

Construction of Derived Polygon Meshes

There are two ways to derive a polygon mesh from existing objects: from closed polylines or from control points of NURBS surfaces. These methods are explored in the sections that follow.

Polygon Mesh from Closed Polyline

In constructing a polygon mesh from a closed polyline, the polyline can be planar or 3D. In the following you will construct a 3D polyline. From the polyline, you will construct a polygon mesh.

1 Start a new file. Use the metric (millimeter) template.
2 With reference to figure 9-73, construct a closed polyline.
3 Select Tools > Polygon Mesh > From Closed Polyline, or the Mesh from Closed Polyline button from the Mesh toolbar.
   Command: MeshPolyline
4 Select the polyline indicated in figure 9-73.

A polygon mesh is constructed. (See figure 9-74.) Save your file as *PolylineMesh.3dm*.

Polygon Mesh from NURBS Surface Control Points

You can also construct a polygon mesh using the control points of a NURBS surface or curves as vertices. To construct a polygon mesh from the control vertices of an extruded surface, perform the following steps.

Use of Polygon Meshes

Fig. 9-73. 3D polyline constructed.

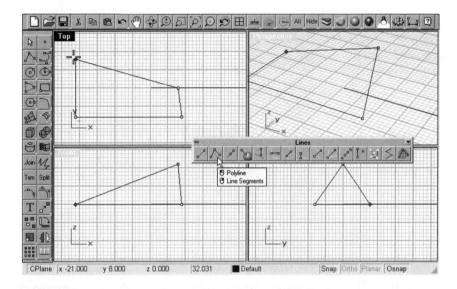

Fig. 9-74. Polygon mesh being constructed from a closed polyline.

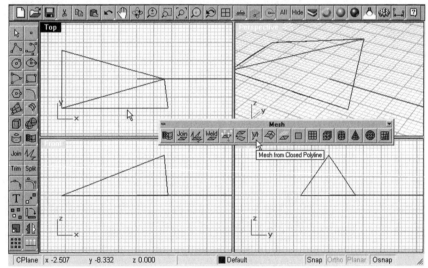

1 Start a new file. Use the metric (millimeter) template.

2 With reference to figure 9-75, construct a curve in the Top viewport and extrude a surface from the curve.

3 Select Polygon Mesh > From NURBS Control Polygon, or the Extract Mesh from NURBS Control Polygon button from the Mesh toolbar.

Command: ExtractControlPolygon

4 Select the surface and press the Enter key.

A polygon mesh is constructed. (See figure 9-76.) Save your file as *ControlVerticeMesh.3dm*.

Fig. 9-75. Curve and extruded surface constructed.

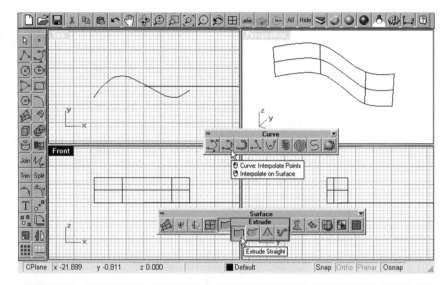

Fig. 9-76. Polygon mesh constructed from the control vertices of the extruded surface.

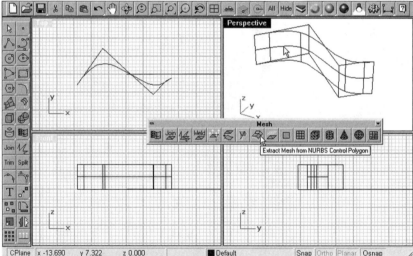

Editing Polygon Meshes

You edit polygon meshes in several ways, including reducing the number of polygons in the mesh, exploding joined polygon meshes into individual polygon meshes, joining contiguous polygon meshes, and welding polygon meshes to form a single polygon mesh object. These operations are explored in the sections that follow.

Use of Polygon Meshes

Reduction of Polygon Density

As explained earlier, a high polygon mesh density results in large file size. If the polygon mesh is overly dense, you can reduce its density. Decreasing the polygon count means simplifying the mesh. To reduce a polygon density, perform the following steps.

NOTE: *This process is nonreversible.*

1   Open the file *MeshPrimitives.3dm*.
2   Select Tools > Polygon Mesh > Reduce, or the Reduce Mesh Polygon Count button from the Mesh toolbar.
    Command: ReduceMesh
3   Select the mesh sphere.
4   In the Reduce Mesh Options dialog box, reduce the polygon count to 8 and click on the OK button. (See figure 9-77.)

The polygon mesh count is reduced. As can be seen, the box, being a set of joined simple planar meshes, has not much effect. Save your file.

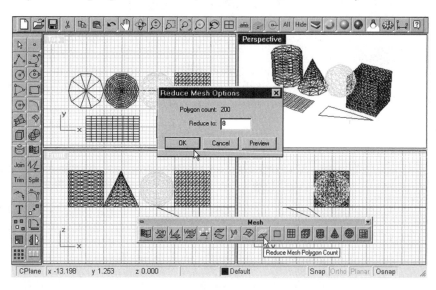

Fig. 9-77. Polygon mesh density being reduced.

Exploding Joined Polygon Meshes

A mesh box, for example, is a set of polygon meshes joined at their edges. To separate the polygon meshes, you explode the mesh box. To construct a mesh box and explode it into separate polygon meshes, perform the following steps.

1   Start a new file. Use the metric (millimeter) template.
2   With reference to figure 9-78, construct a mesh box.

Fig. 9-78. Mesh box being exploded.

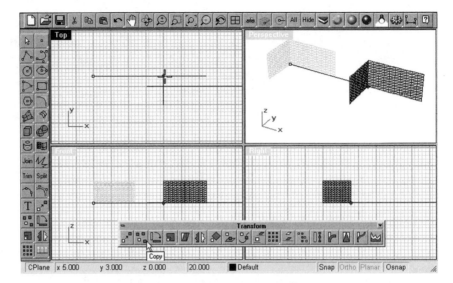

3 Select Polygon Mesh > Explode, or the Explode Polygon Mesh button from the Mesh toolbar.

Command: ExplodeMesh

4 Select the mesh box and press the Enter key.

The mesh box is exploded into six contiguous polygon meshes. To delete some of the polygon meshes and copy the remaining meshes, continue with the following.

5 With reference to figure 9-79, delete four polygon meshes and copy the meshes.

Fig. 9-79. Four polygon meshes deleted and remaining meshes copied.

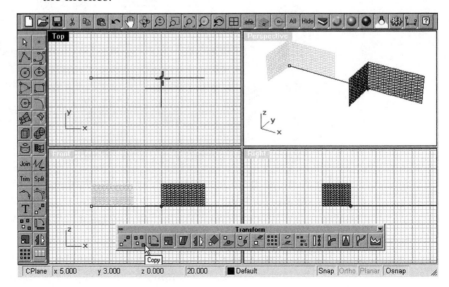

Use of Polygon Meshes

Joining Polygon Meshes

Contrary to exploding, you join contiguous polygon meshes, as follows.

1 Select Polygon Mesh > Join, or the Join Polygon Meshes button from the Mesh toolbar.

Command: JoinMesh

2 Select the polygon meshes indicated in figure 9-80 and press the Enter key.

The selected polygon meshes are joined.

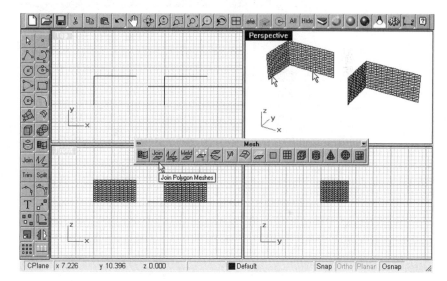

Fig. 9-80. Polygon meshes being joined.

Welding Polygon Meshes

In the following you will weld two contiguous polygon meshes. The difference between joining and welding is subtle. You can explode a joined set of polygon meshes into individual meshes. As for welded polygon meshes, the duplicated contiguous edges are removed and the welded mesh becomes a single mesh.

1 Select Polygon Mesh > Join, or the Join Polygon Meshes button from the Mesh toolbar.

Command: JoinMesh

2 Select the polygon meshes indicated in figure 9-81 and press the Enter key.

Now you have two sets of joined polygon meshes. Continue with the following steps.

3 Select Polygon Mesh > Weld, or the Weld Mesh Vertices button from the Mesh toolbar.

Command: Weld

4 Select the joined polygon meshes indicated in figure 9-81 and press the Enter key.

5 Type *270* at the command line area to specify the angle tolerance between contiguous polygon meshes. Normally, meshes less than or equal to 270 degrees are welded.

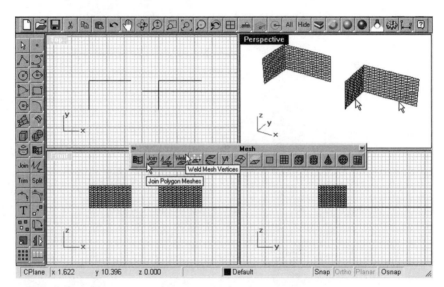

Fig. 9-81. Polygon meshes being welded.

Unifying Mesh Normals

When you join polygon meshes, there is a chance their normal direction is not uniform. To unify the normal directions of the joined meshes, perform the following steps.

1 Select Polygon Mesh > Unify Normals, or the Unify Mesh Normals button from the Mesh toolbar.

Command: UnifyMeshNormals

2 Select the polygon mesh indicated in figure 9-82.

Save your file as *JoinWeldMesh.3dm*.

Construction of Polygon Meshes from NURBS Surfaces

NURBS surfaces and polygon meshes are two different types of objects. A NURBS surface is an exact representation, but the polygon mesh is an ap-

Use of Polygon Meshes

Fig. 9-82. Normal direction being unified.

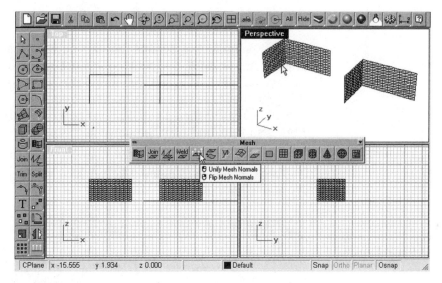

proximated representation of the 3D object. If you already have a NURBS surface model, you can construct a polygon mesh from it, as follows.

1   Start a new file. Use the metric (millimeter) template.

2   With reference to figure 9-83, construct an ellipsoid.

3   Select Polygon Mesh > From NURBS Objects, or the Mesh from NURBS Object button from the Mesh toolbar.

Command: Mesh

4   Select the ellipsoid and press the Enter key.

5   In the Polygon Mesh Options dialog box, set the polygon density by moving the slider bar, and then click on the OK button.

Fig. 9-83. Ellipsoid constructed and polygon mesh being constructed from the mesh.

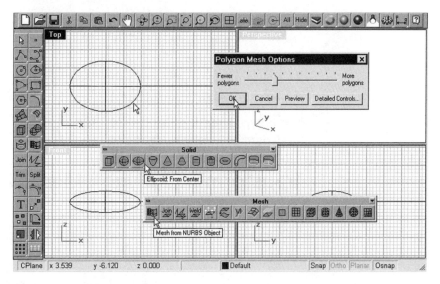

Mapping Polygon Meshes

In many animation programs that use polygon meshes to represent 3D objects, morphing of two different shapes requires the polygon meshes to have an identical vertex count and mesh structure. To construct two different shapes having identical vertex count and mesh structure, you construct two shapes as NURBS surfaces, construct a polygon mesh from a surface, and map the polygon mesh to the other surface, as follows.

1 With reference to figure 9-84, construct a circle and extrude a surface from the circle.

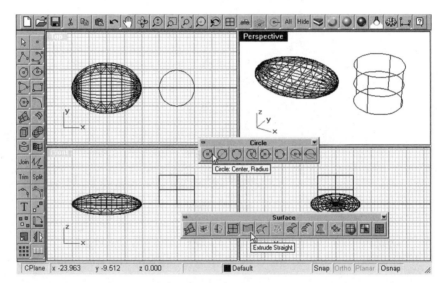

Fig. 9-84. Extruded surface constructed.

2 Select Polygon Mesh > Apply to Surface, or the Apply Mesh to NURBS Surface button from the Mesh toolbar.
Command: ApplyMesh

3 Select the mesh and press the Enter key. (See figure 9-85.)
4 Select the extruded surface.

The polygon mesh is applied to the surface. Save your file as *ApplyMesh.3dm*.

Transforming Polygon Meshes

The set of tools you use on surfaces also applies to polygon meshes. See chapters 4 and 7 to learn how to use the transform tools.

Use of Polygon Meshes

Fig. 9-85. Polygon mesh being applied to an extruded surface.

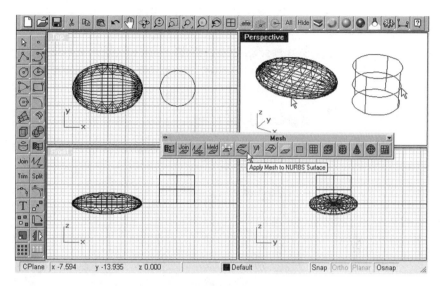

More Analysis Tools

The analysis tools you learned about in previous chapters are applicable to polygon meshes. Table 9-4 outlines more analysis commands and their functions.

Table 9-4: Polygon Mesh Analysis Commands and Their Functions

Command Name	Pull-down Menu	Function
EvaluatePoint	Analyze > Point	For checking the coordinates of selected points
Length	Analyze > Length	For measuring the length of selected curves or edges of surfaces
Distance	Analyze > Distance	For measuring the distance between two selected points
Angle	Analyze > Angle	For measuring the angle between two lines by indicating the end points of the lines
Radius	Analyze > Radius	For measuring the radius of curvature at selected locations along a curve
BoundingBox	Analyze > Bounding Box	For constructing a bounding box to enclose a selected curve, surface, or polysurface
SelBadObjects	Analyze > Diagnostics > Select Bad Objects	For selecting any object that does not pass the check test

Modeling Projects

In this section, you will work on a few modeling projects. These projects are the sports shoe, model car assembly, and medal you have worked on previously.

Sports Shoe

In the following you will construct the surface model of a shoe. Figure 9-86 shows the completed model.

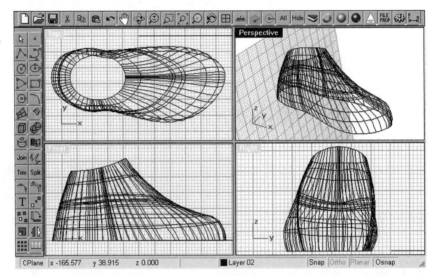

Fig. 9-86. Shoe model.

1 Start a new file. Use the metric (millimeter) template.

2 Maximize the Front viewport.

In the Front viewport, you will construct the curves of the top and front views of the shoe. Constructing the curves for two views of an object in a single viewport helps you realize the relationship among the curves. To construct a closed-loop curve, an ellipse, and a line, continue with the following. The locations of the curves are specified.

3 Set the current layer to *Layer 01*.

To set the grid properties, continue with the following steps.

4 Select File > Properties.
Command: DocumentProperties

5 In the Document Properties dialog box, shown in figure 9-87, select the Grid tab.

6 In the Grid tab, set the grid spacing to 5 units and the grid extents to 200 units, and then click on the OK button.

Modeling Projects

Fig. 9-87. Document Properties dialog box.

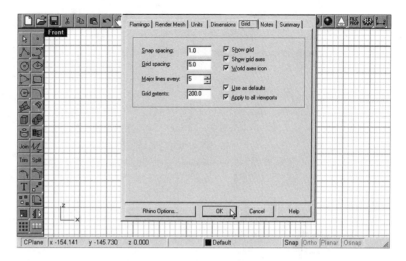

To construct the necessary curve, continue with the following steps.

7 With reference to figure 9-88, construct a free-form curve, an ellipse, and a line in the Front viewport.

Fig. 9-88. Closed-loop curve, ellipse, and line constructed.

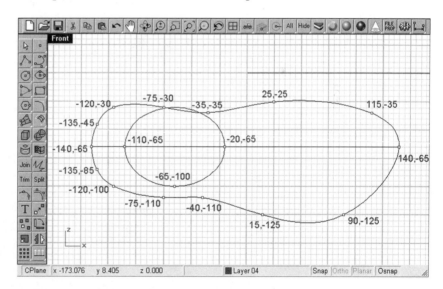

To construct three free-form curves and a line in the Front viewport, continue with the following steps.

8 With reference to figure 9-89, construct three free-form curves and a line in the Front viewport.

To remap the curves representing the top view of the shoe to the construction plane in the Top viewport, continue with the following steps.

474 CHAPTER 9: Consolidation

Fig. 9-89. Curves and line constructed.

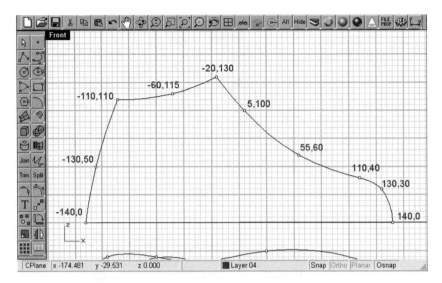

9 Set the display to a four-viewport configuration.

10 Select Transform > Orient > Remap to CPlane, or the Remap to CPlane button from the Transform toolbar.
   Command: RemapCPlan

11 Select the curves indicated in the Front viewport shown in figure 9-90 and press the Enter key.

12 Select the Top viewport.

Fig. 9-90. Curves being remapped.

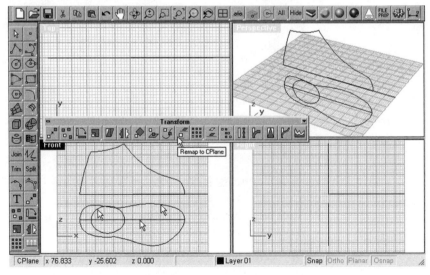

To construct a 3D curve from two planar curves residing in the Top viewport and the Front viewport, continue with the following steps.

Modeling Projects

13 Select Curve > From 2 Views, or the Curve From 2 Views button from the Curve Tools toolbar.

Command: Crv2View

14 Select the curves indicated in the Top and Front viewports shown in figure 9-91.

To move the two curves, continue with the following steps.

15 Maximize the Perspective viewport.

Fig. 9-91. 3D curve constructed from two planar curves.

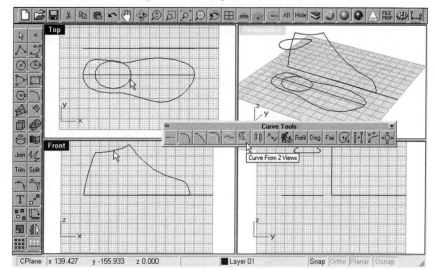

16 Move the curves and line indicated in figure 9-92.

Fig. 9-92. Curves being moved.

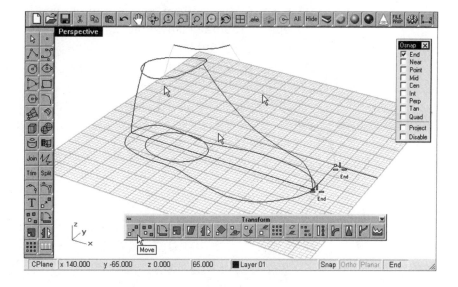

The general framework of the shoe is complete. As can be seen, you need some more curves to define the silhouette and profile. To join the two curves, establish a construction plane, and construct two curves on the plane, continue with the following steps.

17 Set the display to a four-viewport configuration.
18 Select Edit > Join, or the Join button from the Main toolbar.
Command: Join
19 Select curves A and B (see figure 9-93) and press the Enter key.

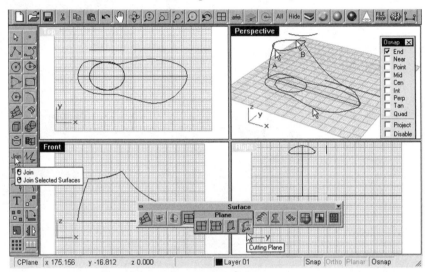

Fig. 9-93. Cross-section rectangle being constructed.

20 In the Join dialog box, click on the OK button.
21 Select Surface > Rectangle > Cutting Plane, or the Cutting Plane button from the Plane toolbar.
Command: CutPlane
22 Select the curves indicated in the Perspective viewport shown in figure 9-93 and press the Enter key.
23 Move the cursor to the Front viewport.
24 Type -75,190 at the command line area to specify the start of the cutting plane.
25 Type -75,0 at the command line area to specify the end of the cutting plane.
26 Press the Enter key.

To construct four points at the intersection of the cross-section rectangle and two curves, continue with the following steps.

27 Maximize the Perspective viewport.

Modeling Projects 477

28 Select Curve > From Objects > Intersection, or the Intersection button from the Curve From Object toolbar.

Command: Intersect

29 Select the objects indicated in figure 9-94 and press the Enter key.

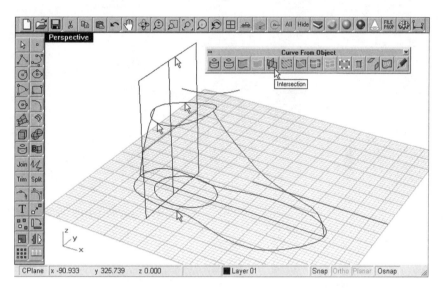

Fig. 9-94. Intersection points constructed.

To set up a new construction plane and hide the rectangular surface, continue with the following steps.

30 Select View > Set CPlane > 3 Points, or the Set CPlane/3 Points button from the Set CPlane toolbar.

Command: CPlane3Pt

31 Select end point A (see figure 9-95) to specify the CPlane origin.

32 Select end point B (see figure 9-95) to specify the X axis direction.

33 Select end point C (see figure 9-95) to specify the CPlane orientation.

34 Select Edit > Visibility > Hide, or the Hide Objects button from the Visibility toolbar.

Command: Hide

35 Select the rectangular surface A (see figure 9-95) and press the Enter key.

To construct two curves on the new construction plane, continue with the following steps.

36 With reference to figure 9-96, construct two free-form curves.

478   **CHAPTER 9: Consolidation**

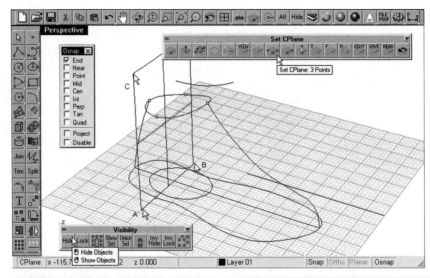

Fig. 9-95 Construction plane being set up.

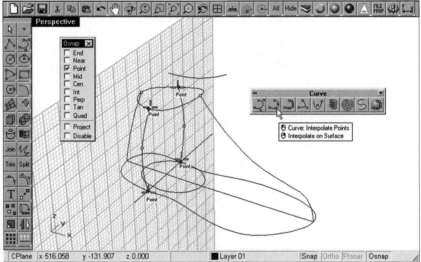

Fig. 9-96. Curves constructed.

Two curves are constructed. Continue with the following.

37 Repeat steps 17 through 36 to construct two move curves. (See figure 9-97.)

The curves for making the surface are complete. To construct a surface, continue with the following steps.

38 Set the current layer to *Layer 02*.

39 Select Surface > Sweep 2 Rails, or the Sweep Along 2 Rails button from the Surface toolbar.

Command: Sweep2

Modeling Projects

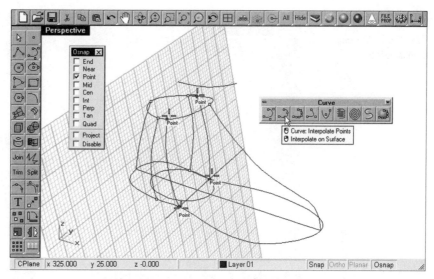

Fig. 9-97. Two curves on a new construction plane constructed.

40 Select curves A and B (see figure 9-98) as the rail curves.

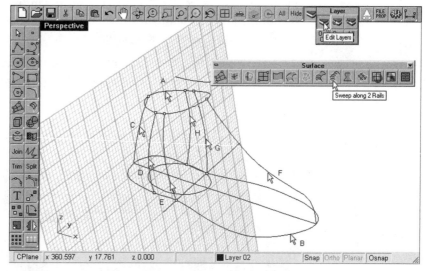

Fig. 9-98. Swept surface being constructed.

41 Select curves C, D, E, F, G, and H (see figure 9-98) as the cross-section curves and press the Enter key.

42 In the Sweep 2 Rails dialog box, click on the OK button.

43 Turn off *Layer 01*. (See figure 9-99.)

The model is complete. Save your file as *Shoe.3dm*.

Fig. 9-99. Surface constructed.

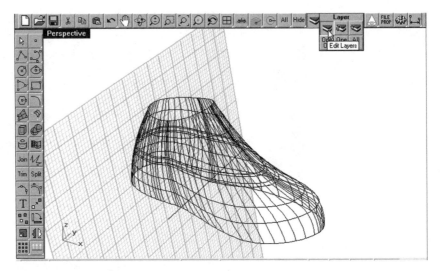

Model Car Assembly

In the following you will construct an assembly of the car chassis by merging the base, shaft, tire, wheel, and car body you constructed in previous chapters into a single file.

1 Start a new file. Use the metric (millimeter) template.

2 Select File > Import/Merge, or the Import/Merge button from the File toolbar.

Command: Import

3 In the Import File dialog box, select the file *Base.3dm* you constructed in Chapter 4.

4 Delete the curves indicated in figure 9-100.

Fig. 9-100. Base imported.

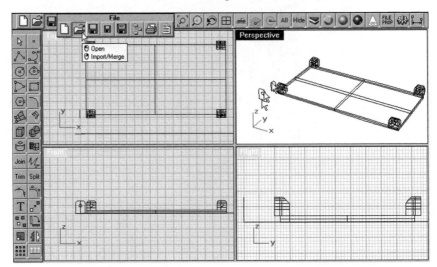

Modeling Projects

5   Repeat the Import command to import the file *Shaft.3dm* you constructed in Chapter 4.
6   Select and drag the file *Shaft.3dm* to the location indicated in figure 9-101.
7   Repeat the Import command to import the file *Wheel.3dm* and drag it to the position shown in figure 9-102.

Fig. 9-101. Shaft imported and relocated.

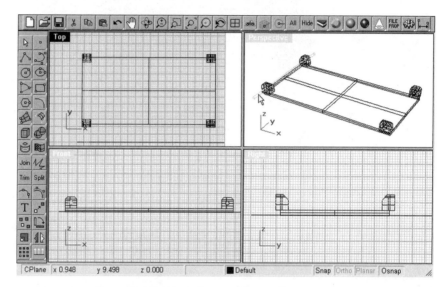

Fig. 9-102. Wheel imported and positioned.

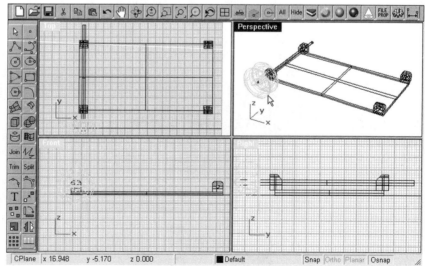

8   Use the Import command again to import the file *Tire.3dm* and drag it to the position shown in figure 9-103.

To mirror the tire and wheel, continue with the following steps.

Fig. 9-103. Tire imported.

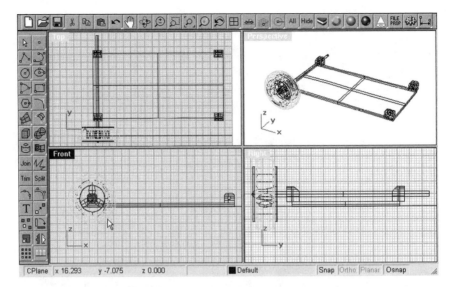

9 Check the Mid box in the Osnap dialog box.

10 With reference to figure 9-104, construct a mirror copy of the tire and wheel.

Fig. 9-104. Tire and wheel being mirrored.

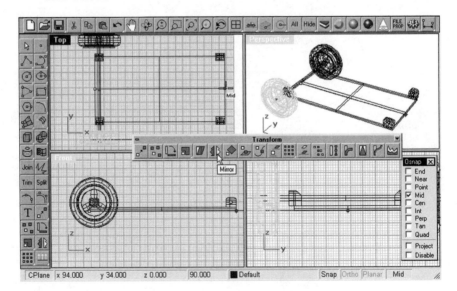

11 Repeat the Mirror command to mirror the tires, wheels, and shaft. (See figure 9-105.)

12 To complete the model, import the file *Beetle.3dm* you constructed in Chapter 3 and position it in accordance with figure 9-106.

13 Turn off *Layer 01*, *Layer 02*, *Layer 03*, and *Layer 04*.

Fig. 9-105. Tires, wheels, and shaft being mirrored.

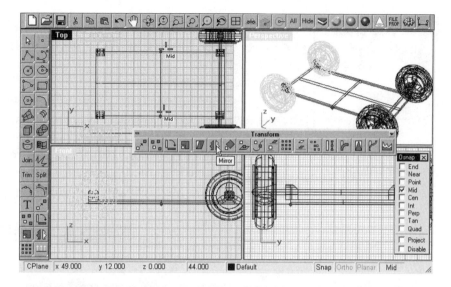

Fig. 9-106. Body imported and positioned.

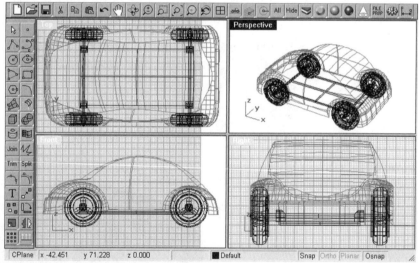

The assembly is complete. Save your file as *Car.3dm*. Note that there is no data associativity between the original solid models and the assembly.

Medal

In the following you will export a solid in STL format for use in rapid prototyping. Before using the STL file, you will open it and inspect it for any gaps or openings.

484 CHAPTER 9: Consolidation

Fig. 9-107. STL Export Options dialog box.

1 Open the file *Heightfield.3dm* you constructed in Chapter 4.
2 Select File > Save As, or the Save As button from the File toolbar.
   Command: SaveAs
3 In the Save As dialog box, select *Stereolithography (*stl)* from the Save As Type box, specify a file name, and click on the OK button.
4 In the STL Export Options dialog box (figure 9-107), select Binary, and click on the OK button.
5 In the Polygon Mesh Options dialog box (figure 9-108), click on the OK button.

Fig. 9-108. Polygon Mesh Options dialog box.

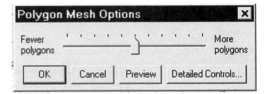

An STL file is exported. To open the STL file (figure 9-109), continue with the following.

6 Click on the STL file to open it.

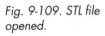

Fig. 9-109. STL file opened.

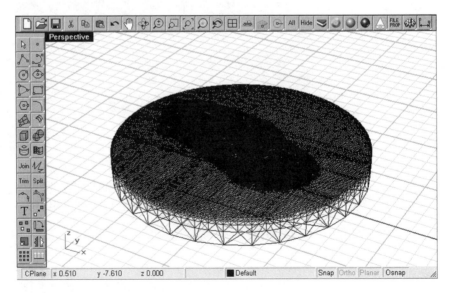

You may now use the polygon mesh tools to make any necessary modifications. If there is no gap or opening, you can use the file to create a rapid prototype. Figures 9-110 and 9-111 show, respectively, the STL file being processed and rapid prototype being constructed.

Summary

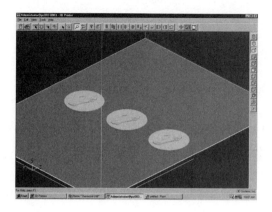

Fig. 9-110. STL file being processed.

Fig. 9-111. Rapid prototype being constructed.

■■ Summary

Although it is logical to construct curves and/or points before making surfaces and polysurfaces, detailing a 3D free-form model needs intertwined construction of curves, points, and surfaces. You construct surfaces from points and curves, and you construct points and curves from surfaces.

To construct objects in 3D space, you input a set of coordinates at the command line area or use the input device to select points in the viewports. Because there is an imaginary construction plane in each of the viewports, selecting a point in a viewport specifies a point in the corresponding viewport construction plane. To construct points and curves in 3D, you manipulate construction planes in various ways.

To reuse model data from upstream applications and export data to downstream operations, you import and export. To import, you open a file. To export, you save the file with a specific file format. Basically, Rhino objects can be categorized into four types: points and curves, NURBS surfaces and polysurfaces, solids, and polygon meshes. You can export a solid as wireframe, surface, solid, or even polygon mesh. On the other hand, you cannot import a polygon mesh to become a NURBS surface. To work on mesh polygons, you use various polygon mesh manipulation methods.

You also learned more advanced techniques in curve and point construction, manipulation of construction planes, and ways to export Rhino objects. You also explored the use of polygon meshes.

Review Questions

1. Outline the methods of deriving curves from existing surface objects.
2. Give a brief account of methods of setting up the construction plane in the viewport.
3. List the file formats that can be opened and saved in Rhino.
4. Briefly explain the tools used to manipulate polygon meshes.

Chapter 10

Presentation

■■ Objectives

The goals of this chapter are to explain the key concepts involved in rendering, to introduce methods of constructing a rendered image, and to explore the process of constructing 2D engineering drawings. After studying this chapter, you should be able to:

- ❐ State the key concepts involved in rendering
- ❐ Construct photorealistic rendered images from 3D models
- ❐ Describe the process of constructing 2D engineering drawings

■■ Overview

In addition to exporting to other computerized applications for downstream operations, you can present your 3D models as rendered images for better visualization and as 2D engineering drawings for communication in a conventional engineering form. This chapter explores methods by which 3D objects are "visualized" in the computer, methods of adding realism to a 3D object visually, and methods of representing 3D objects as 2D drawings.

■■ Shaded and Wireframe Viewport Display

In reality, there are no curves or lines on a 3D free-form smooth surface. Therefore, you should find only edges on the surface's boundaries. However, boundary edges alone do not provide sufficient information to depict the profile and silhouette of the surface. Hence, a set of isoparm curves in two orthogonal directions, color shading, or a set of isoparm curves together with color shading is used to better illustrate a free-form object in the computer display.

By default, objects are displayed in the viewport in a color assigned to the layer at which the objects reside and in wireframe mode with isoparm curves. To change the color of an object, you change the layer's color or the object's color. By shading the object, you distinguish it from the viewport background. To shade the object and display isoparm curves simultaneously, you set the display to shaded mode.

Color

The default color of a Rhino object is determined by the setting *Bylayer*. This means that the color of the object is determined by the color assigned to the layer in which the object resides. Apart from selecting a named color (a preestablished color option), you can set color using the color swatch; by specifying hue, saturation, and value; and by specifying red, green, and blue values (the component mix of a particular color). To set the color of an object using the color swatch, perform the following steps.

1 Open the file *JoyPad.3dm*.

2 Select Edit > Object Properties, or the Object Properties dialog box from the Properties toolbar.

Command: Properties

3 Select the polysurface and press the Enter key.

4 In the Object tab of the Properties dialog box (figure 10-1), select Other from the Color pull-down box to display the Select Color dialog box.

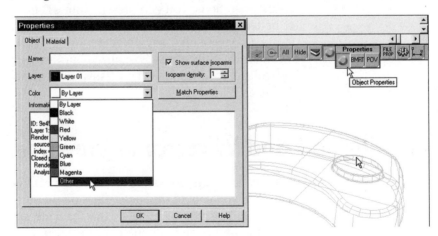

Fig. 10-1. Wireframe color being changed.

5 With reference to figure 10-2, select a color in the circular ring in the color swatch to set the hue, select a color in the square box in the color swatch to set the saturation and value, and click on the OK button.

Shaded and Wireframe Viewport Display

The object's wireframe color is set. In a collaborative design environment, specifying color values (hue, saturation, and value, or red, green, and blue) can accurately convey the color value.

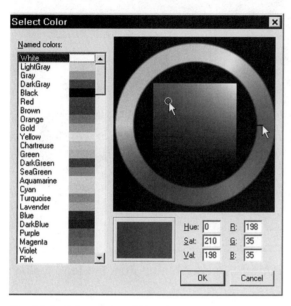

Isoparm Density

In wireframe mode, there are isoparm curves placed on the surface. For a very simple surface, such as a planar surface, one or two isoparm curves are adequate to provide enough information on the curvature of the surface. For more complex surfaces, you need more isoparm curves. To set the isoparm density of an object, perform the following steps.

1 In the Object tab of the Properties dialog box (figure 10-3), set isoparm density to *10* and click on the OK button.

The isoparm density is increased. Although there are more isoparm curves to better represent the profile of the surface, selection of individual objects from a bunch of objects may become difficult.

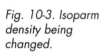

Fig. 10-2. Select Color dialog box.

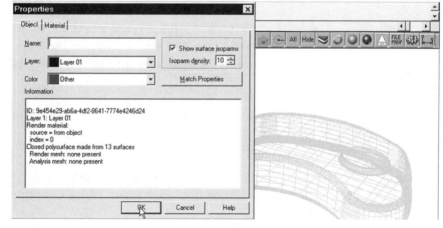

Fig. 10-3. Isoparm density being changed.

Shading and Shade Option

Shading applies a color with highlights to the surface of an object. It lets you see clearly an object's profile and silhouette. The color applied to the surface of an object when it is shaded is determined by the setting in the Shade tab (figure 10-4) of the Options dialog box. To set the color of shading and shade an object, perform the following steps.

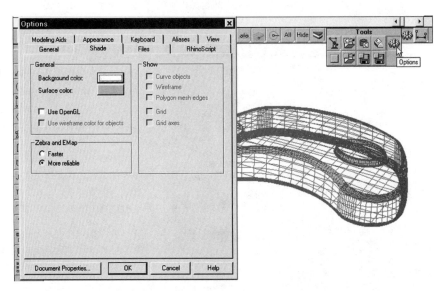

Fig. 10-4. Shade tab of the Options dialog box.

1 Select Tools > Options, or the Options button from the Tools toolbar.
Command: Options

2 In the Shade tab of the Options dialog box, select the background color swatch.

3 In the Select Color dialog box, select a color and click on the OK button.

4 Select the surface color swatch, select a green color from the Select Color dialog box, and click on the OK button.

5 Click on the OK button in the Options dialog box.

Background and surface colors are set. Continue with the following.

6 Select Render > Shade, or the Shade button from the Render toolbar.
Command: Shade

The object is shaded with a color you set in the surface color swatch of the Options dialog box. (See figure 10-5.) Continue with the following steps.

7 Select anywhere on the screen to return to wireframe display mode.

To set the Shade option and shade the object again, continue with the following steps.

Shaded and Wireframe Viewport Display

Fig. 10-5. Object shaded in surface color.

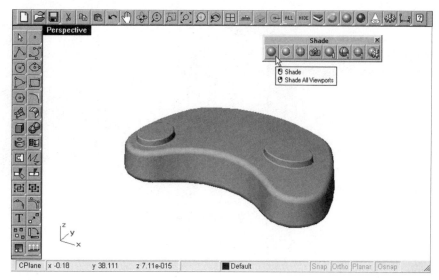

8 Select Tools > Options, or the Options button from the Tools toolbar.
Command: Options

9 In the Shade tab, check the Use OpenGL box and the *Use layer color for objects* box, and click on the OK button.

10 Select Render > Shade, or the Shade button from the Render toolbar.
Command: Shade

The color setting specified in the surface color swatch of the Options dialog box is ignored. The object is shaded with the wireframe color.

Shaded Display Mode

To apply a shaded color and display isoparm curves concurrently, you set the viewport to the Shaded display mode, as follows.

1 Right-click on the viewport title and select Shaded.
Command: ShadedViewport

The viewport is set to Shaded mode. (See figure 10-6.) Note that the model expresses two colors: a wireframe color you assigned to the Properties dialog box and a shading color you assigned in the Options dialog box.

Wireframe Display Mode

To change the display to Wireframe mode, perform the following.

1 Right-click on the viewport title and select Wireframe.
Command: WireframeViewport

*Fig. 10-6.
Viewport set to
Shaded mode.*

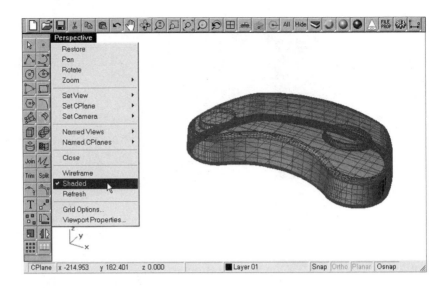

The display is set to Wireframe mode.

Rendering

Rendering is a method of producing a photorealistic image of an object in a 3D scene. Unlike shading, which simply applies a shaded color to the surface of an object, rendering takes into account the material properties (color and texture) assigned to the object and the effect of lighting in the scene.

Material Properties

One of the two crucial elements that contribute to a photorealistic rendered image is the digital material you apply to an object. Rhino provides three methods of assigning material properties: via layer, via plug-in, and via the Basic (property assignment) option. In the first method, you assign material properties to a layer and let the object use the layer's material properties. In the second and third methods, you assign material properties via plug-in or via the Basic (property assignment) option. Strictly speaking, there are only two categories of material assignment in Rhino: plug-in and basic. The "by layer" method makes use of one of these categories.

Assigning Material Properties by Layer

To set the material properties of a group of objects globally, you assign a material (designated in Rhino as either "plug-in" or "basic") to a layer

Rendering

and let the objects residing on the layer use the layer's material. To set material properties by layer, perform the following steps.

1 Select Edit > Object Properties, or the Object Properties dialog box from the Properties toolbar.
Command: Properties

2 Select the polysurface and press the Enter key. (See figure 10-7.)

3 In the Material tab of the Properties dialog box, select Layer in the Assign By box and click on the OK button.

The selected surface will now take on the characteristics of the material assigned to the layer on which the surface resides.

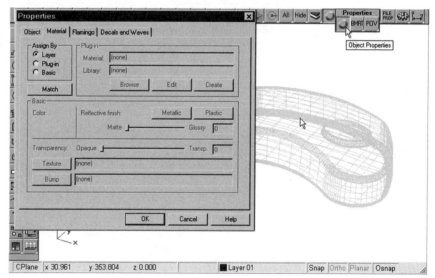

Fig. 10-7. Material tab of the Properties dialog box.

To assign material to the layer, continue with the following steps.

4 Select Edit > Layers > Edit Layers, or the Edit Layers button from the Layer toolbar.
Command: Layer

5 In the Edit Layers dialog box, select a row in the Material column (figure 10-8) to bring out the Material Properties dialog box. (See figure 10-9.)

The Plug-in and Basic options for assigning materials are found in the Material Properties dialog box. These options, explored in the sections that follow, determine the material.

494 CHAPTER 10: Presentation

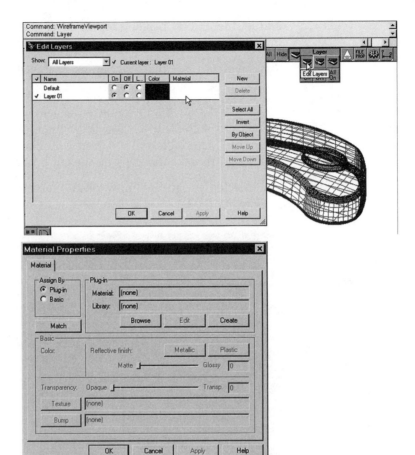

Fig. 10-8. Selecting a row in the Material column.

Fig. 10-9. Material Properties dialog box.

Material Properties Assignment via Plug-in

Assigning material properties via plug-in means that you take a material properties source external to the Rhino program and bring it into a library. You can then assign material properties from that library. To use the default material library or a library you create yourself, you use the Flamingo Photometric or Raytrace renderer.

Using plug-in material is simple. You select material from the plug-in material library, set the parameters of the material if necessary, and apply the material to the object. To use the plug-in, you need to select the Flamingo Photometric renderer or Flamingo Raytrace renderer option. The process works as follows.

1 In the Material Properties dialog box, click on the Browse button. If the Browse button is grayed out, exit the command, select Render > Current Renderer > Flamingo Radiosity (or Flamingo Raytrace), and retry the command.

Rendering

2 In the Material Library dialog box (figure 10-10), select CAR PAINTS in the browser, select Blue Metallic Clear Finish, and click on the OK button.

3 In the Material Properties dialog box, click on the Edit button. In the Material Editor dialog box, change the parameters of the plug-in as required. When you are satisfied with the settings, click on OK. Upon returning to the Material Properties dialog box, click on OK.

The layer's material is set. Because the object uses the layer's material, the layer's material also applies to the object.

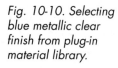

Fig. 10-10. Selecting blue metallic clear finish from plug-in material library.

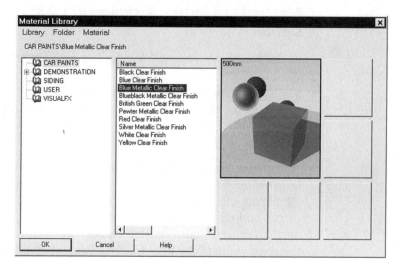

To render the object, continue with the following steps.

4 Select Render > Current Renderer > Flamingo Raytrace to use the Flamingo Raytrace renderer. (You will learn more about the Flamingo Raytrace renderer later in this chapter.)

5 Select Photometric > Render, or the Render button from the Main toolbar.

Command: Render

The object is rendered per the material property assignments, as shown in figure 10-11.

6 To save the rendered image, select File > Save As.

7 In the Save Bitmap dialog box, select a file type, specify a file name, and click on the Save button.

Fig. 10-11. Object with blue metallic clear finish rendered via the Flamingo Raytrace renderer.

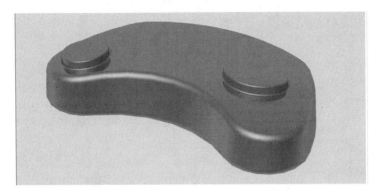

Material Properties Assignment via the Basic Material Option

In the following you will assign material properties via Rhino's Basic option, which incorporates five main settings: basic color, reflective finish, transparency, texture, and bump. Basic color is the main color of the object. Reflective finish concerns how the object reflects light. You set the color and the strength of the reflection and you choose either a metallic reflective finish or plastic reflective finish.

Transparency sets the opacity of the object. You set the object in the range of 100% opaque to 100% transparent. Texture and bump are bitmap mappings applied to the surface. Texture bitmap fundamentally places an image over the surface with the effect of "painting a picture" on the surface. Bump bitmap causes the surface to look bumpy, using the color value of the bitmap. To assign these material properties via the Basic option, perform the following steps.

1 Select Edit > Object Properties, or the Object Properties dialog box from the Properties toolbar.
Command: Properties

2 Select the polysurface and press the Enter key.

3 In the Material tab of the Properties dialog box, select Basic in the Assign By box.

4 In the Reflective Finish box, move the slider bar to the right to set the glossy value to *80*, select the color swatch and a whitish color, and click on the Plastic button.

5 In the Transparency box, move the slider bar to the right to set the transparency value to *32*. (See figure 10-12.)

6 Click on the OK button.

To render the object, continue with the following steps. (See figure 10-13.)

7 Select Render > Current Renderer > Rhino.

Rendering

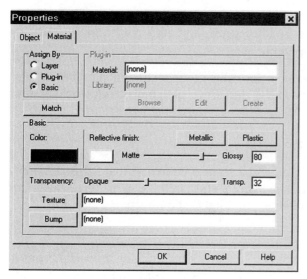

Fig. 10-12. Basic material set.

8 Select Render > Render, or the Render button from the Main toolbar.
Command: Render

9 Select File > Save As to save your image.

To apply a texture bitmap and render the object, continue with the following steps. (See figure 10-14.)

10 Invoke the Properties command and select the polysurface.

11 In the Properties dialog box, select the Texture button, select a bitmap file, and click on the OK button.

12 Invoke the Render command.

In the rendered image shown in figure 10-14, the texture map masks the basic color. To apply a bump bitmap and render the object again, continue with the following steps. (See figure 10-15.)

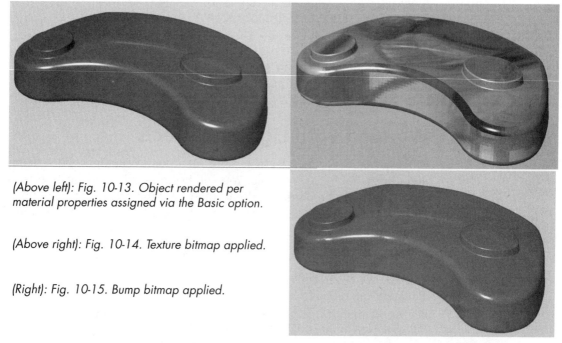

(Above left): Fig. 10-13. Object rendered per material properties assigned via the Basic option.

(Above right): Fig. 10-14. Texture bitmap applied.

(Right): Fig. 10-15. Bump bitmap applied.

13 Invoke the Properties command and select the polysurface.

14 In the Properties dialog box, clear the texture bitmap, click on the Bump button, select a bitmap file, and click on the OK button.

15 Invoke the Render command.

In the rendered image, the color value of the texture map causes the surface to look bumpy. Material properties assignment is complete. Save your file.

Lighting

To better control the lighting in a scene, you add virtual lights and apply sunlight and environmental lighting effects. Rhino offers three categories of virtual lighting: spotlights, point light, and directional light.

Spotlights

A spotlight is characterized by a light source emitting a conical beam of light toward a target. It illuminates a conical volume in the scene. In working with spotlights in Rhino, you specify the light source, the target, and the diameter of the base Main toolbar of the cone.

Point Light

A point light is analogous to an ordinary electric light bulb that emits light in all directions. Point light illuminates an entire scene. In working with point light in Rhino, you specify a location of the light source.

Directional Light

Directional lighting derives from a light source at a distance, such as the sun. This lighting, generally represented as a beam or parallel beams of light, illuminates an entire scene. You specify a direction vector.

Working with Lighting Effects

In this section you will learn how to incorporate spotlight, point light, and directional light in a scene. You will begin by constructing objects to form a scene to be illuminated and then light the scene using the three types of lighting effects. To create the scene to be illuminated, perform the following steps.

1 Open the file *Joypad.3dm* you constructed in Chapter 7.
2 Join the surfaces of the mobile phone to form a single polysurface.
3 With reference to figure 10-16, construct two boxes.
4 Apply material properties to the boxes and the mobile phone. (Use solid colors, cadet, medium, glossy for the boxes, and car paint, yellow clear finish for the mobile phone.)

Rendering

Fig. 10-16. Two boxes constructed.

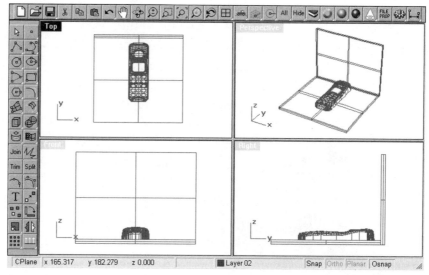

5 Set the current layer to *Layer 03*. You will construct the light-source objects in this layer.

To construct a spotlight and render the scene, continue with the following steps.

6 Select Render > Create Spotlight, or the Create Spotlight button from the Lights toolbar.

Command: Spotlight

7 Type *V* at the command line area to use the Vertical option.

8 Select the point indicated in the Top viewport in figure 10-17 to specify the base of the cone.

Fig. 10-17. Spotlight being constructed.

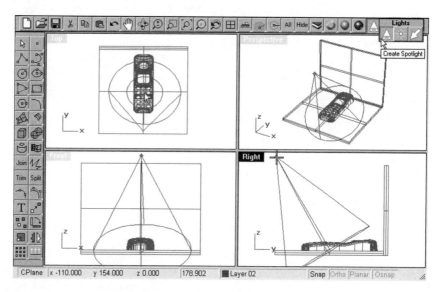

9 Select another point indicated in the Top viewport in figure 10-17 to specify the radius of the cone.

10 Select the point indicated in the Right viewport in figure 10-17 to specify the end of the cone.

11 Maximize the Perspective viewport.

12 Select Render > Current Renderer > Flamingo Raytrace to use the Flamingo Raytrace renderer to render the viewport. (See figure 10-18.)

To adjust the parameters of the light, continue with the following steps.

13 Select Edit > Object Properties, or the Properties button from the Properties toolbar.

Command: Properties

14 Select the spotlight and press the Enter key.

15 In the Light tab of the Properties dialog box (figure 10-19), select the color swatch, set the color to green, and click on the OK button to exit.

16 Render the viewport again. (See figure 10-20.)

Fig. 10-18. Perspective viewport rendered.

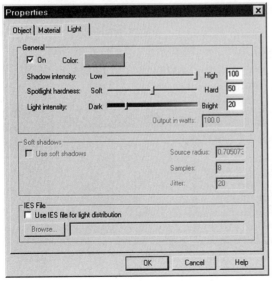

Fig. 10-20. Perspective rendered with a green spotlight.

Fig. 10-19. Color of spotlight being changed.

To turn off the spotlight, construct a point light, and render the scene, continue with the following steps.

17 Apply the Properties command to the spotlight.

Rendering

18 In the Light tab of the Properties dialog box, deselect the On button and click on the OK button.

19 Set the display to a four-viewport configuration.

20 Select Render > Create Point Light, or the Create Point Light button from the Lights toolbar.

Command: PointLight

21 With reference to figure 10-21, select a point in the Front viewport.

Fig. 10-21. Point light being constructed.

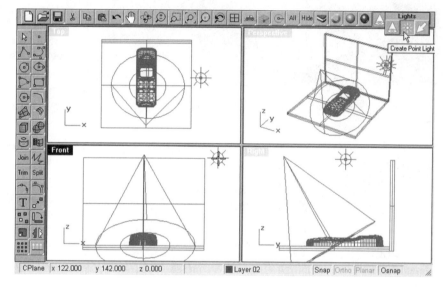

22 Maximize the Perspective viewport.

23 Render the viewport. (See figure 10-22.)

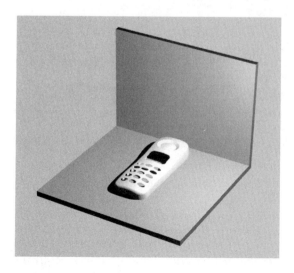

To change the intensity of the point light and render the scene again, continue with the following steps.

24 Apply the Properties command to the point light.

25 In the Light tab of the Properties dialog box, move the slider bar in the Light Intensity box to set the value to *90*, and click on the OK button.

26 Render the scene. (See figure 10-23.)

Fig. 10-22. Viewport rendered.

502 CHAPTER 10: Presentation

Fig. 10-23. Intensity of point light changed to 90.

To turn off the point light, construct a direction light, and render the scene, continue with the following steps.

27 Apply the Properties command to the point light.

28 In the Light tab of the Properties dialog box, deselect the On button and click on the OK button.

29 Return the display to a four-viewport configuration.

30 Select Render > Create Directional Light, or the Create Directional Light button from the Lights toolbar.

Command: DirectionalLight

31 With reference to figure 10-24, select two points in the Right viewport to specify the start and end of the light direction vector.

Fig. 10-24. Directional light being constructed.

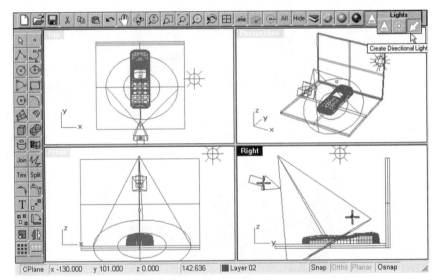

32 Maximize the Perspective viewport and render the viewport. (See figure 10-25.)

Construction of the light source is complete. Save your file.

Renderer

Rhino incorporates the capacity to produce output from three types of renderers: Rhino, Flamingo Photometric, and Flamingo Raytrace. In the sections that follow you will render images using each of these renderers.

Rendering

Fig. 10-25. Viewport rendered with directional light.

Rhino Renderer

The Rhino renderer is the basic renderer. In producing a rendered image, this renderer takes into account the materials assigned to objects and lights included in a scene. To set Rhino renderer options and produce a rendered image, perform the following steps.

1   Open the file *Joypad.3dm*.
2   With reference to figure 10-26, construct a box.
3   Apply material properties to the box. (Use the Wood, Solid, Stained, Dark, and Medium Gloss material options from the plug-in material library.)

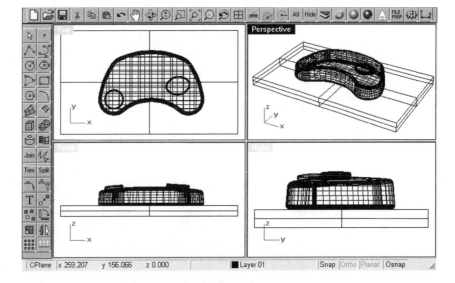

Fig. 10-26. Box constructed and material properties applied.

4   Select Render > Current Renderer > Rhino to use the Rhino renderer.
5   Select Render > Properties. (See figure 10-27.)
    Command: RenderOptions

The Rhino (renderer) tab includes five option fields, the functions of which are summarized in table 10-1.

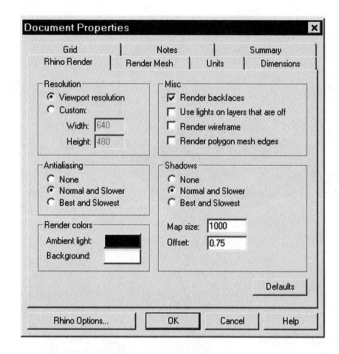

Fig. 10-27. Rhino Render tab of the Document Properties dialog box.

Table 10-1: Rhino Tab Option Fields and Their Functions

Field	Function
Resolution	Controls the resolution of the rendered image.
Antialiasing	Controls how jagged edges in an image are treated. Because each discrete pixel in an image has a unique color value, an inclined edge in a rendered image may look jagged. To minimize this jagged effect, adjacent pixels in the inclined edges are averaged. The process is called antialiasing. Antialiasing produces a blurred edge to mask the jagged effect. For example, in a red box on a white background image, pixels are either red or white. Antialiasing produces a set of pinkish pixels between the red and white pixels.
Misc	Controls the rendering of backfaces, the use of lights in regard to layers that are turned off, the display of wireframe in a rendering, and the display of mesh edges in a rendering.
Shadows	Controls the treatment of shadows.
Render Colors	Controls ambient light color and background color. Ambient light is an approximation of all indirect light in the scene. Indirect light is normally referred to as radiosity, which is the effect of reflection from all objects in a scene. To accurately simulate indirect light, you use the Flamingo Photometric renderer.

Continue with the following steps.

Rendering

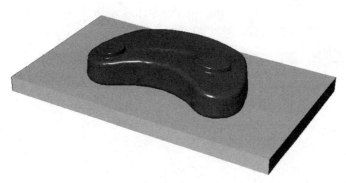

Fig. 10-28. Antialiasing turned off.

6 In the Rhino (renderer) tab of the Document Properties dialog box, select None in the Antialiasing box, select the background color swatch and set the color to white, and click on the OK button.

7 Maximize the Perspective viewport.

8 Select Render > Render to render the viewport using the Rhino renderer. (See figure 10-28.)

Note in figure 10-28 that the plug-in materials applied to the box and the joypad do not yet show up. Continue with the following steps.

Fig. 10-29. Best antialiasing effect.

9 Select Render > Properties to invoke the RenderOptions command again.

10 In the Rhino (renderer) tab, select Best and Slowest in the Antialiasing box and click on the OK button.

11 Select Render > Render to render the viewport. (See figure 10-29.)

Compare figures 10-28 and 10-29. The edges of the antialiased image are smoother.

Flamingo Raytrace Renderer

Raytracing is a technique that incorporates the effect of the path of light rays from all light sources in a scene to the viewer's eye. Raytracing takes into account the intensity of the light source (or sources), and the transparency and reflectivity of objects between the light source and the viewer's eye. The Flamingo Raytrace renderer uses raytracing techniques to compute the combined effect of all direct lighting. To employ the Flamingo Raytrace renderer, perform the following steps.

1 Select Render > Current Renderer > Flamingo Raytrace to use the Flamingo Raytrace renderer.

2 Select Render > Properties. (See figure 10-30.)

Command: RenderOptions

3 In the Flamingo tab of the Document Properties dialog box, click on the Environment button. (See figure 10-31.)

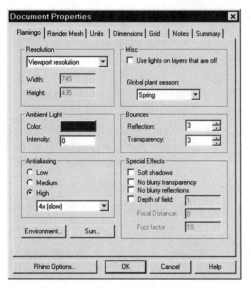

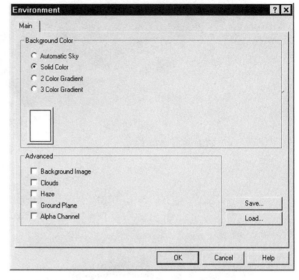

Fig. 10-30. Flamingo Raytrace renderer options.

Fig. 10-31. Environment dialog box.

4 In the Environment dialog box, select the color swatch in the Background color box, select a white color, and click on the OK button to exit.

5 Return to the Flamingo tab of the Document Properties dialog box and click on the Sun button.

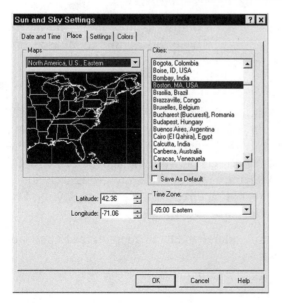

6 The Sun and Sky Settings dialog box (see figure 10-32) includes four tabs containing options for specifying date and time, place, setting, and colors. Select the Place tab and select the place where you live to specify the location of the sun. Click on the OK button.

7 Return to the Document Properties dialog box and click on the OK button.

8 Select Raytrace > Render to use the Flamingo Raytrace renderer to render the viewport. (See figure 10-33.)

Fig. 10-32. Sun and Sky Settings dialog box.

Rendering

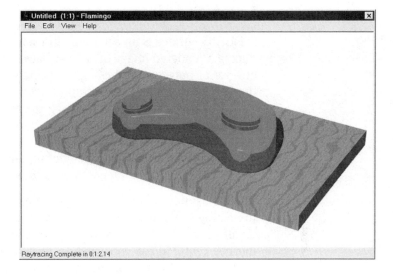

Fig. 10-33. Image rendered using the Flamingo Raytrace renderer.

Flamingo Photometric Renderer

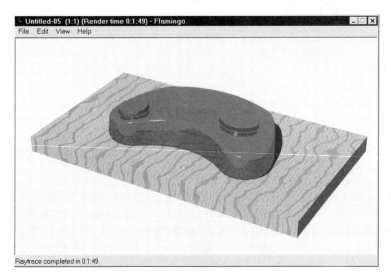

Fig. 10-34. Raytraced radiosity image.

Radiosity refers to the indirect lighting in a scene. Unlike rendered ambient light, which approximates indirect lighting, radiosity is a more accurate representation of the light emanating from all directions. You use radiosity for scenes in which the indirect lighting effect is significant. The Flamingo Photometric renderer adds to the rendered image subtle radiosity lighting effects produced by the combination of all indirect lighting sources. To employ the Flamingo Photometric renderer, perform the following steps.

1 Select Render > Current Renderer > Flamingo Photometric to use the Flamingo Photometric renderer.

2 Render the viewport.

3 Leave the default setting of Sun.

An image is generated. (See figure 10-34.)

Save Small

A NURBS surface is an accurate representation of a smooth surface. Producing a rendered image of such surfaces typically requires the existence of a set of polygon meshes representing the surfaces. When you invoke the Render command, the computer constructs a set of polygon meshes and uses the meshes for rendering purposes. Normally, these polygon meshes are preserved when you save your file. The next time you render the object (provided it is not modified), the computer will use the saved set of meshes for rendering. Rendering takes less time this way because polygon meshes do not have to be reconstructed.

The inclusion of polygon mesh data makes file sizes larger. To minimize the storage requirement for polygon mesh data, you can use the Save-Small command, which saves the NURBS surfaces but not the polygon meshes required for construction of a rendered image. However, as a result, the next time you open the file and want to render the object, you will have to reconstruct the set of polygon meshes.

1 To save small, select File > Save Small.
Command: SaveSmall

■■ 2D Drawing

2D orthographic engineering drawings are the conventional means of communication among engineering personnel. Although the advent of computer-aided design applications replaced some of the uses of 2D drawings, there remain many situations or purposes for which 2D drawings and/or 2D drawing output is useful or necessary.

One important function of 2D drawings is to specify precisely the dimensions of the objects they represent, along with annotations that convey other information about the object or objects represented. In Rhino, you use the Dimension toolbar (shown in figure 10-35) to create 2D drawings and incorporate dimensions and annotations.

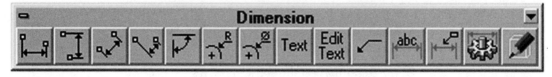

Fig. 10-35. Dimension toolbar.

The process of outputting a 3D model as an engineering (2D) drawing is fairly simple using Rhino. Basically, you select a command and let the computer do all of the 2D drawing construction work, as follows.

1 Open the file *Tire.3dm* you constructed in Chapter 8.

2D Drawing

2 Select Dimension > Make 2-D Drawing or the Make 2-D button from the Dimension toolbar.
Command: Make2d

3 Select the solid indicated in figure 10-36 and press the Enter key.

4 In the Make2-D Options dialog box, click on the OK button.

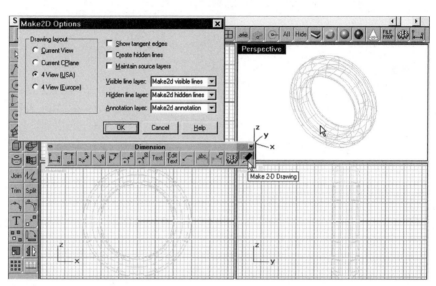

Fig. 10-36. 2D drawing being produced.

Visible lines and hidden lines of the front, top, side, and isometric views of the selected object are constructed. Continue with the following steps.

5 Set the current layer to *Make2dvisiblelines* and turn off the default layer.

6 Maximize the Top viewport. (See figure 10-37.)

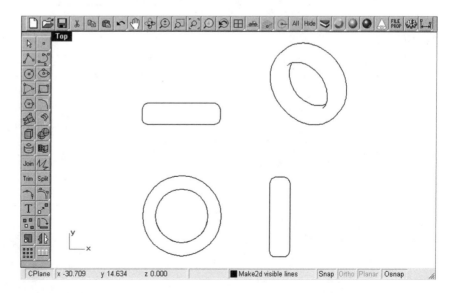

Fig. 10-37. Front, side, top, and isometric views constructed.

■■ Dimensioning and Annotation

2D drawings typically indicate the dimensions of the 3D object or objects they represent, and typically incorporate informative annotations. To add dimensions and annotations to a 2D Rhino drawing, perform the following steps.

1   Select Dimension > Properties, or the Dimension button from the Dimension toolbar.

    Command: DimOptions

2   In the Dimensions tab of the Document Properties dialog box, set text height to 3 units and arrow length to 3 units, and click on the OK button. (See figure 10-38.)

3   With reference to figure 10-39, add dimensions and annotations.

The 2D drawing is complete. Save your file.

Fig. 10-38. Dimension options.

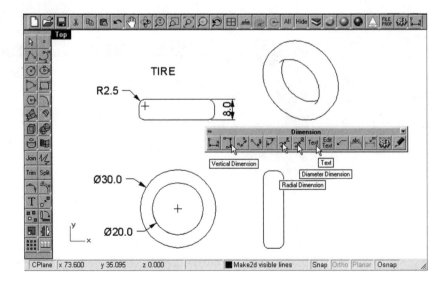

Fig. 10-39. Dimensions and annotations added.

■■ Presentation Projects

In the following sections you will work with the models you have developed in other chapters. These include the joystick, the medal, and the toy car.

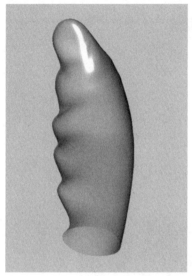

Fig. 10-40. Joystick rendered.

Joystick

To apply material properties to the joystick you constructed in Chapter 8 and create a rendered image, perform the following steps.

1 Open the file *Joystick.3dm* you created in Chapter 8.
2 Apply material properties to the joystick.
3 Access the Basic material option in the Material tab of the Properties dialog box.
4 Select a green color.
5 Select a white reflective color.
6 Move the slider bar in the Reflective Finish box to the right to set the reflective value to *80*.
7 Select the Plastic finish.
8 Move the slider bar in the Transparency box to the right to set the transparency value to *50*.
9 Use the Rhino renderer to render the viewport. (See figure 10-40.)

Medal

To apply material properties to the medal you constructed in Chapter 8, perform the following steps.

1 Open the file *Heightfield.3dm*.
2 With reference to figure 10-41, construct a box and a directional light.
3 Maximize the Perspective viewport.
4 Access the Flamingo Raytrace renderer.
5 Apply plug-in material. Select the Metals, Gold, Polished, Plain options for the medal, and the Wood, Pine, Yellow, Natural, and Polished options for the box.
6 Render the viewport. (See figure 10-42.)

Fig. 10-41. Box and directional light constructed.

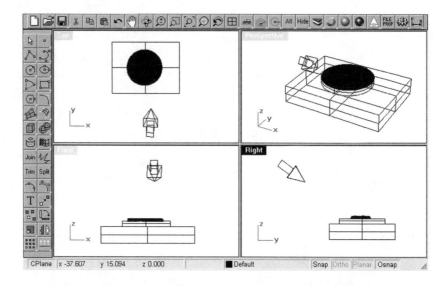

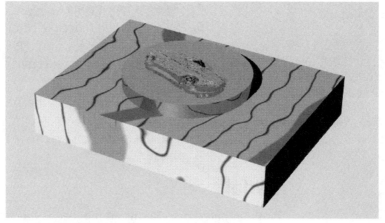

Fig. 10-42. Rendered viewport.

Toy Car Model

To apply material properties to the toy car model you constructed in Chapter 9 and output a 2D engineering drawing of it, perform the following steps.

1 Open the file *Car.3dm*.
2 With reference to figure 10-43, construct two boxes and a spotlight.
3 Maximize the Perspective viewport.
4 Access the Flamingo Raytrace renderer.
5 Apply material from the plug-in material library to various component parts of the car in accordance with table 10-2.
6 Render the viewport. (See figure 10-44.)

Presentation Projects

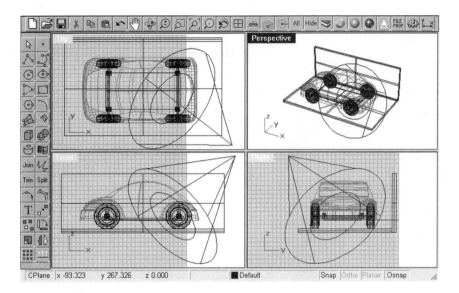

Fig. 10-43. Boxes and spotlight constructed.

Table 10-2: Library Material Characteristics to Apply to Model Car Parts

Model Car Part	Material Characteristics
Car body	Car Paints, Yellow Clear Finish
Base	Plastics, Green, Forest, Medium, Smooth
Shaft	Metals, Stainless Steel, Polish, Plain
Tire	Plastics, Blues, Cadet, Dark, Smooth
Wheel	Plastics, Reds and Oranges, Red, Light, Smooth
Horizontal box	Concrete, Exposed Aggregate, Tan
Vertical box	Concrete, Non-Uniform, Gray, Reflective

Fig. 10-44. Rendered viewport.

7 Output an engineering drawing. (See figure 10-45.)

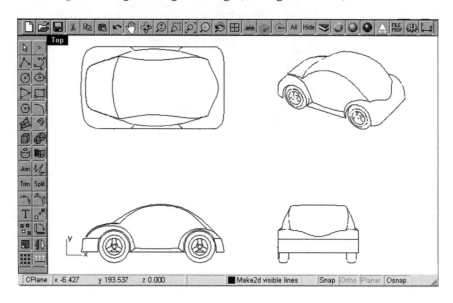

Fig. 10-45.
Engineering
drawing.

■■ Summary

Shading is a means of enhancing the visual representation of a 3D object by applying a color shade to the surface of the object to contrast it from its surrounding background. Because you have the option of shading and displaying a surface or shading a surface and displaying its isoparm curves, there are two color settings: wireframe color and surface color.

Wireframe color is applied to isoparm curves. It is the color you assign to the object in the Properties dialog box. If the color is Bylayer, the wireframe color is the same color as the color assigned to the layer on which the object resides. Hence, for example, an object with a Bylayer color moved from a layer of yellow color to a layer of red color will change from yellow to red.

On the other hand, an object having a designated color will retain its color in any layer. Surface color applies to the surface. You set the surface color in the Options dialog box. When you shade the surface, the shading color can be either the surface color or the wireframe color, depending on the setting you made in the Options dialog box.

To add realism to an object, you apply material properties. Strictly speaking, there are two option categories of material properties: Plug-in and Basic. A huge repertoire of material is available in the plug-in material library. In this process, you choose a material, establish the parameters of application of the material, and apply the material to the object. The other category of material is Basic. In this process, you specify the basic color, reflective finish, transparency, texture, and bumpiness.

To make a scene more authentic, you can add lighting effects. Lighting options include spotlight, point light, and directional light. Lighting effects are applied via a choice of three renderers: Rhino, Flamingo Photometric, and Flamingo Raytrace. The Rhino renderer is the basic renderer. The Flamingo Raytrace renderer traces beams of light from light sources and the Flamingo Photometric renderer accurately computes the combined effect of all indirect light sources.

Polygon meshes are constructed and incorporated in a file whenever a rendered image is constructed. To reduce file sizes, you can "save small" by not saving the polygon meshes. However, this requires that meshes be reconstructed if you subsequently want to render images saved this way.

2D orthographic engineering drawings are a standard means of communication among engineers. In Rhino, you create 2D drawings by selecting an object and letting the computer generate the drawing views. A Rhino 2D drawing can incorporate information on the dimensions of the object or objects the drawing represents and other annotation (textual information).

■■ Review Questions

1 Explain the methods by which an object is shaded in the viewport.
2 State the methods by which material properties are applied to an object.
3 Describe the procedures of setting up spotlight, point light, and directional light.
4 Explain the concepts of raytrace and radiosity.
5 Explain how 2D engineering drawings are produced from computer models of 3D objects.

Index

Numerics

2D drawings 508–509
3D curves, deriving from planar 94–96
3D drawings, converting to 2D 508
3D faces, polygon meshes 461
3D free-form surfaces 214–218, 226–228
3D objects
 as wireframe models 28
 illustration 6
3DFace command 456, 461
4-point lines 42–44

A

Analyze toolbar 151, 304
Analyze, menu option 9
analyzing
 curves 151–161
 points 152–154
 polygon meshes 471
 solid models 384–386
 surface models. *See* surface models, analyzing
Angle command 471
 curve and point analysis 152, 155
angles, measuring 152, 155
annotations 510
application window 8
ApplyCrv command 101, 435
ApplyMesh command 456, 470
Arc command 66, 67
Arc toolbar 66
Arc3Pt command 66
ArcDir command 66, 67
arcs 66–67
ArcSer command 66
ArcTTR command 66
area centroid 306–307
Area command 304, 306–307
area moments 306–307
AreaCentroid command 304, 306
AreaMoments command 304, 306
Array command 126
ArrayCrv command 126, 138
arraying curves 136–140
ArrayPolar command 126, 137
ArraySrf command 126, 139
assembly models 3

B

Background Bitmap toolbar 85
basic free-form surfaces 213–240
Bend command 127, 144
bending curves 144
bisecting lines 45
Bisector command 38, 45
bitmaps
 Background Bitmap toolbar 85
 HideBackgroundBitmap command 86
 PlaceBackgroundBitmap command 85
Blend command 94
blending curves 93–94
Boolean commands 375
BooleanDifference command 375, 376
BooleanIntersection command 375, 377
BooleanUnion command 375, 376
boundary representation 353
bounding boxes, constructing 156–157, 305–306
BoundingBox command 471
 curve and point analysis 152, 156
 surface modeling 304, 305
Box command 356
Box3pt command 356
boxes
 polygon meshes 460
 solid model 357–358
B-rep 353
bump. *See* material properties

C

Cap command 373, 375
capping planar holes 373–375
car body project
 curves 184–198
 rendering 512–514
 solid modeling 387–406
 surface models 337–349
cellular deconstruction 353
Chamfer command 91
chamfering curves 91
ChangeDegree command 116, 121
ChangeLayer command 52
Check command 152, 161

517

Circle command 61, 63, 421
Circle toolbar 60
Circle3Pt command 61, 62
CircleD command 61, 62
circles, constructing 60–66
　tangent circles 64–66
CircleTTR command 61, 64
CircleTTT command 61, 65
closed polylines 462
ClosestPt command 36
color
　objects, setting 488–489
　user interface, setting 25
　See also material properties
command line area 10
commands
　aliases 24
　arcs 66
　Boolean 375
　circles 61
　conics 72
　construction plane 437–438
　curve analysis 152
　curve editing, advanced 116
　curve editing, basic 102
　curve editing, points 108
　derived 100–101
　digitizer 242
　ellipses 68
　extend 87
　extrude solids 370
　fillet 90
　fillet solids 370
　free-form curves 48
　free-form surfaces 213–214
　from 2 curves 95
　helixes 74
　line construction 38–39
　parabolas 70
　point analysis 152
　point construction 36–37
　point editing 108
　polygon mesh analysis 471
　polygon meshes 456–457
　polygons 79
　primitive solids 356–357
　primitive surfaces 208–209
　rectangles 81
　solid analysis 384
　solid models 373, 379
　spirals 76
　surface analysis 304–305
　surface models 242
　transform 125–127
　See also entries for specific commands
commands, running from
　command line 10
　menus 9
　toolbars 12
computer modeling 1–2
Cone command 357, 364
cones
　polygon meshes 458–459
　solid models 363–365
Conic command 72
conics 72–73
constructing
　bounding boxes 156–157
　boxes 357–358
　circles 60–66, 158
　cones 363–365
　curvature circles 158
　curves, NURBS surface models 416–421
　cylinders 365–366
　derived surfaces. See deriving surfaces
　ellipsoids 360–362
　frameworks 202
　lines 38–47
　mesh primitives 457–462
　NURBS surface models 201
　paraboloids 363
　pipes 367–368
　points 36–37, 54–57
　polygon meshes 462–464, 468–469
　solid models. See solid models
　spheres 359–360
　surface models. See surface models, constructing
　torus 367
　truncated cones 364–365
　tubes 366
　wireframe models 28–29
constructing shapes. See entries for specific shapes
construction plane
　commands 437–438
　definition 14–16
　elevation, setting 441–443
　next 441
　object construction 443
　orientation, specifying 439
　origin 440–441
　origin, specifying 439
　perpendicular to a curve 447–448
　previous 441
　restricting to last selected point 20
　right-click menu 447
　rotating 443
　vertical 444
　viewing 446–447
　world front 445
　world right 445–446
　world top 445
　X axis, specifying 439, 449
　Z axis, specifying 449
construction plane configuration
　editing 451–452
　reading from a file 452–453
　retrieving 450–451
　saving 450
constructive solid geometry (CSG) 354–355
continuity curves 103
Contour command 101, 423
control point curves 50
control points
　displaying 53
　editing 110–111
　extracting 426–427
　location 32
　setting. See setting control points
　weight 32
　See also points
ConvertToPolyline 39
Coons surfaces 214–218
　See also surface-from-curve network surfaces
coordinate systems 16–17
Copy command 125
copying
　borders 428–430
　curves 127–128
　edges 428–430
CPlane3Pt command 438, 439

Index

CPlaneElevation command 214, 238, 438, 442
CPlaneFront command 438, 445
CPlaneNext command 437, 441
CPlaneOrigin command 437, 440
CPlanePerpToCrv command 438, 448
CPlanePrev command 437
CPlaneRight command 438, 446
CPlaneToObject command 438, 443
CPlaneTop command 438, 445
CPlaneToView command 438, 447
CPlaneV command 438, 444
CPlaneX command 438, 449
CPlaneZ command 438, 449
CreateUVCrv command 101, 434
cross-sections
 curve profiles 96–100
 digitizing 247–251
Crv2View command 95, 220
CrvDeviation command 152, 159
CrvEnd command 37, 55
CrvSeam command 116, 124
CrvStart command 36, 55
Csec command 97, 98
CSG (constructive solid geometry) 354–355
cursor movement, restricting 20
curvature arcs 310
curvature circles 158
Curvature command 152, 158
curvature graphs 156–157
curvature graphs circles, constructing 158
curvature rendering 307–308
CurvatureAnalysis command 304, 307
CurvatureGraphOff command 152
CurvatureGraphOn command 152, 157
CurvatureSrf command 304, 310
Curve command 48, 50
Curve from Objects toolbar 100
Curve menu 34
Curve toolbar 47
Curve Tools toolbar
 advanced editing tools 116
 Blend button 93
 Chamfer button 91

Csec Profiles button 97
Fillet button 90
Offset button 92
Curve, menu option 9
curves
 4-point lines 42–44
 analyzing 151–161
 arcs 66–67
 changing object layers 52
 conics 72–73
 continuity 103, 158
 control point curves 50
 control point location 32
 control point weight 32
 control points, displaying 53
 curvature radius, measuring 156
 deviation between 159
 direction, displaying 160
 ellipses 68–70
 error checking 160–161
 free-form 47–57
 handlebar editor 34
 helixes 72–76
 interpolating points 51
 kink points 33
 knots 32
 length, measuring 153
 line bisector 45
 line construction 38–47
 line segments and polylines 39–42
 lines, perpendicular to 57–58
 lines, tangent to 59–60
 matching 117–118
 NURBS 31
 parabolas 70–71
 periodic 33
 point construction 36–37, 54–57
 polygons around 79–81
 polylines and free-form curves 53–55
 polylines and line segments 39–42
 polylines through points 46–47
 polynomial degree 31
 polynomial splines 30
 projects. See projects, curves
 Rhino 31–34
 spirals 76–78
 surface models 203–204

See also NURBS surface models, curves; solid models; surface models; wireframe models
curves, derived
 3D from planar 94–96
 blending 93–94
 chamfering 91
 commands for 100–101
 cross-section profiles 96–100
 extending 87–89
 filleting 89
 from objects 100–101
 offset curves 92
curves, editing 101–125
 advanced editing 115–125
 commands for 102, 108, 116
 control points 110–111
 edit points 33, 109–110
 end bulge 123
 exploding 105
 fairing 121–122
 handlebar editor 114–115
 joining 102–103
 kink points, inserting 111–113
 kink points, removing 113–114
 knot points, inserting 111
 knot points, removing 113–114
 making periodic 124
 matching two curves 117–118
 point editing 107–115
 polylines, fitting 119–121
 polynomial degree, changing 121
 seam location 124
 simplifying 121–122
 splitting in two 106
 trimming 106–107
curves, transforming
 arraying 136–140
 bending 144
 commands for 125–127
 copying 127–128
 flowing along a curve 147–148
 mirroring 127–128
 moving 127–128
 orienting 130–136
 projection 142
 rotating 127–128
 scaling 128–129
 setting control points 141

shearing 130
smoothing 146–147
tapering 144–146
twisting 143–144
customizing Rhino. *See* system settings
CutPlane command 209, 211
Cylinder command 139, 357, 365
cylinders
   polygon meshes 458
   solid models 365–366

D

databases of objects, listing 159, 160
deconstruction, surface models 202–203
deriving
   3D curves from planar 94–96
   curves from objects 421–436
   periodic curves 124
   points from surface objects 412–414
   polygon meshes 462–464
   See also constructing; digitizing
differences, solid models 376
Dig command 242, 243
DigCalibrate command 242, 244
DigDisconnect command 242
digitizer 241–251
digitizing
   cross-sections 247–251
   surface models 204, 240–251
   See also constructing; deriving
DigPause command 242
DigScale command 242
DigSection command 242, 247
DigSketch command 242, 244
Dimension, menu option 9
dimensioning 510
Dir command 152, 160, 305
directional lights 498, 502
DirectionalLight command 502
displaying
   control points 53
   curve direction 160
   edges 312–314, 318–319
Distance command 471
   curve and point analysis 152, 154

Divide command 37, 57
DivideByLength command 37, 56
draft angle rendering 308
DraftAngleAnalysis command 304, 308
DrapePt command 36, 413
drawings
   2D 508–509
   3D, converting to 2D 508
   engineering 3–4, 508–509
DupBorder command 101, 429
DupEdge command 101, 428
duplicating. *See* copying

E

edges
   broken 318
   displaying 312–314, 318
   joining 317–318
   manipulating 312–319
   merging 316–317
   naked 313–314
   rebuilding 318
   splitting 315–316
EdgeSrf command 213, 216, 217
edit points 33, 109–110
Edit, menu option 9
editing
   construction plane configuration 451–452
   control points 110–111
   curves. *See* curves, editing
   end bulges 116, 123
   handlebar editor 34, 114–115
   points 107–115
   polygon meshes 464–468
   polysurfaces 378–383
   surface models. *See* surface models, editing
EditPtOn command 108
elevator mode 20
Ellipse command 68
Ellipse fromfoci command 68
Ellipse toolbar 68
EllipseD command 68
ellipses 68–70
Ellipsoid command 356, 361
Ellipsoid FromFoci command 356, 361
ellipsoids, constructing 360–362

EMap command 304, 308
end bulges 116, 123
EndBulge command 116, 123
engineering drawings 3–4, 508–509
environment. *See* user interface
environment map rendering 308–309
error checking, objects and curves 160–161
EvaluatePoint command 471
   curve and point analysis 152, 153
EvaluateUVPt command 304, 312
Explode command 102, 105, 318
ExplodeMesh command 456, 466
exploding
   curves 105
   polygon meshes 465–466
   polysurfaces 378
exporting files 7, 453–455
Extend command 87, 88
Extend toolbar 87
ExtendByArc command 87, 121
ExtendByArctoPt command 87
ExtendByLine command 87
ExtendCrvOnSrf command 87, 415
extending curves 87–89, 414–415
ExtractControlPolygon command 456, 463
extracting
   control points 426–427
   isoparms 425
   surfaces 378–380
   wireframes 432–433
ExtractIsoparm command 101, 425
ExtractPt command 101, 427
ExtractSrf command 379
ExtractWireframe command 101, 432
Extrude command
   solid models 370
   surface models 213, 221
ExtrudeAlongCrv command 213, 221
ExtrudeSrf command 370, 372
ExtrudeToPt command 223
extruding
   solid models 370–373
   surfaces 220–224

Index

F

Fair command 116, 121
fairing curves 121–122
FAQs (frequently asked questions) 21–23
File, menu option 9
files
  exporting 7, 453–455
  formats supported 7, 455
  importing 7, 453–455
  locations, setting 24
  reducing size 508
Fillet command 90
FilletEdge command 370, 372
filleting
  curves 89
  solid models 370–373
FitCrv command 116, 119
Flamingo Photometric renderer 507
Flamingo Raytrace renderer 505–507
Flow command 127, 148
flowing curves along a curve 147–148
free-form curves 47–57
free-form sketch curves 418–420
frequently asked questions (FAQs) 21–23
from-curve-network surface 226–228

G

Gcon command 152, 158
graphics area 10–11
Grid command 108, 116
grids 19–20

H

handlebar editor 34, 114–115
Hbar command 108, 114
Helix command 74
helixes 72–76
help system 21–23
Help, menu option 9
Hide command 217, 432
HideBackgroundBitmap command 86
hybrid system, solid modeling 355

hydrostatic values 386
Hydrostatics command 384, 386

I

importing files 7, 453–455
InsertEditPoint command 108
inserting
  kink points 111–113
  knot points 111
InsertKink command 108, 112
InsertKnot command 108, 111
InterpCrv command 48, 51
InterpCrvOnSrf command 48, 417
interpolating points 51
InterpPolyline command 48, 53
Intersect command 101, 431
intersections
  curves 101
  solid 377–378
  surfaces 101
isoparm curves 201
isoparms
  density 489
  extracting from surfaces 425

J

Join command 102, 118
JoinEdge command 305, 317
joining
  curves 102–103
  edges 317–318
  polygon meshes 467
  polysurfaces 378
JoinMesh command 456, 467
joypad project
  curves 162–169
  solid models 406–407
  surface models 320–329
joystick project
  curves 169–177
  rendering 511
  solid models 407
  surface models 329–330

K

kink points
  definition 33
  inserting 111–113
  removing 113–114

knot points
  definition 32
  inserting 111
  removing 113–114

L

Layer command 52
layers
  changing 52
  definition 18–19
Length command 471
  curve and point analysis 152, 153
lighting effects 498–502
Line command 38, 39
Line4Pt command 38, 42
LineAngle command 38
LinePerp command 38, 57
LinePP command 38, 58
lines
  4-point 42–44
  angles, measuring 155
  at an angle to another line 44–45
  constructing 38–47
  construction commands 38–39
  normal to a surface 416–417
  perpendicular to curves 57–58
  segments 39–42
  tangent to curves 59–60
  See also polylines
Lines command 38
Lines toolbar 38
LineTan command 38, 59
LineTT command 38, 60
LineV command 39, 40
List command 152, 160
Loft command 213, 225
lofted surfaces 224–226

M

Main toolbar 12
MakeCrvPeriodic command 116, 124
mapping polygon meshes 470
Match command 116, 117, 217
matching curves 117–118
material properties, assigning by
  Basic option 496–498
  layer 492–494
  plug-in 494–496

measuring
  angles 155
  curvature radius 156
  curve length 153
  distance between points 154
  units of measure, setting 24
medal project
  rendering 511-512
  solid modeling 407-410
menus
  pull-down 8
  running commands 9
MergeEdge command 305, 316
merging edges 316-317
mesh boxes 460
Mesh command 456, 469
mesh cones 458-459
mesh cylinders 458
mesh density, setting 457
mesh primitives, constructing 457-462
mesh rectangular planes 460-461
mesh spheres 459-460
MeshBox command 456, 460
MeshCone command 457, 458
MeshCylinder command 457, 458
MeshDensity command 457
MeshPlane command 456, 460
MeshPolyline command 456, 462
MeshSphere command 457, 459
MicroScribe digitizer 241-251
MicroScribe toolbar 242
Mirror command 126
mobile phone project
  curves 177-184
  surface models 330-336
model car. See car body project
model tolerances, setting 24
mouse, left vs. right click 13
Move command 125
MoveUVN command 108, 278
moving curves 127-128

N

NamedCPlane command 438, 451
NetworkSrf command 226
non-uniform rational B-splines.
  See NURBS
Normal command 39, 416-417

NURBS
  definition 31
  polygon meshes, from surface control points 462-464
  polygon meshes, from surfaces 468-469
NURBS surface models
  constructing 201
  definition 201
  description 412
  lines, normal to a surface 416-417
  points, deriving from surface objects 412-414
  polyline, on polygon mesh 420-421
  types of 207-208
  See also surface models
NURBS surface models, curves
  borders, duplicating 428-430
  constructing on objects 416-421
  contour lines 423-424
  control points, extracting 426-427
  deriving from objects 421-436
  extending to a surface 414-415
  free-form on a surface 418
  free-form sketch on a surface 418
  free-form sketch on polygon mesh 418-420
  isoparms, extracting from surfaces 425
  projecting to a surface 421-423
  pulling to a surface 421-423
  reference U and V, on the X-Y plane 433-436
  sections 423-424
  silhouette of a surface 425-426
  surface edge, duplicating 428-430
  surface intersections 430-431
  wireframes, extracting 432-433
  See also curves

O

object snap 21
objects
  error checking 160-161
  listing database of 160

Offset command 92
offset curves 92
Orient command 126, 131, 230
Orient3Pt command 126, 132
orienting
  construction planes 439
  curves 130-136
OrientOnSrf command 126, 135
OrientPerpToCrv command 126, 133
ortho mode 20
orthographic drawings 508-509

P

Parabola command 70
Parabola Vertex command 70
parabolas 70-71
Paraboloid command 356, 363
Paraboloid Vertex command 357, 363, 419
paraboloids, constructing 363
Patch command 214, 234
patch surfaces 233-237
periodic curves 33, 124
Pipe command 357, 367
pipes, constructing 367-368
PlaceBackgroundBitmap command 85
planar holes, capping 373-375
planar mode 20
PlanarSrf command 213, 219
Plane command 209, 210
Plane3Pt command 209, 210
PlaneThroughPt command 209, 212
PlaneV command 209, 211
Point command 36
point construction commands 36-37
point editing 107-115
Point Editing toolbar 108
point lights 498, 501-502
Point toolbar 36
PointDeviation command 305, 312
PointFromUV command 304
PointLight command 501
points
  analyzing 152-154
  at naked edges 314
  constructing 36-37, 54-57

Index

deriving from surface objects 412–414
deviation from surface 312
distance between, measuring 154
edit points 33
kink points 33
on a surface 310–311
surface models 203
U and V coordinates 311–312
See also control points
Points command 36
PointsAtNakedEdges command 305, 314
PointsFromUV command 310
Polygon command 79
polygon meshes
analysis tools 471
commands 456–457, 471
deriving from closed polylines 462
deriving from NURBS surface control points 462–464
description 455–456
editing 464–468
exploding 465–466
free-form sketch curves 418–420
from NURBS surfaces 468–469
joining 467
mapping 470
mesh boxes 460
mesh cones 458–459
mesh cylinders 458
mesh density, setting 457
mesh primitives, constructing 457–462
mesh rectangular planes 460–461
mesh spheres 459–460
polylines 420–421
reducing density 465
surface models 200
transforming 470–471
unifying normals 468
welding 467–468
PolygonEdge command 79
polygons, around curves 79–81
Polyline command 38, 40, 119
PolylineOnMesh command 39, 420–421
polylines
and free-form curves 53–55

and line segments 39–42
fitting to curves 119–121
on polygon mesh 420–421
polygon meshes 420–421
through points 46–47
PolylineThroughPt command 39, 47
polynomial degree 31, 121
polynomial splines 30
polysurfaces
capping planar holes 373–375
editing 378–383
exploding 378
extracting surfaces from 378–380
joining 378
splitting 382–383
transforming 384
trimming 380–382
preferences. *See* system settings
primitives
mesh 457–462
solids, commands 356–357
solids, instancing 352
surfaces 208–212
See also entries for specific shapes
Project command 100, 422
projecting curves 142, 421–423
projects, curves
car body 184–198
joypad 162–169
joystick 169–177
mobile phone 177–184
projects, rendering
car body 512–514
joystick 511
medal 511–512
projects, solid models
car parts 387–406, 480–483
joypad 406–407
joystick 407
medal 407–410, 483–485
projects, surface models
car body 337–349
joypad 320–329
joystick 329–330
mobile phone 330–336
sports shoe 472–480
ProjectToCPlane command 126, 142
Properties command 488

PtOff command 53, 108, 110
PtOffSelected command 108
PtOn command 52, 108, 110, 111, 120
Pull command 100, 423

Q

quadrilateral planar surfaces 209

R

Radius command 471
curve and point analysis 152, 156
RailRevolve command 214, 233
ReadNamedCPlanesFromFile command 438, 452
Rebuild command 116, 118
RebuildEdges command 305, 318
rebuilding edges 318
Rectangle command 81
Rectangle toolbar 81
Rectangle3Pt command 81
RectangleCen command 81
rectangles 81
RectangleV command 81
rectangular planes 460–461
rectangular surfaces 210–212
ReduceMesh command 456, 465
reducing file size 508
reflective finish. *See* material properties
RemapCPlane command 126, 134, 220
RemoveKnot command 108, 113
removing
kink points 113–114
knot points 113–114
Render, menu option 9
renderers
Flamingo Photometric 507
Flamingo Raytrace 505–507
Rhino 503–505
rendering
assigning material properties. *See* material properties, assigning
curvature 307–308
directional lights 498, 502
draft angle 308
environment map 308–309
illustration 6

524   Index

lighting effects 498–502
   point lights 498, 501–502
   projects. *See* projects, rendering
   reducing file size 508
   spotlights 498
   zebra stripes 309–310
RestoreCPlane command 438, 450
Revolve command 214, 232
revolved surfaces 232–233
Rhino curves 31–34
Rhino environment. *See* user interface
Rhino renderer 503–505
Ribbon command 213, 223
Rotate command 125
Rotate3D command 125, 128
RotateCPlane command 438, 443
rotating
   construction planes 443
   curves 127–128

S

SaveCPlane command 438, 450
SaveSmall command 508
Scale command 126, 129
scale model car body. *See* car body
Scale1D command 126
Scale2D command 126
ScaleNU command 126
scaling curves 128–129
screen descriptions. *See* user interface
seam location, changing 124
Section command 101, 424
SelBadObjects command 471
   curve and point analysis 152, 161
SetPt command 126, 141
setting control points, curves 141
Shade command 433, 490, 491
shaded display mode 491
shaded viewport display 487–492
ShadedViewport command 491
shading 489–491
Shear command 126, 130
shearing curves 130
shortcut keys, setting 23
Show command 217, 429
ShowBrokenEdges command 305, 318

ShowEdges command 305, 313
ShowNakedEdges command 305, 313
Silhouette command 101, 425
SimplifyCrv command 116, 122
simplifying curves 121–122
Sketch command 48, 86, 117
sketching, tracing images 84–86
SketchOnMesh command 48, 419
SketchOnSrf command 418
Smooth command 127, 146
smoothing curves 146–147
smoothness, analyzing 308–309
snap to grid 20
snap to objects 21
solid analysis commands 384
solid models
   analyzing 384–386
   B-rep 353
   cellular deconstruction 353
   creating with curves 29
   CSG (constructive solid geometry) 354–355
   definition 3, 352
   differences 376
   extracting surfaces 378–380
   extruding 370–373
   filleting 370–373
   generalized 352
   hybrid system 355
   hydrostatic value 386
   intersections 377–378
   planar holes, capping 373–375
   polysurfaces. *See* polysurfaces
   primitive instancing 352
   primitive solids, creating. *See* entries for specific shapes
   projects. *See* projects, solid models
   solid objects, combining 375–378
   spatial occupancy enumeration 353
   unions 375–376
   See also curves; surface models; wireframe models
Solid toolbar 356
spatial occupancy enumeration 353
Sphere command 135, 356, 359
Sphere3Pt command 356, 359

SphereD command 356, 359
spheres
   polygon meshes 459–460
   solid models 359–360
Spiral command 76
spirals 76–78
spline curves 30–31
Split command 383
   definition 102
   splitting curves 106
SplitEdge command 305, 315
splitting
   curves 106
   edges 315–316
   polysurfaces 382–383
Spotlight command 499
spotlights 498
SrfPt command 208, 209
SrfPtGrid command 214, 239
standard toolbar 12
starting Rhino 8
status bar 12
surface direction 319
surface models
   advantages 205–206
   approaches to 201
   deconstruction 202–203
   definition 2, 200
   derived surfaces. *See* deriving surfaces
   digitizing 204
   frameworks, constructing 202
   isoparm curves 201
   limitations 206
   modifications 205
   points 203
   polygon mesh 200
   prerequisites 205
   projects. *See* projects, surface models
   See also curves; NURBS surface models; solid models; wireframe models
surface models, analyzing
   area 306–307
   area centroid 306–307
   area moments 306–307
   bounding boxes 305–306
   curvature arcs 310
   curvature rendering 307–308

Index

draft angle rendering 308
edge manipulation 312–319
edges, broken 318
edges, displaying 312–314, 318–319
edges, joining 317–318
edges, merging 316–317
edges, naked 313–314
edges, rebuilding 318
edges, splitting 315–316
environment map rendering 308–309
points on a surface 310–311
points, at naked edges 314
points, deviation from surface 312
points, U and V coordinates 311–312
smoothness 308–309
surface direction 319
surface profiles 307–312
zebra stripes rendering 309–310
surface models, constructing
   3D free-form surfaces 214–218, 226–228
   basic free-form surfaces 213–240
   commands 208–209, 213–214, 242
   digitizing 240–251
   extruding surfaces 220–224
   from point grids 238–240
   from-curve-network surface 226–228
   lofted surfaces 224–226
   patch surfaces 233–237
   primitive surfaces 208–212
   quadrilateral planar surfaces 209
   rectangular surfaces 210–212
   revolved surfaces 232–233
   sweep surfaces 228–232
   trimmed planar surfaces 218–219
   with curves 29
surface profiles 307–312
Surface, menu option 9
Sweep1 command 213, 228
Sweep2 command 213, 231

sweeping
   solid models 352
   surfaces 228–232
system settings
   appearance 25
   file locations 24
   general settings 23
   model tolerances 24
   shortcut keys 23
   units of measure 24
   user interface, color and appearance 24

T

tangent circles 64–66
Taper command 127, 145
tapering curves 144–146
TCone command 357, 365, 434
text objects 368
TextObject command 357, 369
texture. *See* material properties
toolbars 12
   See also entries for specific toolbars
Toolbars dialog box 13
Tools, menu option 9
Torus command 357, 367
torus, constructing 367
tracing images 84–86
Transform toolbar 125
Transform, menu option 9
transforming
   curves. *See* curves, transforming
   polygon meshes 470–471
   polysurfaces 384
   surface models. *See* surface models, transforming
transparency. *See* material properties
Trim command 428
   definition 102
   trimming curves 106
   trimming polysurfaces 380–382
trimmed planar surfaces 218–219
trimming
   curves 106–107
   polysurfaces 380–382

truncated cones, constructing 364–365
Tube command 357, 366
tubes, constructing 366
Twist command 127, 143
twisting curves 143–144

U

UnifyMeshNormals command 456, 468
unions, solid models 375–376
units of measure, setting 24
user interface
   application window 8
   command line area 10
   graphics area 10–11
   pull-down menu 8
   setting color and appearance 24
   status bar 12
   Windows title bar 8
   See also toolbars

V

View, menu option 9
Volume command 384, 385
VolumeCentroid command 384, 385
VolumeMoments command 384, 385

W

Weight command 108, 110
Weld command 456, 468
welding polygon meshes 467–468
Windows title bar 8
wireframe display mode 491
wireframe models
   concepts 28–29
   constructing 28–29
   curves. *See* curves
   definition 2
   limitations 29
   representing 3D objects 28
   See also solid models; surface models
wireframe viewport display 487–492
wireframes, extracting 432–433
WireframeViewport command 491

Z

Zebra command 304, 309
zebra stripes rendering 309–310